CBRN and HAZMAT Incidents at Major Public Events

CBRN and HAZMAT Incidents at Major Public Events

Planning and Response

Second Edition

Daniel J. Kaszeta

Registered Office
John Wiley & Sons, Inc., 111 River Street, Hoboken, NJ 07030, USA

For details of our global editorial offices, customer services, and more information about Wiley products visit us at www.wiley.com.

Wiley also publishes its books in a variety of electronic formats and by print-on-demand. Some content that appears in standard print versions of this book may not be available in other formats.

Library of Congress Cataloging-in-Publication Data
Names: Kaszeta, Dan, author. | John Wiley & Sons, publisher.
Title: CBRN and HAZMAT incidents at major public events : planning and
    response / Daniel J. Kaszeta.
Description: Second edition. | Hoboken, NJ : Wiley, 2023.
Identifiers: LCCN 2022023838 (print) | LCCN 2022023839 (ebook) | ISBN
    9781119742999 (cloth) | ISBN 9781119743002 (adobe pdf) | ISBN
    9781119743040 (epub)
Subjects: LCSH: Emergency management. | Special events–Safety measures. |
    Public safety. | Hazardous substances. | Weapons of mass destruction.
Classification: LCC HV551.2 .K387 2023 (print) | LCC HV551.2 (ebook) |
    DDC 363.34/8068–dc23/eng/20220711
LC record available at https://lccn.loc.gov/2022023838
LC ebook record available at https://lccn.loc.gov/2022023839

Cover image: © badahos/Shutterstock
Cover design by Wiley

Set in 9/12pt MinionPro by Straive, Pondicherry, India

Contents

PART I

The Operational Environment

PART III
Response

PART IV
Practical Scenarios

Appendices

List of Figures

Preface to the First Edition (2012)

This book is about protecting large, high visibility events and public gatherings from accidents or incidents involving hazardous substances. By this, I mean chemical weapons, biological weapons, radioactive substances, nuclear devices, and the whole spectrum of toxic, flammable, and otherwise dangerous commercial and industrial hazardous materials. Collectively, this is the *CBRN/HAZMAT* threat.

WHO NEEDS TO READ THIS BOOK?

This book is really designed for anyone involved in preparation of safety and security plans for large events. I am writing this book for two groups of readers. First, I am writing this book for the many people who plan, manage, and provide emergency services support to major events. Few of these people will be subject matter experts in CBRN/HAZMAT but need to know about how to correctly consider the CBRN/HAZMAT threat in their plans. The second part of the readership is the CBRN/HAZMAT practitioner who may be tasked to support a major event. Many practitioners have great expertise in response, but supporting a major event can be quite different than normal operations. It will be difficult to address both categories of reader equally and consistently throughout the book. Wherever possible, if I think something is very useful one or the other, I will highlight it.

As the reader will soon see, this book cuts across many different disciplines. Because I try to connect some ideas and practices from many different sources, I hope that this is useful to the security, safety, or emergency planning generalist as well as specialists.

WHY?

I am writing this book because I have spent 12 years working on security and safety arrangements for major public events, both in specific CBRN/HAZMAT roles and in less specialized antiterrorism roles. I have wanted to write something like this since January 2005. I was sitting in an assistant fire chief's car in downtown Washington DC as part of a "joint hazard assessment team" for the second inauguration of President George W. Bush. I was sitting with a command officer from the DC Fire Department, an agent from the Washington field office of the Federal Bureau of Investigations, a sergeant from the District of Columbia Metropolitan Police Department, and a military officer from one of the multitude of US military units supporting the event. I was sitting in the car in a capacity as CBRN specialist from the US Secret Service. If we had more room in the car, we could have added at least a dozen others with a valid need to be there.

Because we had the whole day to sit in the car, we talked about many things. However, as we all were CBRN/HAZMAT specialists to one degree or another, we engaged in rather a lot of "shop talk" and airing of grievances, as one does in such a circumstance. I realized that a lot of people cared about doing the right thing, but no one person or department had the whole answer on how to prepare for a large event. This particular day in 2005 was certainly not my first or my last major event. However, certain thoughts started to crystallize in my head. The discussions with my comrades from other agencies and backgrounds reinforced what I already suspected. I realized several important points, which I should describe.

Major Events Take a Lot of Planning and Involve a Lot of Agencies

Any major security and safety planning effort will involve many people and many agencies. This means that there will be many different agendas, varying levels of knowledge and experience, and different philosophical approaches. Because CBRN/HAZMAT incidents do not happen at every event, sometimes the planning for them is lost in the bureaucratic noise or is not given the emphasis that it should rightly have. More often than not, this is not because of deliberate decisions but because of the bureaucratic nature of the process.

The Wheel Gets Reinvented Every Single Time

The planning and execution of major events is complicated. Major events do not occur every day or even every year in some places. Conditions change. Organizations have personnel turnover and attrition. Even over the course of a six month planning effort, the individuals who are assigned to turn up to planning meetings are likely to change. This results in planning efforts that repeat themselves year after year, often without much effort made to capture lessons learned. Some departments and agencies are better than others, but after about ten years, it really did feel that I was starting from scratch each time. It seemed to me that even a minimal effort to capture and write down some "best practices" would help manage the next event.

The Workers and the Bosses Have Different Perspectives

There are dedicated and knowledgeable people in CBRN/HAZMAT. There are certainly many more people working in the field now than when I started out in the early 1990s. However, if you go far enough up the chain of command, everyone has a boss somewhere who is not a CBRN/HAZMAT specialist and has a variety of concerns broader than the CBRN/HAZMAT niche. And these are the people likely to be making the big decisions about planning and response at the large events. In 20 years in the CBRN/HAZMAT business, I spent a lot of time working for a boss who was not a specialist in the field, but I also worked in various capacities under bosses who were specialists as well. But once I left the US Army Chemical School, I never once had a second level supervisor, a boss's boss, who was a CBRN specialist. This was certainly the case with just about everyone I worked with. And I suspect that this situation is prevalent around the world.

People from Different Occupational Backgrounds Will Work on the Problem in Different Ways

Firefighters, paramedics, police officers, environmental specialists, public health specialists, physicians, soldiers, and scientists (to name only a few) will all look at the problem of CBRN/HAZMAT in different ways. This is because their training and experience are different in important ways. All of these perspectives have valid things to say about how to make the public safe and secure, but none of them have a monopoly on the truth. I have literally seen fights break out between police officers and firefighters. The efforts involved in CBRN/HAZMAT planning and response for a major event tend to be somewhat outside the parameters of everyone's normal day-to-day roles.

There is no Default Script to Fall Back on

In 2004, Ronald Reagan died. The State Funeral for President Reagan required a planning effort that was crushed into a few days rather than the months or years normally allotted for planning for major events in Washington DC. The US Army, at the Military District of Washington, had a standing default plan for state funerals. However, we did not have much to work with in the CBRN/HAZMAT arena except for our common sense and experience. If ever there were a day we could have used a "CBRN at Special Events" for dummies manual, 11 June 2004 was that day. I thought then and continue to believe now that there is a requirement for a good planning basis that can be picked up and used in a hurry.

All of these realizations mean that I feel that there is ample scope for someone to try to cut through all of these problems and put a useful common body of knowledge onto paper.

WHY AM I QUALIFIED TO WRITE THIS BOOK?

First, by definition I am more qualified than anyone else who has written a book on this subject, because nobody else has a book in print on this specific subject. However, that's just a technicality. I have been working for 20 years in the field of CBRN defense and HAZMAT response. My career has taken me on a grand tour through the whole sphere of CBRN/HAZMAT, while also giving me experience in other related sectors such as emergency medicine, military operations, law enforcement, and emergency planning. As we have already discussed, planning for major events sits at the nexus of several different important operational disciplines, and I feel particularly privileged to have worked precisely in that nexus. I was originally trained as a Chemical Corps officer in the US Army, but my career path forced me to receive training and experience in many other disciplines, including protective security, emergency management, intelligence, radiation safety and health physics, incident command, explosives/demolition, fire safety, hazardous materials, and physical security.

In particular, I had three different assignments in government service that put me squarely in the line of fire for multidisciplinary CBRN/HAZMAT planning and response:

- Disaster Preparedness Advisor, White House Military Office 1996–2002—I was the CBRN subject matter expert in the Defense Department office at the White House that does emergency planning for the office of the President.

- CBRN Specialist, Technical Security Division, US Secret Service 2002–2008—I served in CBRN countermeasures in direct support of the White House and the US President, in a wide variety of assignments.

- Operations Officer, 32nd Civil Support Team, Maryland National Guard. 1998–2003. As with most National Guard jobs, this was a part time job. I worked to establish a state-level emergency response team for CBRN incidents in Maryland.

Because Washington DC and the White House are central to a disproportionately large portion of political major events in the United States and because the US Secret Service, by law, is the lead agency in the US government for planning and executing security at "National Special Security Events," I was in a privileged position to take part in many events as a CBRN specialist.

The following list details the various major events in which I had some involvement. In all of these situations, I was involved in at least some CBRN/HAZMAT aspects of the operation.

Three Presidential Inaugurations (1997, 2001, 2005)
Ten State of the Union addresses (1997–2006)
NATO 50th Anniversary Summit, Washington (1999)
World Trade Organization Summit (1999)
Y2K Transition (31 December 1999)
World Bank Summit (2000)
G8 Summit, Genoa (2000)
Democratic national convention (2000)
IMF Spring Meetings (2003)
Two UN General Assemblies (2003 and 2004)
Million Woman March, Washington DC (2004)
State funeral of Ronald Reagan (2004)
Republican national convention (2004)
G8 Summit, Scotland (2005)
Eastern European Summit, Lithuania (2006)

Every one of these events provided some insight and perspective on what to do and what not to do.

One thing that I wish to make clear is that this book is not a platform for me to advocate that the rest of the world should always imitate the United States or the practices of the various agencies where I worked. One of the many things that I learned over the course of my career is that the American way, when even such a thing could be defined at all, has not always been the best. There is good and bad to be found everywhere, and no agency or country has a monopoly on best practices. Some serious mistakes were made during some of the events listed above. I made some of them. I was a witness to many others. In several places in this book, I am very critical of the way things have evolved in the United States. So, please do not throw down the book thinking that I am yet another man from Washington who thinks that the US government way is the best.

Acknowledgments

A number of people have helped me with this new, revised second edition. First, it is important that I acknowledge that dozens of people across a number of disciplines read the first edition and provided valuable feedback to me. This showed me that there was demand for an updated version. Special thanks are owed to Dr. Holly Carter, who opened my eyes on behavioral science aspects of CBRN incidents; to Abigail Tattersfield, who reviewed my chapter on forensics; and to Professor John Drury, for significant assistance on numerous psychological and behavioral aspects. I would also like to thank Chris Aldous, Ian Munoz, Robert Wagner, Richard Mead, and David Crouch. Gwyn Winfield and Steve Johnson encouraged me to start this process years ago with the first edition, and I would not be where I am now without the boost that they gave me at the time. Special thanks to the team at Wiley who commissioned this second edition and who were generous enough to grant me several extensions during the writing process. The staff at the British Library, the Oriental Club, and the Cask Pub and Kitchen, all in London, are worthy of thanks for help in several different ways. Finally, special thanks to my wife, Sophie Tyler, for putting up with me during this process.

Introduction to the Second Edition

The purpose of this book is to help the reader prepare for accidents and incidents involving chemical, biological, radiological, or nuclear (CBRN) materials at major public events, whether from a terrorist origin or as the result of an accident. I have written this book in a way which should benefit both the serious CBRN practitioner or hazardous materials (HAZMAT) specialist, as well as newcomers to the emergency services and emergency planning fields. Although I did not write the first edition with the intention of this book becoming a general introductory primer on CBRN/HAZMAT planning and response, many readers have used it as such. I hope that it will be useful in that secondary role, even if it is not my primary objective. Others have used the book as a guide to major events' security and safety planning in general. I am comfortable with that as well, as most of the broad principles and general advice are certainly applicable. Wherever possible I will attempt to make clear where people using the book in these off-label usages may find additional resources and information.

WHAT IS A "MAJOR PUBLIC EVENT?"

My career field is plagued with jargon and technical terms. It is a minefield of shifting definitions and confusing words, abbreviations, and acronyms. But if I am going to write a book, I must pin a few definitions down. We cannot get far in a discussion of "major public events" without defining the term.

When is an event "major?" Is it a high visibility event, a significant eve, or a "national security special event?" (The latter is an official Federal government US term.) Other terms are in use elsewhere in the world. By "major event," I really mean any large gathering of people that is sufficiently high profile enough to warrant special planning for security and public safety. The threshold for declaring something to be a major event is tricky and will vary greatly. Political party conventions, large sporting events, royal weddings, state funerals, large concerts, state visits by important foreign dignitaries, inaugurations of US presidents, and large festivals all certainly count as major events. I do not think that a lengthy discussion of what is or is not a "Major Event" will really be helpful, as the local definitions will vary so widely. If it is "major" to the reader or the local officials, then perhaps this book can be helpful. If it is something out of the ordinary, and you are going to do planning and preparedness above and beyond what you usually do, then, by all means it is "major." As such, the term "major event" is used throughout.

Some Conventions Used in this Book

As this book is written for an American publisher, it uses American spelling and writing conventions. While Chapter 1 describes my understanding of the threat in some level of detail, I refer to the threat environment and the general subject area covered by this book as CBRN/HAZMAT. CBRN is normally taken to refer to military chemical warfare agents, biological warfare agents, and radiological/nuclear materials. HAZMAT generally is used to refer to hazardous substances used in commerce and industry. In order to avoid arguments, I will often use this collective term CBRN/HAZMAT rather than get into the argument of where CBRN ends and HAZMAT begins. That is a lengthy discussion that is beyond the scope of the book. I realize that there are numerous definitions of both these terms in different sectors and in different countries around the world. When in doubt, please assume that I am taking a broad and inclusive interpretation. As you will see as you read this book, I often draw no distinction between the two, as the effects are often identical from a response viewpoint.

The book aims to avoid excessively technical jargon, because much of the jargon in existence is not universally recognized and some people who are reading this will not be specialists in the field. I make an earnest effort to define any acronyms or technical terms as I go along. The glossary that appeared in the first edition is not included as few readers found it helpful or necessary.

Wherever it is necessary, I cite useful source materials and include them in the bibliography at the end. There are many resources available on the internet. The various hyperlinks listed in various chapters or in the end notes were current at the time of the completion of the book, but web addresses do change over time.

What's new? Readers are assured that there is a fair bit of new material here. First of all, there are new chapters on social/behavioral aspects and crisis communication. The older edition's work on CBRN forensics has been augmented and spun off into a whole new chapter. The illustrations are, by and large, different. A few things that were not viewed as particularly interesting or helpful have been reduced or cut out. The various chapters of the book end with one or more "lessons learned" from practical experience to try to apply the theoretical material in real life. Finally, the practical scenarios in the latter part of the book are all new material.

This book is organized into several parts. The first edition of this book had a several page introduction to each part. Lengthy discussions with readers led me to the conclusion that nobody read them, so I will summarize here. Part I talks about the general operational environment for major events. It sets the scene. It also describes general planning provisions that apply to most, if not all, major events, and it lays out a few possible general tools and techniques that can be broadly applied. This part is as applicable to general planning and emergency preparedness as it is to CBRN/HAZMAT incidents.

Part II addresses preparedness, training, and readiness BEFORE an incident occurs. This part is aligned according to the major response disciplines and I break years of practice, prejudice, and precedent by treating the private sector as part of the solution, and not just some sort of thing running along in the background. Building managers, facility staff, and private security staff are part of the emergency response to terrorism and accidents, and we need to treat them as part of the solution and not just as part of the landscape.

Part III talks about how to actually deal with an incident or attack if and when it occurs. I certainly cannot tell you what to do in any circumstance. Every incident or terrorist attack will be different. In this book the focus is on broad guidance, best practices, lessons learned in the past, and crucially, what NOT to do. It is very difficult to prescribe exactly what to do at some point in the future. The entire rationale for this book is to create an environment and instill processes and ways of thinking that will lead you to make the right decisions when the time comes.

Part IV is, in many ways, the most important part of the book. It ties together parts I–III in a series of practical scenarios. These scenarios, and their accompanying guidance, are useful readings on their own. But they are also intended to be the basis for training and exercises. Readers of the first edition consistently informed me that they were the most helpful part of the book. Readers of the first book can be assured that this part contains a high percentage of new material. Readers new to this book can seek out the previous edition for the older training scenarios, which still have significant value for those seeing them for the first time.

PART I

The Operational Environment

CHAPTER 1

The CBRN and Hazardous Materials Threat

The major event environment faces threats from both deliberate and accidental dispersal of hazardous substances. Safety and security efforts for major events are aimed at preventing, responding, or mitigating the adverse effects of such acts or situations, whether these situations are deliberate or accidental. Chemical, biological, radiological, and nuclear (CBRN) and hazardous materials (HAZMAT) can provide serious disruption, injury, illness, or fatalities at major public events.

There are many reasons why you might be reading this book. However, you have probably not taken the effort to locate this book if you did not already take the threat of CBRN and HAZMAT seriously. An encyclopedic inventory of possible threats is out of the scope of this book, as is a discussion about whether or not to take the threat seriously. If you've got so far as to open this book, I am assuming that you take the threat seriously. As for detailed technical deeper dives into individual types of threats, there is adequate literature already in existence.

ADVERSE EFFECTS

Much of the existing literature talks about "combating" or "responding to" CBRN threats or attack, or HAZMAT incidents or situations. The reality is that we are surrounded by chemical products, biological background materials, and cosmic radiation every day. It is actually much more helpful if we look at the whole planning and response field from the standpoint of preventing, responding to, and mitigating the adverse effects of CBRN/HAZMAT situations. We aren't fighting, for example, chlorine gas. Chlorine gas exists. You do not fight it. However, we COULD be confronting the adverse effects of an accident or deliberate incident involving chlorine. When one starts to analyze how to plan for and respond to problems in this CBRN/HAZMAT field, I find it more useful and simpler to look at the whole subject from this perspective of adverse

CBRN and HAZMAT Incidents at Major Public Events: Planning and Response, Second Edition. Daniel J. Kaszeta.
© 2023 John Wiley & Sons, Inc. Published 2023 by John Wiley & Sons, Inc.

FIGURE 1.1 Traditional chemical munitions are only a small part of the threat spectrum.
Source: Photo: US Army, public domain image.

FIGURE 1.2 Hazardous materials accidents are a significant fraction of potential threats to major events.
Source: Photo: US Air Force, public domain image.

effects. It is far more effective to plan for dealing with large numbers of sick and injured people than it is to conduct planning for specific categories of chemical substance. It is a better use of resources and intellectual capital to have one very good general purpose plan for sick, injured, and contaminated people than a number of specific plans for specific chemical substances.

Planners and responders are better served by considering the end-state of CBRN/ HAZMAT scenarios and working backwards from them. One could write literally 6000 scenarios and work through them to their conclusion, but this is really a waste of precious time as there are really only a handful of outcomes. The thousands of available CBRN/ HAZMAT substances can cause six categories of damage. The adverse effects of CBRN/ HAZMAT incidents or accidents include any combination of the six categories of damage described below.

Death (Immediate or Delayed)

CBRN/HAZMAT materials may cause immediate or delayed death. The overwhelming operational imperative of major event planning will likely be to reduce or eliminate death. It is important to understand that most CBRN/HAZMAT materials do not have much potential to cause instant lethality. While the small number of CBRN terrorist incidents in modern times has caused deaths, they have caused only a handful of immediate fatalities. Even the fastest acting biological warfare agents (BWAs) cause death hours or days after exposure. Even the most radioactive "dirty bomb" is likely to cause fatalities only through an explosive dissemination. Many chemical substances, including chemical warfare agents (CWAs), are theoretically capable of rapidly killing exposed individuals, but field conditions, especially in terrorist or accident settings, rarely allow for the necessary concentrations to be present.

Physical Injury and Illness (Immediate or Delayed)

Illness and injury, which may or may not lead to eventual fatalities, are a more significant planning consideration than fatalities. The vast majority of the CWAs and toxic industrial chemicals are far more effective at causing illness, injury, incapacitation, or serious discomfort than they are at killing people outright. BWAs, such as pathogens and toxins, are designed by natural evolution to cause sickness, but not necessarily death. Indeed, rapid death of a host does not serve a useful evolutionary purpose for a disease-producing microbe. Some of the mental effects, discussed below, can lead to physical injuries. For example, people moving quickly, out of fear and anxiety, can injure others or themselves.

Latency is an issue. Despite what Hollywood may have you believe, many CBRN/ HAZMAT threat materials do not cause immediate injury or illness. Many materials have delayed effects because the chemical or biological mechanisms that cause bodily harm take some time. In most conceivable radiation exposure scenarios, radiation sickness and other effects will take hours, days, or a week or longer. In many radiological situations, the long-term delayed effects are statistical in nature and may take years if ever to become apparent. Aside from a handful of fast-acting toxins, BWAs tend to have delayed effects, as there are incubation periods involved. While many chemicals are fast acting, some are not. One example is phosgene.

It is a dangerous industrial chemical and CWA responsible for the majority of "gas warfare" fatalities in the First World War, but it takes many hours for its effects to appear.

In the context of emergency planning and emergency response, ill and injured people provide a far greater burden than dead victims. While dead people must be taken care of, the urgency is far less than with living victims who need rescue, decontamination, immediate first aid, and/or transport to definitive medical care.

Mental Injury and Illness (Immediate or Delayed)

CBRN/HAZMAT may cause many psychological and social effects. The psychological and emotional effects of CBRN warfare and terrorism have generally been less studied than the physical effects. For the most part, CBRN weapons are invisible. For many people, a threat that you cannot see produces far more fear and anxiety than a well-known or highly visible danger. Fear can seem to be "contagious."[1] In addition, there is the possibility of psychogenic effects, where fear and anxiety may produce physical symptoms not unlike exposure to some of the threat materials. In other circumstances, people with existing mundane illnesses may mistake their symptoms for exposure. People with nausea may mistake it for acute radiation sickness and people with respiratory infections will think that they have anthrax. The term "worried well" is often used, and this phenomenon will be discussed in more detail in Chapter 8. Psychological, social, and behavioral aspects of CBRN/HAZMAT incidents are discussed in greater detail in Chapter 3.

Damage to Property

In many situations and scenarios, property may be contaminated and rendered unusable for its intended purpose. Sometimes actual contamination is not necessary for people to imagine that contamination might be present. If people think that an area or a building still poses a threat, they will not go there, causing businesses to suffer. The psychological taint may prove harder to remove than any physical taint. Because many major events occur in sites of unique cultural importance, the damage to property may assume more dimensions than merely economic.

Damage to the Environment

The most CBRN/HAZMAT would qualify as environmental contaminants. The long-term environmental effects of dispersal of such materials could cause problems for decades. As with damage to property, some threats such as radiation may induce such fear that people will assume that they are present long after they have decayed or been decontaminated.

Economic Damage

Both CBRN terrorism and HAZMAT accidents can be responsible for vast economic damage. Response to disasters costs money. The economic impact of conventional terrorism is well documented[2] and effects of CBRN terrorism could be significantly higher due to contamination of property. It may be more expensive to deal with contaminated property than the mere demolition of property destroyed in conventional terrorist attacks. The cost of decontaminating property will only serve to extend the economic impact of terrorist attacks. In many scenarios, it is possible that the indirect costs of CBRN/HAZMAT incidents will greatly exceed the direct costs. The loss of property and/or the extensive efforts to restore property to usable condition could be very expensive. Recovery efforts, including decontamination, could take a very long period of time and many resources. Businesses could lose revenue from closure or go out of business. Buildings and areas of cities may be isolated or abandoned for periods of time, having adverse effects on the economy. Indirect effects are possible too, as financial markets will react to terrorist events and major accidents.

CATEGORIES OF THREAT MATERIALS

It is counterproductive to attempt a vast catalog of possible materials that can cause death, illness, or injury by design or accident. A catalog of death and destruction is beyond the scope of this book, and such an attempt would fill a hundred redundant pages. There are many excellent references that can serve as the A–Z of CBRN and HAZMAT threats, and it will serve little purpose for me to try to walk in the steps of others. US Army Field Manual 3-11-9,[3] available widely on the internet as a PDF, is a classic reference on chemical weapons, as are Jan Medema's *Basic Principles of Chemical Defense*[4] and the various works of Frederick Sidell, particularly *Chemical Warfare Agents: Toxicology and Treatment*.[5] For biological weapons, *Medical Aspects of Biological Warfare*[6] is recommended, as is its sister volume, *Medical Consequences of Radiological and Nuclear Weapons*,[7] for radiological threats. Both are provided online, free, by the US Army's Borden Institute.

In order to understand the rest of the book, it is necessary to have a basic overall understanding of the categories of threat materials we are worried about.

Chemical Warfare Agents (CWAs)

The term "chemical warfare agent" (CWA) broadly means the use of chemical substances that have been developed for use in warfare because of their toxic properties. The CWAs are widely known in the US and NATO countries by their "digraph"—a two letter symbol (e.g., VX) originally developed many decades ago in the NATO countries. The world's various militaries historically classified the CWAs according to their mechanism of physiological action (in some cases a very dated approach to classification), and they include the following families of chemicals:

Nerve Agents: These are organophosphate compounds that cause lethal damage to the nervous system. In particular, they interfere with the operation of an enzyme called acetylcholinesterase, causing a chemical imbalance in the nervous system. GA ("Tabun"), GB ("Sarin"), GD ("Soman"), GF, and VX are the primary examples. There's also a newer family, developed in the 1970 and 1980s in the Soviet Union, which is broadly termed "the Novichoks." However, all nerve agents operate using the same biochemical mechanisms in the human body.

The nerve agents are one of the more dangerous chemicals produced by man and are lethal in very small quantities. They are also noted for being lethal through dermal exposure (absorption of liquid agent through the skin), as well as through inhalation. The primary difference between the various nerve agents is not their method of action, but the rate at which they evaporate. The term "nerve gas" is not really appropriate as the nerve agents are all liquids at normal temperatures.

Nerve agents have been of particular interest in recent years. Nerve agents were success-fully employed by the Aum cult. They have been used in the Iran–Iraq war, the civil war in Syria, and in both attempted and successful assassinations by the Russian and North Korean state security services. All of these uses are discussed in my 2020 book *Toxic* if readers want more information.[8]

Blister Agents: Also known as vesicants, this family of agents includes the mustard agents (HD—so-called mustard gas even though it is a liquid, HN, H, etc.) as well as Lewisite (L). These compounds provide significant damage to exposed skin, eyes, and respiratory tract. The mustard agents have effects that are delayed. While mustard was used to great effect in the First World War to cause casualties, it caused relatively few battlefield deaths and most of the injured victims that eventually died did so because they succumbed to complications that would have been treatable with access to modern medicine.

Cyanides—"Blood Agents": Hydrogen cyanide (AC) and cyanogen chloride (CK) were referred to historically and incorrectly as "blood agents." Arsine may also be considered in this category, but arsine never achieved much result as a warfare agent. It should be noted that hydrogen cyanide is a rarity among the CWAs in that it is lighter than air. Despite the fact that hydrogen cyanide has been used for executions and genocide, it has a poor history of use in actual warfare. AC and CK need a very high concentration in order to cause lethal-ity, which makes their use as weapons problematic. Hydrogen cyanide is used in industry as well.

Pulmonary Agents: This category of substances is also known under the older and some-what less accurate term of "choking agents." This family includes chlorine, phosgene, diphos-gene, and other respiratory irritants. Such agents are often considered "first-generation" CWAs as they were the first to be used in modern warfare. These kill by inducing pulmonary edema. The effects of phosgene and diphosgene are delayed acting. A majority percentage of the chem-ical warfare fatalities in the First World War were due to this category of agents,[9] principally phosgene. These chemicals, because of their use in industry, are often considered under the category of industrial hazardous materials. Many industrial chemicals have characteristics nearly identical to the military choking/pulmonary agents.

Chlorine often gets special attention in this category because of its widespread use in industry. In some instances, it is available in very large quantities for water purification. While its usefulness as a battlefield weapon was very poor, the near global availability of chlorine in large quantities gives some people concern. The scope of danger posed by chlorine in industrial quantities is well established in industry publications.[10] Chlorine has been extensively used in the civil war in Syria as a weapon.

"Nonlethal agents"—incapacitants and riot control agents: This family comprises a wide variety of compounds that cause (generally) less-than-lethal effects. Some examples include the vomiting agent Adamsite (DM), tear gasses (like OC, CS, CR, CN), BZ (an unpredictable hallucinogen), and derivatives of fentanyl, a narcotic tranquilizer. While riot control agents are not viewed as a likely terrorist agent, they must still figure into planning, as they may still be of significance in a major event setting. Some major events, such as political summits, may attract demonstrations and civil disorder. Police may use riot control agents in public order opera-tions, and the dispersal of such agents may have unexpected effects. In general, the toxicology of riot control agents and the various decomposition products and filler material associated with it is poorly understood by the typical users of such chemicals.[11]

Riot control agents could also be used as a distraction accompanying some other form of terrorist incident. For example, tear gas could be used to start a panic and move large numbers of people in a certain direction for other more lethal attacks to take place. While "nonlethal" agents are designed to be incapacitating rather than lethal, there have been circumstances when serious injury or death have occurred from their use.

Industrial Chemicals/HAZMAT

In a way, chemical warfare is basically just the produce of the chemical industry put to work in warfare. However, the amounts of chemicals produced and used in industry are vastly greater and are often more dangerous in many ways than traditional warfare agents. Industrial HAZMAT can pose serious threats as weapons if used by governments, terrorists, or lone individuals. Toxic, flammable, explosive, or corrosive materials may be stolen from industry. Attackers could target industrial facilities for the purpose of causing a release of toxic materials. For example, during the wars that accompanied the breakup of Yugoslavia in the early 1990s, the chemical industry in Croatia was deliberately targeted.[12]

Accidents are a risk to the major event environment as well. Many dangerous chemicals are routinely used in commerce and industry and are transported on the world's railroads, highways, and waterways in large quantity. Accidents involving such substances can pose as much a threat to a major event as a deliberate terrorist incident. We need only to look at the Union Carbide incident in Bhopal, India, to understand the potential for harm represented by industrial chemicals.

Many industrial and commercial chemicals have characteristics that make them more dangerous to the general public than some of the CWAs. Industrial chemicals can include any of the following characteristics:

- Flammability
- Explosion risk
- Corrosiveness
- Reactivity to air or water
- Toxicity

Because commercial and industrial chemicals are much more freely available for purchase or theft in large quantities, I consider them to be much more accessible to the modern terrorist than the CWAs.

The number of chemicals considered dangerous to life, health, property, or the environment numbers in the many thousands. We can fill feet of shelf space merely cataloging the potential threat chemicals. The point here is that we cannot afford to get fixated on individual substances, as there are far too many. Because the options are simply too many, response plans against chemical threats cannot be too tied to the individual characteristics of single materials.

Many of the HAZMAT accidents that have occurred in the past arise from transportation accidents. The US Department of Transportation (DOT) provides useful statistics on hazardous materials accidents, particularly those involved in transportation accidents. US DOT provides many rank-ordered lists, using a variety of schemes for weighting applicable variables, to illustrate what substances are most commonly involved in accidents and what mechanisms caused the accident.[13]

There are literally thousands of additional substances that we cannot ignore.[14] As only one example, the bureaucrats of the United States have been busy. They not only have lists of chemicals, but lists of lists. For example, the US Environmental Protection agency has an excellent "meta-list."[15] Europe also has its lists, with over 100,000 substances being listed in European Union regulations and directives.

Biological Warfare Agents (BWAs)

Biological warfare has been described as "public health in reverse."[16] It is the use of naturally occurring microbes and nature's own poisons to cause illness or death. Biological warfare is actually much older than chemical warfare. While many BWAs are lethal, much of the biological warfare research conducted in the Cold War was aimed at the development of incapacitating agents. It was widely believed that making people ill for a long period of time would have had a greater strategic impact on war-fighting capability than merely killing people. This is because a sick soldier was felt to be more of a logistical burden than a dead soldier. Of course, this logic probably does not necessarily apply to terrorist selection of potential weapons.

Pathogens: BWAs include pathogens and toxins. Pathogens are disease-producing organisms. They include bacteria, viruses, fungi, rickettsia, and various parasites. Again, many good references exist.[17] Some examples of pathogens that have been cited as having potential as biological weapons include:

Anthrax
Smallpox
Plague
Tularemia
Q-Fever
Glanders
Brucellosis

Pathogens may be developed in powder or liquid forms. It should be noted that pathogens are living organisms and they are often quite fragile. For example, UV radiation in normal sunlight is very destructive to most pathogens.

Toxins: Biological toxins are chemical poisons that are produced by plants and animals. Snake venom would be one example, albeit one not really suited for widespread terrorist usage. The toxins most cited as useful by terrorists are ricin, a poison extracted from castor beans, and botulinum toxin, produced by botulism bacteria. Ricin has been used for assassinations. The efficiency of toxins as area weapons, other than poisoning food or water, is somewhat speculative. In general terms, toxins are chemicals, and a situation involving toxins may more closely resemble a chemical incident rather than a pathogen incident. When encountered in liquid form, it is important to remember that toxins are usually large, complex, and heavy molecules that lack volatility. In other words, a puddle of toxin will not evaporate into a vapor form and pose a respiratory hazard.

Radioactive Materials

Radioactive isotopes can provide unhealthy levels of ionizing radiation to persons. Depending on the exact isotope, alpha particles, beta particles, gamma rays, and/or neutrons may be emitted. There are only a finite number of isotopes, and only a subset of them has properties and/

or availability to make them viable as threats. Some materials exist for only fractions of a second in a laboratory or are available in such minute quantities that their threat is academic. While authorities vary, the Argonne National Laboratory of the United States has developed useful list of radioactive materials thought to be useful for terrorism.[18]

Radioactive materials may be found as part of the nuclear fuel cycle, or in a wide variety of scientific, medical, and industrial applications. The US government has published an excellent primer on what types of radiation sources are likely to be encountered around the world called Technical Guide 238,[19] which is an excellent reference.

The primary concern about terrorist use of radioactive materials is that they would be disseminated in an explosive device, a so-called dirty bomb or radiation dispersal device (RDD). A secondary concern with radioactive materials is that they may be covertly placed in a location that would expose people nearby to dangerous amounts of radiation.

Except in extremely large quantities, radioactive materials typically produce a level of health hazard that is occupational or environmental rather than acute. It is difficult for a terrorist to cause acute radiation sickness with commercial or industrial materials except in some fairly constrained circumstances. However, the mere presence or suspicion of radioactive material is sufficient to cause panic.

Nuclear Devices: Nuclear devices are nuclear bombs, producing massive destruction through fission or fusion. Nuclear devices could be state-produced nuclear weapons or improvised nuclear devices (INDs). This is the least likely threat, due to the technical barriers involved. A poorly constructed IND may not function as intended and become a crude RDD, dispersing much plutonium or uranium. However, a communicated threat or a hoax involving a nuclear device is certainly a credible possibility.

MEANS OF DISSEMINATION

CBRN weapons and industrial HAZMAT rarely pose a widespread threat merely by their presence. Many of us live in cities surrounded by chemicals every day. For CBRN/HAZMAT to be a threat to major events, they require a means of dissemination in order to cause adverse effects. Different means of dissemination have differing operational impacts. This section is designed to provide an overview of the methods of dissemination that I believe is of relevance to the major event environment.

Efficiency

With deliberate employment of CBRN/HAZMAT substances, there is a useful concept known as "munition efficiency." Munition efficiency was originally derived as a concept when nations were manufacturing chemical weapons for use on the battlefield. Munition efficiency is usually expressed as a percentage and is a measure of how much chemical or biological agent is actually disseminated in a form that is going to achieve the desired tactical effect. For example, it does not take much engineering knowledge to estimate that using explosive dissemination (such as in an artillery shell) of a highly volatile and highly flammable liquid is probably not going to result in effective dispersal, as most or all of the agent will burn up. In a similar example, the Soviet Union discovered that aerial sprays of hydrogen cyanide had little effectiveness, since the hydrogen cyanide vapor is lighter than air, and a chemical agent needs to be heavier than air to reach troops on the ground if dispersed at altitude.[20]

When evaluating potential threats, developing a planning threshold, or when estimating the future course and harm of an incident, the munition efficiency of the dispersal means should be considered. Rarely, if ever, an accident or incident provides a mechanism that results in 100% effective dissemination of a threat material. When discussing planning scenarios, I have often seen emergency planners make poor assumptions about terrorist device efficiency. I have encountered training exercises where if 25 kg of chemical agent is in a device, planners assume that all 25 kg of agent will be disseminated in a useful form. This is rarely, if ever, the case, even with sophisticated chemical munitions designed by the research and development program of a large nation-state, let alone an improvised device using homemade components.

Primitive or Bulk Dissemination

The most primitive methods of dissemination are generally to dump bulk agent or to have some kind of leak or drip. Such dissemination generally relies on the volatility of chemicals to evaporate from liquid into vapor state or on gravity to allow dangerous liquids to flow from the point of dissemination. The 1995 Tokyo subway attack was a very crude dissemination. The 2001 Anthrax attacks had no real method of dissemination. Fine powder was simply put into envelopes, yet death, illness, and contamination of property was the result. It is also possible that bulk or primitive dissemination may happen.

Covert Emplacement of a Radiation Source

Radioactive materials do not need to be inhaled or ingested. Radioactivity, particularly gamma rays and neutrons, can penetrate through many types of material and travel a long way in air. A radiation source, such as a commercial or medical source, could be placed in a location where people could be exposed to harmful amounts of radiation. Since radiation exposure is cumulative over a period of time, such an attack might be effective. Major events often force people to congregate in the same area for a long period of time, such as stadium seating or entry queues.

Contamination of Food or Water

Food and water could be contaminated by deliberate action by terrorists. An accident could expose food or water to industrial chemicals. Major events often have catering restricted to a handful of sources. While history has few actual incidents of bioterrorism, but one that actually occurred was an incident in Oregon in 1984 when a religious cult made 751 ill with a food-borne pathogen, *Salmonella typhimurium.*[21]

Spraying

Both chemical and biological weapons are most effective as respiratory threats. Radiological particles that can be inhaled could also pose a threat. Devices and mechanisms that spray gas, vapor, or fine mists of droplets can be used as methods of dissemination by terrorists. Some accidents may involve leaks to pressurized containers, which is effectively the same thing as a spray device. The earliest military chemical weapons were chlorine cylinders that were taken to frontline trenches in the First World War and opened when the wind was in the correct direction. More sophisticated techniques such as aerial spray tanks were developed later. Correctly constructed spray devices can achieve high munition efficiency.

Explosive Dispersal

Most chemical weapons designed by the major nation-state participants in chemical weapons development were explosive dissemination devices, such as bombs, artillery shells, rockets, and missile warheads. Chemicals, biological agents, and radioactive material could all be disseminated with some sort of explosive charge. Munition efficiency can vary widely using this technique. Homemade or improvised explosive dispersal is likely to result in an incorrect charge-to-agent ratio. History tells us that effective explosive dispersal devices are the result of rather a lot of experimentation, trial, and error.

Binary Devices

A device could be constructed to combine precursor chemicals in such a manner as to formulate a chemical weapon. Binary devices are an established military technology, and a few attempts have been made to use binary devices as terrorist weapons.[22] As a rule of thumb, binary devices are actually less efficient than "unitary" devices of the same size. They sacrifice munition efficiency for safety among the handlers, which is not always a concern for terrorists. Binary devices are, historically, far more difficult to create in practice than in principle. They require more testing, trials, and validation than nonbinary devices.

Vectors

Some pathogens are spread in nature by "vectors" such as insects or parasites. For example, mosquitoes spread yellow fever virus. In practice, using vectors as a method of dissemination is likely to be very inefficient. However, the presence of unusual numbers or types of vermin at a major event bears examination.

Commercial, Industrial, and Transportation Accidents

Hazardous chemicals are used in industry every day and are transported around roads, railways, and waterways in large quantities. Accidents involving such materials near a major event are more likely than deliberate terrorism. Such scenarios can have more potential for harm (due to quantities involved) than many of the other threat mechanisms that I have discussed. A wide variety of mechanisms could be present during a transportation accident that might cause dispersal of hazardous substances.

The Bhopal disaster in India in 1984, where large quantities of a toxic methyl isocyanate vapor leaked, demonstrated firmly that industrial accidents in factories can kill thousands outside the factory perimeter. Industrial accidents or deliberate sabotage is a possible mechanism of dissemination. Major event planners need to consider just what exactly is used, stored, or produced in significant quantities upwind of major event venues.

Another concern pertinent to major event security is the possibility of an accident or disaster involving nuclear power facilities. Three Mile Island, Chernobyl, and Fukushima are all in the public consciousness. Such major accidents are blessedly rare. But the risk cannot be overlooked. While major events in the immediate precincts of large power stations are not likely, it is certain major events will be held within the downwind hazard area of commercial or military reactor facilities. Accidents or sabotage scenarios at such facilities is a possible scenario.

THE CAUSE OF THE PROBLEM: THE PERPETRATOR

The other part of the equation is the perpetrator. Who or what has caused the materials to be released?

Accidents, Natural Disasters, Bad Luck, and Incompetence

Not every incident is the result of terrorist activity. Incompetence has killed more people than terrorism. The largest dispersals of radioactive material and toxic industrial chemicals have been accidents brought about by natural or man-made disaster, not terrorism. By their very nature, the effects of such causative agents will have a lot of randomness. It is very important to understand that a CBRN/HAZMAT incident could, in fact, be caused by some sort of other natural or man-made disaster. A natural disaster such as a hurricane, flood, or a tornado could affect an industrial facility or railroad line and cause a serious release of dangerous materials. A large structural collapse, like that of the World Trade Center after the 9/11 attacks, could cause release of substances with long-term health and environmental effects. Indeed, one can make the argument that the World Trade Center collapse should be classified as a HAZMAT incident given the amount and type of long-term fatalities that have occurred from exposure to dangerous substances.

The Lone Perpetrator

Many violent or destructive acts have been perpetrated by a single individual. Not every lone perpetrator is necessarily a terrorist in the classic sense, as some incidents have been the result of mental illness rather than political, religious, or ideological motivation. A major event can easily present a logical target to someone with real or imagined grievances. There are practical limits to what a single person can accomplish, but we can look at some perpetrators in history (such as Bruce Ivins, the likely perpetrator of the 2001 anthrax attacks in the USA) to see that the limit is still quite high.

The Organized Group

An organized group of perpetrators can generally accomplish far more than a single individual. A terrorist group is likely to have more resources than an individual, and CBRN terrorism is a resource-intensive proposition. Much of the "process of CBRN terrorism" (discussed in Chapter 9) is labor-intensive, and there is a finite limit to what a single perpetrator or even a small group can hope to accomplish. On the other hand, a group can leak its intentions more easily than a single individual.

The Nation-State

It is possible that a large and coherent conspiracy may be mounted by the efforts of a nation-state. While this is less likely than individuals or groups, the support of a nation-state greatly increases the resources available to a terrorist plot. However, it should be noted that the overwhelming recent history of CWA usage has been by states, e.g., Russia, Syria, and North Korea.

NUISANCES, HOAXES, AND COMMUNICATED THREATS

An important category of the threat spectrum is the category of "apparent and possible threats." This category includes a wide variety of nuisances, hoaxes, and the traditional "communicated threat," which I will address individually. Any major event safety and security effort is likely to spend more time and effort addressing this category of threat than any actual employment of CBRN agents.

Nuisances

I define nuisances as relatively harmless situations that are frequently mistaken for a CBRN incident or HAZMAT accident. Depending on the parameters of the event, the nuisance category may include such events and situations as:

- Malfunctions by sensors
- Overzealous reporting or well-meaning bystander reporting of innocuous phenomena
- Abandoned parcels or items containing liquids or powders
- Substances mistaken for potentially hazardous materials or reported as such by well-meaning individuals
- Chemical sensors operating correctly but detecting cleaning products
- Detection of benign radiation sources, such as legitimate nuclear medicine sources or procedures
- Changes in natural background phenomenon that cause alarms on biological or radiological sensors

Nuisances will increase in frequency commensurate with the effort to detect CBRN substances. Every sensor has some theoretical false-positive rate, even if quite low. Therefore, with an increase in the number of sensors that are employed, there will be a higher number of false alarms. Nuisances will also increase in number after an actual incident, as was shown in 2001 in the United States after the anthrax incidents. If an actual incident involves a powder as was the case in 2001, more powders will be deemed suspicious by more people.

Hoaxes

The hoax is a situation contrived to imitate an actual CBRN/HAZMAT situation, without the actual presence of CBRN materials. Some examples from recent years include:

- Benign powders purporting to be anthrax
- Liquids marked as containing HIV or another infectious disease
- Liquids thrown onto VIPs that turn out to be harmless
- Highly odorous gases or vapors used as "stink bombs" to disrupt an event
- Small, harmless radiation sources placed deliberately to cause an alarm

An important subcategory of the hoax is a modified conventional device. This would be a conventional hazard or hoax, such as a real or fake IED that is accompanied by hoax materials, in order to increase the psychological effect of the device or to complicate conventional explosive ordnance disposal (EOD) procedures. An example would be a pipe bomb attached to a barrel with hazardous materials marking (thus mimicking a chemical IED), or a car bomb that is paired with small amounts of radioactive material, thus mimicking an RDD.

A Case Study: Furries and Chlorine

On 7 December 2014, a strong smell of chlorine permeated the Hyatt hotel in Rosemont, Illinois, not far from Chicago O'Hare Airport. The hotel was hosting the Midwest FurFest convention—a meeting of the "furry" community. "Furries" are enthusiasts who dress in non-human costumes. It is a subculture unto itself. Approximately 4000 "furries" congregated for this particular convention.

The strong odor and irritating effects of chlorine initiated an emergency response. Nineteen people were taken to hospital (all but one of them were quickly released) and more were treated at the scene. The building was evacuated and thousands of furries were, ironically, given refuge in a neighboring dog show. It was determined that the source of the irritating vapors was a quantity of a powdered swimming pool chemical in a stairwell on the 9th floor.

A large number of people were interviewed by police and the FBI was involved in the investigation. Although it is presumed that this was a deliberate act, no arrests were ever made. To this day, the incident remains a mystery.

The mere fact that this event happened at all points out several things relevant to this book. First, a relatively small amount of material can have a disproportionate effect. In this case, it was the evacuation of thousands of people and 19 taken to hospital. This illustrates the disproportionate effect that hazardous materials can have. The second point is that you don't have to seriously hurt someone to cause major disruption. The event was temporarily disrupted. The third point is that motives can be idiosyncratic. If, before the event, you had asked me to assess the likelihood of CBRN terrorism against a furry convention, I'd have laughed. Not a lot of people were laughing that day.

Hoaxes may be relatively harmless gestures by people seeking thrills or attention. Alternatively, they may be more serious efforts to cause undesirable effects, such as panic or confusion. Planning considerations should include the possibility that hoaxes may be designed to deliberately confuse emergency response efforts, waste resources, or demonstrate to observers what the response capability for a certain type of incident may be. Hoax perpetrators could conceivably range from disaffected individuals with specific grievances (either outwardly rational or irrational) all the way to serious terrorists with ill intent, who merely lack the resources to conduct a more potent attack.

Communicated Threat

This category of threat includes situations where some form of direct or implied threat of employment of hazardous substances has been communicated through some means, such as email, a phone message, or a letter. The so-called bomb threat is a (generally) non-CBRN example. A telephonic claim to have parked a vehicle with a chemical or radiological device in a certain area is one example.

Communicated threats cannot be ignored as obvious hoaxes. It has been an established practice with some terrorist groups to give warning after placing a dangerous device. This has been a particular practice of Irish Republican extremist groups. While most communicated threats are nonsense or provocations, authorities cannot rule out the possibility that a real hazard lies behind the threat.

The communicated threat may be linked to a nuisance situation, often a "target of opportunity" scenario. For example, a strange smell may appear in a public area and someone could call the police to say that they spilled a chemical agent there. Likewise, a threat can be included with a hoax device or substance, such as a threatening letter with an unidentified powder.

Lesson Learned: Tear Gas. Sometimes Police Action Can Cause a CBRN Incident

Overzealous, inept, or disproportionate use of tear gas in law enforcement operations can backfire and cause a serious incident at a major public event. Tear gas used to disperse protestors can enter a building or change direction based on weather and seriously disrupt a major event. Similarly, police riot control agents can inadvertently use excessive quantities due to unfamiliarity or poor training. There have been incidents where excessive tear gas was used for purposes of brutality. Use of tear gas at the WTO summit in Seattle in 1999 and at the Genoa G8 summit in 2001 had the potential to become an issue for the event participants and not just protesters outside.

REFERENCES

1. Palmer, I. (2004). The psychological dimension of chemical, biological, radiological, and nuclear terrorism. *Journal of the Royal Army Medical Corps* 150: 3–9.
2. London Chamber of Commerce and Industry. The economic effects of terrorism on London: Experiences of firms in London's business community. London: August 2005.
3. United States Army (2005). *Field Manual 3-11-9: Potential Military Chemical/Biological Agents and Compounds*. Headquarters, Dept. of the Army.
4. Medema, J. (2010). *Principles of Chemical Defense*. Netherlands: Self published.
5. Marrs, T., Maynard, R., and Sidell, F. (2007). *Chemical Warfare Agents: Toxicology and Treatment*, 2nde. Chichester (UK): Wiley.
6. Dembek, Z.F. (ed.) (2008). *Medical aspects of biological warfare*, Textbooks of Military Medicine. Office of the Surgeon General, United States Army.
7. Defense Department (2013). *Medical Consequences of Radiological and Nuclear Weapons*. US Government Printing Office.
8. Kaszeta, D. (2020). *Toxic: A History of Nerve Agents*. C. Hurst & Co (UK).
9. Medema, J. (2010). *Principles of Chemical Defense*, 25. Netherlands: Self published.
10. The Chlorine Institute (2006). *Guidance on Complying with EPA Requirements Under the Clean Air Act by Estimating the Area Affected by a Chlorine Release*. Arlington (VA): Chlorine Institute, Inc.; Edition 4, rev 1.
11. Kaszeta, D. (2019). Restrict use of riot-control chemicals. *Nature* 573 (7772): 27–30.
12. United States Army (2002). *Field Manual 3-06.11 Combined Arms Operations in Urban Terrain*. Washington: Headquarters, Dept. of the Army; US Government.
13. United States Department of Transportation, Pipeline and Hazardous Materials Safety Administration (2011). *Top Consequence Hazardous Materials by Commodities and Failure Modes 2005–2009*. Washington (DC): US Government.

14. Hincal, F. and Erkekoglu, P. (2006). Toxic industrial chemicals – chemical warfare without chemical weapons. *FABAD Journal of Pharmaceutical Sciences* 31: 220–229.

15. US Environmental Protection Agency (2001). *List of Lists - EPA 550-B-01-003*. Washington (DC): US Government.

16. US Department of Health, Education, and Welfare (1959). *Effects of Biological Warfare Agents*. Washington (DC): US Government.

17. Sidell, F., Takafuji, E., and Franz, D. (ed.) (1997). *Medical Aspects of Chemical and Biological Warfare*, Textbook of military medicine. Washington (DC): Office of the Surgeon General, US Army.

18. Argonne National Laboratory. *Radiological and Chemical Fact Sheets to Support Health Risk Analyses for Contaminated Areas*. March 2007. https://remm.hhs.gov/ANL_ContaminantFactSheets_All_070418.pdf.

19. United States Army Center for Health and Preventive Medicine (1999). *Technical Guide 238: Radiological Sources of Potential Exposure and/or Contamination*. Edgewood (MD): US Army.

20. Medema, J. (2010). *Principles of Chemical Defense*, 21. Netherlands: Self published.

21. Torok, T. et al. (1997). A large community outbreak of salmonellosis caused by intentional contamination of restaurant salad bars. *Journal of the American Medical Association* 278 (5): 389–395.

22. Campbell, L.M. (2009). *A Technological Countermeasure for Chemical Terrorism Against Public Transportation Systems: A Case Study of the "Protect" Program*. Monterrey (CA): US Naval Postgraduate School Available from the Homeland Security Digital Library.

The Major Events Operating Environment

Major events occur in a setting. They happen at a time and a place. This setting, the surroundings, the people that are at an event, and the duration of the event can all be considered to be the "operating environment." The operating environment will have a considerable effect on how well you can manage the safety and security of your event, both from a generic standpoint and from the standpoint of protecting against CBRN/HAZMAT threats. You will have to adapt your planning to accommodate a wide variety of variables, and it is only really possible to generalize in this chapter.

ASPECTS OF EVENTS

Size of Events

Size matters. Geographic size, the actual physical footprint of an event, is significant. So too is the demographic size, the number of people involved in an event. The size, both in terms of physical and attendance.

Duration of Events

Major events could range from a single hour, an example being a funeral of a public figure, to a series of large events spread over across nearly a month, such as the Olympic Games. Both size and duration will greatly affect the number of people you will need to do the necessary work.

The Location

Where the event is held is so important that I devote an entire chapter to it. Chapter 6 discusses the physical aspects of event venues.

CBRN and HAZMAT Incidents at Major Public Events: Planning and Response, Second Edition. Daniel J. Kaszeta.
© 2023 John Wiley & Sons, Inc. Published 2023 by John Wiley & Sons, Inc.

Complexity of Events

Some events are very simple in conception, while others have numerous inter-related components and lots of movements. If there are lots of movements of celebrities, VIPS, differing groups of the public coming in and out, changes of performers, different activities with different logistical footprints, media arriving and departing, and similar features. Some events have a linear progression where something happens, then something else happens in a different place. But many are more complex, with lots of events simultaneously or overlapping.

Type and Diversity of Attendees

The nature of the attendees at a major event can affect the type and nature of support that will be needed. A largely fit, adult audience will not need the same type of emergency support as a group that is full of elderly or otherwise vulnerable attendees larger than the number of attendees.

ATTENDEES—THE COMMONPLACE "CAST OF CHARACTERS"

Events will have a wide variety of people at them. While I might be accused of engaging in stereotyping, it is useful to look at some basic archetypes of who might be at a major event. These include but are not limited to the following categories:

Paid/Ticketed Attendee

Many events will have ticketing or some sort of registration. This allows some degree of understanding of numbers. This also means that you will have to devote security or stewarding staff to collect tickets. Attendees at a paid event may have different expectations at an event than attendees at free events. There may be ways to use ticket sales as an intelligence tool, such as checking names against watchlists.

Free Attendee

Some events are either open to the public or so difficult to enforce entry that you will get attendees turning up without tickets or registration. You may often have a mix of both. For example, major sporting events may have tens of thousands of ticketed attendees inside, but thousands of more fans outside the event and in local bars and pubs. One reason why I draw a distinction between free and paid attendees is that people who have paid for a ticket, or at least registered for it, can be made subject to various terms and conditions. A private venue is well within its rights, for example, to make searching of bags a condition of entry as part of the terms of selling the ticket to an event. Another reason is that it may be reasonably possible to estimate the number of paid/ticketed attendees who might attend. Free attendees are, in effect, members of the general public. At best, you might be able to estimate their numbers ahead of time based on previous experience.

Vulnerable Attendee

With large events, you will inevitably get attendees with vulnerabilities or special needs that might require significant additional planning in the event of an emergency. Children, the elderly, people with disabilities, people with service animals, and others with special needs may require consideration in emergency plans. The US National Fire Protection Association has an excellent guide available online for dealing with people with disabilities.[1] While this resource is oriented toward fire situations, much of its content is immediately applicable to CBRN/ HAZMAT scenarios.

Foreign Attendee

Not everyone will be able to speak, read, or understand the prevailing local language. Some people may not be able to understand security or safety procedures. They may have difficulty understanding questions and instructions from event staff, both in routine situations and in the event of an emergency. Language issues are discussed in more detail in Chapter 13.

Neighbors/Onlookers/Passersby

No matter how big the event is, there will be some outer perimeter. Unless it is Burning Man in remote Nevada, there will likely be people living or working outside the outer perimeter. There may be onlookers, people who merely live or work adjacent to the event, people passing by in traffic, and all of the rest of modern life. A serious incident at a major event may be large enough to affect adjacent properties and people, so you should take some steps to understand who else is in the area. By similar logic, something that is close to a major event, such as an embassy, may even be a target in its own right. It should also be noted that neighbors can be a useful source of intelligence. The people who live or work in an area every day will be in a better position to notice things that are different. Further, they have a vested interest in bad things not happening in their neighborhood. The routine occupants of the area should be treated as an ally, not as a nuisance.

Opportunists and Criminals

Large events will almost certainly attract a variety of borderline or actual criminals. This could include unlicensed vendors, ticket scalpers, drug dealers, pickpockets, car thieves, and worse. The presence of such people could complicate emergency response in many ways. Is a drug dealer going to give up his supply in order to be decontaminated?

Aggrieved Locals

Not everyone who lives or works near a major event is happy at the prospect of the event. Major events can cause serious disruption to business or daily life. Noise, inconvenience, property damage, and myriad annoyances may arise from a major event and bother local businesses and/or residents. You must take this into account when planning. Some protective measures that you might wish to adopt may create a lot of animosity, thus creating problems. An example would be street closures. Some are no doubt necessary for various reasons. However, excessive street closures that inconvenience locals might provoke petty acts of revenge or pressure on local government that might make your job harder rather than easier.

Demonstrators/Protestors

Some people and some organizations will use the visibility and media presence at a major event to magnify whatever message they might wish to send. Demonstrators may range from peaceful citizens with legitimate and relevant grievances all the way to insane ranting extremists. Tactics may range from the polite to the completely disruptive. It is important to note that it is quite possible to have demonstrators turn up for reasons totally unrelated to the event itself, due to the amount of free publicity that they would get. It is not impossible that demonstrations could be used to distract police and security staff or that a demonstration might be useful in that way as a target of opportunity by terrorists. Finally, some demonstrators might be a target for an incident in their own right.

Press/Media

Every type of media outlet can be expected to cover major events. The very nature of major events is such that we can expect media coverage to occur. This coverage can be of every type and every possible quantity, ranging from a single reporter tapping onto a smartphone all the way up to hundreds of journalists from radio, print, television, and online media outlets. The immediate operational ramification of this is that there will be instantaneous visibility of your event, worldwide, that you cannot control.

Performer/Entertainer/Sportspersons

Many events are of an entertainment or sporting nature. There will be people there whose role is to provide the actual content of the event. This may range from a single performer to vast casts of hundreds of performers or athletes.

VIP/Celebrity Attendee

Major events will have prominent people attending, whether they are political figures, religious figures, or celebrities. Depending on the event, they may be the center of attention and integral to the event, such as at a presidential inauguration or a royal wedding. In other situations, such people may merely be a special class of attendee at the event, such as a politician or Hollywood star attending a concert or sporting event.

Fans/Paparazzi

Some celebrities will attract fans, some of whom are obsessive. There is an entire subculture that stalks some celebrities. Indeed, some celebrities may have problems with fixated individuals who stalk them. This could, in theory, cause problems at a major event:

Venue Workers (Permanent)

The event venues themselves will have their own employees, who work there. They will be necessary for the event, but they also present their own issues. Workers might take umbrage at having to go through additional security screening in their habitual workplace or even take measures to obstruct security measures if such measures are not coordinated well. For example, if the workers in a kitchen routinely leave a particular door open, they may continue to leave it open even if, for the purposes of a safe and secure event, it needs to be closed.

Volunteers

Many types of events rely upon volunteer staff to assist in various functions. These can pose different issues than paid staff. It can, in fact, be more difficult to direct volunteers who have got used to a particular way of working. It is considered an axiom that it is difficult to fire a volunteer. Given the nature of some events, you might get volunteers turning up who are not needed. Advanced planning can alleviate most of these issues.

Venue Workers (Temporary)

Major events often require additional staff such as additional cleaners, security guards, or concession vendors. It is not unheard of for terrorist groups to try to use temporary employment to gain access to venues or to impersonate a worker. Because temporary staff are likely to be unfamiliar to the permanent venue staff, this poses some vulnerability. Temporary employment can be used to gain credentials and keys, for example. It is not unknown to have venue staff who primarily speak a different language. For example, cleaning and grounds staff who speak primarily Spanish.

Tradesmen and Deliveries

Goods need to come to event venues. Waste needs to be removed. People arrive to fix broken items or conduct necessary maintenance. All of the routine business of life continues through special events. As with other categories, people intent on committing violent acts could use cover of routine visits by delivery drivers or similar service staff to enter an event site. Or, at a minimum, unexpected arrivals can cause difficulties at a major event. At one presidential visit, I was confronted by the unexpected arrival of a contractor to service the vending machines. Everyone had forgotten about his visits, and he was used to unaccompanied access to the site.

The Amateur Helper

In the event of an incident, people may seek to help. Indeed, in many scenarios, the response will be aided by people helping each other out. However, untrained, unequipped, and/or unaffiliated responders pose a variety of issues. This is discussed in more detail in Chapter 11.

PUBLIC TRANSPORTATION AND TRANSPORT INFRASTRUCTURE

Many major public events will be attended by large numbers of the public, such as sporting events. While it is generally not seen as an obvious component of security or antiterrorism planning, transportation planning has an important role to play in security and safety plans. Major events are notorious for causing congestion of transportation routes and public transportation networks. Many transportation networks in major cities are at or near their full capacity on a normal day. Injecting tens or hundreds of thousands of visitors to a city will cause chaos and delays, even without the threat of terrorism. Of particular concern to the CBRN/HAZMAT planner is that transport chaos may create large accumulations of people at major choke points, such as public transport hubs and interchanges. Congestion of roads may make emergency response actions more difficult as well.

Such planning considerations are significant because congestion, delays, and diversions may serve to create an easy target for terrorists to attack. Transport hubs and interchanges are already significant targets for terrorism, but if they are more crowded than normal, this may increase their appeal to terrorists. Terrorists, who may seek to exploit such soft targets as well as hard targets, do not necessarily need to directly attack an event venue to gain the symbolic value for their deeds. Why go to great lengths to attack a crowd in a secured site? An unsecured, but equally crowded bus station may serve terrorist objectives. An attack on people at the airport in the same city as the Olympics may carry as much symbolic value as attacking a sporting event directly.

Keeping a city moving and keeping crowd sizes low are an unglamorous but necessary part of antiterrorism planning. One lesson from the COVID-19 pandemic is that many people can, in fact, work from home for periods of time. A tactic that could be used during a major public event, particularly one in or near a major business district, is to encourage working from home. Nobody has yet to my knowledge started an "antiterrorism through telecommuting" campaign, but if such a campaign reduced crowding by ten percent at transport hubs around an event, that would be significant.

EFFECTS OF WEATHER, PARTICULARLY IN URBAN ENVIRONMENTS

Your major event environment literally includes the environment—the air around you. CBRN and HAZMAT, particularly in gas, vapor, and aerosol forms, are highly dependent on weather and movement of air. In fact, the variability of weather and its major impact on chemical and biological warfare agents is one of the reasons why chemical and biological warfare agents did not become more prevalent in warfare. Knowing which way the air may be flowing is very important to understand how CBRN/HAZMAT incidents may unfold at your event. (Air flow inside buildings is discussed in Chapter 6.)

The study of weather is still largely a "macro" phenomenon, based on broad generalizations applicable over a large area. A responder should consider himself very lucky if a CBRN/HAZMAT incident occurs exactly at the spot on the map where there is a weather station. The nature of built-up environments makes prediction of wind direction much more complex and has made efforts to conduct dispersion modeling of CBRN/HAZMAT incidents more complicated.

Generally speaking, weather and air flow at a particular event is poorly understood on the ground. It rarely exceeds the level of folk wisdom and things like "well, the wind is usually from the west this time of year" or similar sayings. To make things worse, the actual movement of air at a particular spot in the urban environment can be counterintuitive. You can look up the weather data on your smart phone, but that may be taken from a weather station on a rooftop three miles away. It might say "wind SE 7mph," but I can almost guarantee you that it will be something different at your exact location. You can look up in the sky and see the clouds moving one way, yet a flag or banner at street level is flapping in a different direction.

The reason why this is the case is that urban environments have lots of features that create local differences and even microclimates. Modern cities, which are the location of most major events, provide for a large proliferation of microclimates. The varying topography of populated areas provides for ducting and channeling of winds, including the so-called urban wind canyon

FIGURE 2.1   The flow of air in urban environments is a complex phenomenon. *Source: Photo: NASA, public domain image.*

effect.[2] The bottom line is that you should take some effort to understand the movement of air at the hyper-local level. You could, for example, put some small windspeed and direction sensors around your venue.

MAJOR EVENTS SERVE AS A MULTIPLIER FOR TERRORIST ATTACKS

Securing major public events from terrorist threats is necessary because of the very nature of terrorism. Broadly defined, terrorism is criminal violence committed to advance some political, ideological, or religious objective. The nature of terrorism is such that small groups with small budgets can commit acts that are small (in overall warfare terms) yet of great symbolic value to their cause. However, small acts by a terrorist group have little or no value if nobody notices them. A bomb that goes off in a shed in the woods and kills nobody rates little coverage.

Major public events serve to make the terrorist's job easier. First, the terrorist does not have to do any target selection. The major event becomes a high profile by default. By holding a major event, society says "Look at this event! This occurrence is something which is of value to us." The event schedule is published, making target analysis easier. Many public events are, by definition, held in public spaces and venues, making it less difficult for an extremist or a group to conduct reconnaissance, particularly if the names of venues and locations are announced well in advance, as it often is.

Major events serve to concentrate targets in one location at a specified time and place. Of even greater importance, a major public event concentrates media and public attention. A terrorist attack at a major event will have hundreds if not thousands of people posting about it live on social media. Some events may have literally hundreds of media feeds for radio, television, and online media broadcasting in real time. The "audience" for a terrorist attack is already formed, and the media strategy is already in place. Such a concentration of targets and attention may be irresistible. Terrorist attacks unfold live on social media in this modern era. Major public events serve to concentrate the cameras and social media accounts at the right time and place for exploitation by terrorists.

The concentration of media, particularly live media, increases the attractiveness of an event to possible attackers in net terms. However, such a conclusion does not always lead to a conclusion that use of CBRN materials are made more likely than conventional means.

An argument can be made that most CBRN materials may be less appealing for immediate use. Many CBRN materials have delayed effects and lack a certain "Hollywood Drama" appeal for live news coverage, unlike cheaper, more visibly gory, and more commonplace methods like shooting, stabbing, and bombing. Because the effects of most CBRN materials take hours or even days to appear, and then often appear in ways that often aren't terribly photogenic, terrorists want to make a strong visible statement may feel more inclined to use the more visible methods.

Another aspect of major events is that queues and crowds in and around the event venues themselves may become soft targets appealing to a terrorist. Likewise, large collections and congregations of people outside of venues may be vulnerable to a HAZMAT incident. Stringent security measures at perimeters and entry points may have value and may indeed be necessary. But you should bear in mind that if you create a lengthy queue of people who are not well protected, you have managed to create an easier target for terrorists.

THE ORGANIZATIONAL AND BUREAUCRATIC ENVIRONMENT

CBRN and HAZMAT, while important to people reading this book, are well down on the list of planning concerns for the upper management. Safety and security, while certainly not ignored, is one of at least a dozen planning areas for the overall management of the event. Housing, transport, catering, architecture, water, electricity, media, sales of tickets, and many other areas of concern will be fighting for a place on the table at planning meetings. It is important to understand this fact. At the end of the day, political events will be about politics and sporting events will be about sports.

Few managers are likely to publicly state that security and safety are not important. However, this doesn't mean that security and safety will get a high priority compared with other factors. Historically, within many bureaucracies, the safety and security people are the ones who often say "no" to plans, often based on legitimate concerns. This has often given the safety and security planning professionals a reputation of being perennial naysayers. Some major events, such as many sporting events, are large sources of commercial revenue, and support functions, such as security, are "nonrevenue" lines of business that may take second place to activities that make money. Any security planner who has tried to shut down a revenue-producing business temporarily, even for very valid reasons, will have run afoul of this factor.

We can look at the London 2012 Olympics as an example of priorities. London 2012's Olympic Delivery Authority describes their work as falling into six "Priority Themes"[3]:

- Design and Accessibility
- Employment and Skills
- Equality and Inclusion
- Health, Safety, and Security
- Legacy
- Sustainability

Within this simple framework, we can deduce that CBRN/HAZMAT concerns are a subset of a subset of an overall event framework.

A significant obstacle for the CBRN/HAZMAT planner may not be terrorists or accidents. It might be one's own bureaucracy. The phrase "death by committee" is relevant. The planning structures for major events continue to grow geometrically. We can look to the US federal government as an interesting case study in how this works out in practice. The US government has an official planning structure for major events. Events that are of national significance can be designated as a "National Special Security Event" (NSSE) by the federal government. Let's look at how the US government suggests that we proceed.

The US government provides a clear example of this, in the form of planning guidance promulgated by the US Department of Justice in 2007 called *Planning and Managing Security for Major Special Events: Guidelines for Law Enforcement.*[4] Despite its age, this is still a highly relevant and thoughtful document. This document breaks down law enforcement and security for major events into 14 functional areas:

- Determining and Acquiring the Security Workforce
- Communications and Communication Technology
- Access Control: Screening and Physical Security
- Transportation/Traffic
- Intelligence
- Credentialing
- Administrative and Logistical Support
- Protecting Critical Infrastructure and Utilities
- Fire/EMS/Hospitals/Public Health
- WMD/HAZMAT Detection, Response, and Management
- Tactical Support and Crisis Management
- Public Information and Media Relations
- Training
- Planning for and Managing Demonstrations

This framework is comprehensive, but it can lead to a stovepipe mentality, segregating CBRN/HAZMAT issues into one of only 14 functional areas, even though the other 13 functional areas need to worry about CBRN/HAZMAT to some extent.

However, the US government being what it is has reinvented the wheel and issued a different document in 2013, more or less purporting to do the same thing as the 2007 document, without rescinding the older document. *Managing Large-Scale Security Events: A Planning Primer for Local Law Enforcement Agencies*[5] also by the Department of Justice, lays out 18 "Core Operational Areas" for structuring the planning for major events:

- Administrative and Logistics Support
- Command and Control
- Credentialing
- Crowd Management
- Dignitary/VIP Protection
- Financial/Grant Management

- Fire/Emergency Medical Services/Hospitals/ Public Health
- Intelligence/Counterterrorism/Counter surveillance
- Interagency Communications and Technology
- Legal Affairs
- Nonevent Patrol
- Prisoner Processing
- Protecting Critical Infrastructure and Utilities
- Public Information and Media Relations
- Screening and Physical Security
- Tactical Support and Explosive Device Response/Hazardous Materials
- Training
- Transportation and Traffic Management

This list overlaps the 2007 list considerably. But it also relegates CBRN/HAZMAT issues into a subset, lumped in with bomb squads and SWAT teams. Both are useful lists, however, as an overall guide.

A Case Study: 2004 Republican National Convention

It is useful to look at historical examples of major event security planning. One that I have extensive familiarity with was the 2004 Republican National Convention. It was held in New York City, at the Madison Square Garden arena. Security for this event was an enormous undertaking, colored by the 9/11 attacks and overshadowed by wars in Iraq and Afghanistan, and complicated by the complex nature of the city itself. Rather early in the planning for this event, an executive steering committee was established with the following members:

- Commissioner, New York City Police Department
- Commissioner, New York City Fire Department
- Commissioner, New York City Office of Emergency Management
- Director, New York State Office of Homeland Security
- U.S. Attorney, Southern District of New York
- Assistant Director and Special Agent in Charge, FBI New York Field Office
- Regional Director, FEMA Region 2
- Special Agent in Charge, US Secret Service New York Field Office
- Port Authority of New York and New Jersey
- Commander, Coast Guard Activities New York

This is certainly one group that would have come to the table with a large number of institutional and personal rivalries. It is inconceivable that this steering group did not have serious arguments. In a later chapter, when I discuss operations centers, it will be apparent that some interesting things came out of this management structure.

You can look at basically any example of a major event planning and coordinating structure and see that things sprawl a bit. This is by no means an exclusively American phenomenon. The 2004 Athens Olympics involved approximately 70 agencies and organizations.[6] A simple list of agencies involved in the security arrangements for the London 2012 Olympics exceeds 100 organizations.[7] Organizational diagrams for the 2012 Olympics are too perverse to adequately describe and were often obsolete even before the documents were disseminated.

The point here is not whether this is right or wrong. This is the environment in which you will need to advocate for CBRN/HAZMAT preparedness. Clearly, an argument can be made that there are examples that represent a worst-case scenario of modern bureaucracy running amok. There are also perfectly valid arguments that large events require delegation, subdivision of tasks, and manageable span of control for middle managers. Both sides of the argument have valid points. I am writing this to describe the sea in which we have to swim. The acronyms and agencies will change from country-to-country, but the point is that there will be a large bureaucracy, both formal and ad-hoc, in which the CBRN/HAZMAT planning and response function must exist.

As you can see, in the several examples we have examined, CBRN/HAZMAT is well down the list. We can learn three things from these examples. First, it is highly unlikely that CBRN/HAZMAT considerations will be central or foremost in major events. Second, safety and security at a large event is a broad subject with many different concerns, all of which are valid and will draw on resources. Third, the various lists of functional areas are actually inter-related. All of the other functional areas actually have both direct and indirect bearing on how well we can handle CBRN incidents. For example, can we really have CBRN/HAZMAT planning without health and medical plans?

Many of us in CBRN and HAZMAT take it as a given that the threat is real and growing and that CBRN/HAZMAT is obviously a prime concern for major events like Olympic games. But the main body of the security profession does not always see it that way. A 252-page major study of Olympic security issues published in 2011[8] did not mention CBRN/HAZMAT even once. Often, our work is cut out for us.

The reason why I have written all of this is to show that the CBRN/HAZMAT planner will be swimming in a sea of other fish. Some of these fish are large and aggressive, with institutional objectives and agendas. The appropriate emergency plans will not get developed if you do not learn to swim in this sea.

Lesson Learned: Not Every Major Event is an Officially Designated Event

Most of this book is about large public events that are usually known about well ahead of time and planned for significantly in advance. However, it is important to remember that there are a number of situations that effectively constitute a "major public event" without much, if any, advanced notice or planning. It is easier to explain this by example.

The death of Princess Diana in 1997 saw an outpouring of public grief in the UK, as well as other countries. Thousands of people gathered around Buckingham Palace and other venues associated with the royal family. Another example is large demonstrations, such as "Stop the war," "Black Lives Matter," and "Extinction Rebellion," events. Such events have brought thousands of people onto city streets and into public venues. Not only that, disruptions of traffic and access, both planned and unplanned, will cause congestion of people and/or vehicles.

The large assemblages of people, both in the demonstrations and in the inevitable buildup of congestion around such demonstrations, can become a target. Individuals or groups with an opposing agenda to the demonstration could easily make such an event their target. Or an opportunistic group could simply exploit the chaos and disorder to commit an attack for other reasons. In some cities, police activity to confront civil disturbances may involve use of riot control agents. Historically, such incidents have sometimes got out of hand. For all of these reasons, we must consider large gatherings such as these to be major public events just like sporting events or other traditional "major events."

Of course, such situations inevitably mean that planning and preparedness will be a bit different. There will be variables that you cannot control and you will likely be operating with less information. You will, in effect, be planning for an event that isn't "yours"—it is either spontaneous (and therefore unplanned) or planned without your input. If you are major municipality, you should consider having a standing plan that you can put into place on short notice. It won't be ideal or as detailed as a specific plan for a specific event, but even a generic document is better than making it up on the day. This I know, because I did a lot of making it up on the day.

REFERENCES

1. National Fire Protection Association (2016). *Emergency Evacuation Planning Guide for People with Disabilities*. National Fire Protection Association https://www.nfpa.org/-/media/Files/Public-Education/By-topic/Disabilities/EvacuationGuidePDF.ashx.
2. Gayev, Y. (2004). *Flow and Transport Processes with Complex Obstructions: Applications to Cities, Vegetative Canopies, and Industry*, 15. Dordrecht (NL): Springer.
3. Olympic Delivery Authority (UK) (2011). *Learning Legacy Lessons Learned from the London 2012 Games Construction Project*. London: Olympic Delivery Authority.
4. United States Department of Justice (2007). *Planning and Managing Security for Major Special Events: Guidelines for Law Enforcement*. Washington (DC): U.S. Dept. of Justice, Office of Community Oriented Policing Services.
5. United States Department of Justice (2013). *Managing Large-Scale Security Events: A Planning Primer for Local Law Enforcement Agencies*. Department of Justice.
6. Brig Gen. Ioannis Galatas (Army of Greece ret.) Interview. Conducted by Dan Kaszeta, 8 November 2011.
7. Richards, A., Russey, P., and Silke, A. (2011). *Terrorism and the Olympics*, 203–204. London: Routledge.
8. Ibid

CHAPTER 3

Social, Behavioral, and Psychological Issues

Because we are dealing with "public events," any CBRN/HAZMAT incident is going to have some effect on the "public." How people behave, react, and think will have both direct and indirect impacts on planning and response. In both the planning and response phases of an incident, outcomes will be much better if knowledge about human behavior is adequately incorporated. The material in this chapter should influence all of the subjects covered in later chapters.

In the past, the social and behavioral aspects of CBRN/HAZMAT response have not been adequately considered. Barely a year had passed since the publication of the first edition of this book when I attended a presentation by Dr. Holly Carter, a psychologist who has done excellent work in this area. I started to realize that I had not adequately considered behavioral sciences. Since the point at which I wrote the first edition of this book, many papers and articles on this subject have been published, and a number of relevant specialists have started to examine the subject at some length. It is important for us to incorporate this body of knowledge into our practices. Readers are strongly advised to seek out papers and articles from Dr. Holly Carter, and Professors John Drury and Richard Amlôt, many of which are listed in the endnotes.

In earlier decades of my career, public behavior was largely disregarded in planning or training. Or if it was addressed at all, it was either "the public will do what we tell them to do because we are the authorities" or "the public will panic and do stupid things." Both are crass over-simplifications. Another important aspect of this discussion is that, for decades, serious writing about CBRN/HAZMAT response has been largely the province of technical specialists from the "hard sciences" or the emergency service disciplines. This has served the field poorly because there are psychologists, sociologists, anthropologists, and others in the social sciences, who can shed some much-needed light on the subject. The point of this chapter is to give some

CBRN and HAZMAT Incidents at Major Public Events: Planning and Response, Second Edition. Daniel J. Kaszeta.
© 2023 John Wiley & Sons, Inc. Published 2023 by John Wiley & Sons, Inc.

Mass panic prerequisites

| A real threat | Everyone knows the threat is present | Presence of a crowd | Limited means of escape |

FIGURE 3.1 The prerequisites for "mass panic" (Author's own work).

FIGURE 3.2 Crowds can behave in various ways.
Source: Photo: US Dept. of Defense, public domain image.

background from the behavioral sciences that can assist with both the planning and response phases of an incident, which are detailed later in the book.

CBRN MATERIALS, FEAR, AND ANXIETY

One of the most important factors in understanding social and behavioral aspects of response at major public events is that most people experience uncertainty, fear, and anxiety when confronted by the prospect of CBRN materials or weapons.[1] The psychosocial aspects of possible CBRN use make them potentially lucrative weapons by terrorists, who can use human behavior as a force multiplier for the effects of otherwise lackluster weapons and devices.

Most people are unfamiliar with the relevant technical, operational, and medical aspects of CBRN materials. At best they know little, and at worst, they have developed inaccurate understandings based on films, television, and popular culture. Most people have little direct experience or relevant education to allow them to understand and evaluate risks involving CBRN materials. To make matters worse, the hazards posed by CBRN materials are often unseen. Some are totally beyond human senses. Because a hazard may be invisible or, worse, look innocuous (like a clear liquid), people may discount the risks involved. Underestimating risks is a reason why people die in emergency situations.

Invisible perils are unlike known, visible ones. For example, people have developed some sense of fear of a person with a knife or a gun. A policeman with a gun in his holster is not seen as a threat. But if a stranger is running down the street pointing a gun at people and yelling, most people have some sense of the hazards to our safety and health that this represents. But when confronted with something like anthrax spores or invisible vapors of a toxic chemical, most people simply do not have the facts or experience to understand the risk and know what actions to take to protect themselves. Lack of understanding can combine with anxiety and/or fear to reduce the likelihood that people know how to act in ways to protect themselves.[2]

An additional aspect is that, generally, most people have difficulty accurately perceiving relative risks. Studies of both general terrorism risk perception[3] and perception of various CBRN threats[4] indicate a wide range of perception of risk that often differs from objective rational calculations by experts. This is not to say that CBRN materials are not dangerous. They are, or there would be no point in my writing this book. Because people often do not have sufficient knowledge or experience to objectively understand how to behave in a CBRN incident, both in terms of reacting to the threat itself and reacting to the emergency services response, planners and responders need to account for this situation. Responders will need to assess human behavior and interact with people in a way that helps the situation.

ASSESSING GROUP BEHAVIOR

Groups of people can behave in a lot of different ways. When confronted by the uncertainty, fear, and/or anxiety produced by a CBRN/HAZMAT situation, people will react in various ways. These behaviors can be helpful or harmful, either to themselves or to the overall emergency management situation. Often, group or crowd behavior is discussed in terms of whether or not the crowd is in a state of "mass panic." Lumping mass behavior into binary categories of "panic" and "not panic" is not actually a distinction that is useful for this book. Professor John Drury, a British social psychologist at the University of Sussex, points out that, in emergency situations, there are several reasons why "panic" is an inaccurate word to use.[5]

In doing the research for this edition, I made the mistake of asking about panic. The word panic is troublesome. For the purposes of this book, I will use the word "panic" more carefully as we need to dig a bit deeper into the subject and simplifications do not help us much. It is also important, in the context of major events, to discern between individual behavior and group behavior. Dr. Drury explained several things to me about "panic."

First, panic implies irrational behavior. But sometimes there really is a rational need to act in ways that look to the outside observer as though they might be irrational or a result of "panic." Moving quickly out of fear is actually people acting in a reasonably way, given their perception of a threat. If a building is on fire, fleeing the fire is not irrational. If people simply do not have sufficient information on what to do, as may be the case in CBRN incidents, it is actually hard to judge behavior as irrational. In order to claim some behavior is irrational, there needs to be some kind of reference point as to what the rational behavior would be in a given scenario.[6] This is not always clear in novel incidents involving threats that are unknown (at least at the time of discovery) to the participants.

Second, the word "panic" is applied inconsistently. It can mean different things to different people and in different contexts and often carries a bad connotation. Someone who has "panicked" is often deemed to have done the wrong thing. Arguably, in some situations, while behavior may look like panic to an outside observer, the fear and anxiety induced by an incident may have saved more lives. I speculate that if more people fled in fear from the building during the dreadful Grenfell Tower fire in London in 2017, more people would have survived this incident. Sadly, people were given poor information in this circumstance.[7]

Third, a number of documented situations that provided all the "usual" conditions for mass panic (a real threat, the threat being known to all, presence of a crowd, and limited means of escape) did not cause widespread mass panic. Some incidents that were claimed to have caused "mass panic" come into question once they are put under serious scrutiny. In many disaster situations, people have acted in the exact opposite way than classic "mass panic" by giving aid to those around them. Panic in disaster situations is not commonplace, when the phenomenon is rigorously studied.[8]

Dr. Drury pointed me toward several aspects of behavior that can be observed, assessed, and measured.[9] Instead of "is the public panicking?" you should look at the following three factors:

Flight

Are the people actively fleeing from some real or perceived harm? People who are actively fleeing are likely to have had a fight-or-flight reaction and are fleeing from some set of circumstances. They may be reacting to actual threats or imagined threats, but their fear will be real. If someone shouts fire in a theater, causing people to flee, it is because people believe that there might be a fire. Sometimes you want people to take flight. Other times you do not. People already taking flight from imminent danger are not to be handled the same way as people who are not actually taking flight. The latter might be more receptive to communication efforts. With people actually taking flight from something, sometimes the best response may be to provide direction or deflection, not resistance. "Hey, go over that way" is more useful than "hey, you, stop!" once people have made the decision to flee.

Anxiety

Are people experiencing anxiety? There are many things that could cause anxiety to be felt throughout a group. Anxiety could be caused by many things. It is a natural human reaction that a threat to life or health, whether it is actual or perceived, can cause anxiety. CBRN/HAZMAT situations can certainly cause anxiety, even anxiety out of direct proportion to the actual threat.

It is important to understand that emergency response actions may also themselves cause anxiety. Decontamination, in particular, can entail physical discomfort and loss of dignity. Subjecting people to decontamination is certainly anxiety-producing. But other emergency response actions can also increase anxiety as well.[10] Orders to shelter-in-place, evacuate places of work or residence, and other such actions can all raise anxiety.

In addition, uncertainty and lack of information can lead to anxiety. Not understanding a situation will naturally increase anxiety among people. Calm and clear communications with information content, as opposed to platitudes, might reduce anxiety.

Individual Versus Collective Behavior

Is the group acting as one? Or is it a bunch of people acting independently as individuals? A crowd all acting independently isn't really much of a crowd. A group with a shared identity, even only a temporary one, can be treated and communicated with as a group. A disparate group of people acting quite differently from each other will be harder to communicate with.

WHAT BEHAVIOR DO YOU EXPECT OR WANT?

In an ideal situation, you do not want members of public anywhere near a CBRN/HAZMAT incident. However, this entire book is given over to situations and scenarios where the public are involved and may even be the targets of hostile use of such materials. Emergency responders will have to interact with the public while working to protect them and manage the overall situations. There are several types of behavior that emergency responders may want and/or expect from the public. In general, you will want the public to behave in several general ways, described below.

Not Injuring Themselves or Others

You do not want the public to behave in ways that cause injury to themselves, make existing injuries worse, or cause injuries to others. Some scenarios in this category are self-evident, such as competitive behavior involving pushing that harms other through physical traumas. Others are harmful in less obvious ways. If a hazard is undetectable to human senses, such as radiation, you do not want people going from an area of relative safety into an area of hazard.

Not Making the Problem Worse

People may unintentionally behave in ways that make the overall situation harder to manage rather than easier. This can manifest itself in numerous ways. People could re-enter a contaminated area to help a friend and wind up as a casualty. Well-meaning people could do things that destroy valuable forensic evidence. Uninformed attempts at first aid could do the wrong thing and increase the severity of injuries. There are two particular sub-categories of acute interest in CBRN/HAZMAT situations.

Contamination Limitation

You do not want a crowd of people behaving in ways that transfer hazardous substances unnecessarily. Contamination control is a large part of incident management. Keeping dangerous substances from spreading is a key motivation behind many emergency response measures. However, victims and bystanders may end up behaving in ways that spread contamination. People with chemicals or radioactive particles on their clothing may flee an incident scene, thus taking the hazards with them. Left to their own devices and judgment at an incident, people might inadvertently spread the hazard.

Responder Legitimacy

Compliance and cooperation will be far easier to obtain if people view responder actions as legitimate, such as being told to disrobe and decontaminate. If people believe that emergency responders are legitimately trying to help in a situation, it is easier to execute necessary emergency response tasks. Overly antagonistic or forceful actions or lack of clear communications may reduce the perceived legitimacy of the responders. In some communities, different emergency response organizations may come into a situation with preexisting legitimacy issues. In some places, relationships between the community and local police are more antagonistic than relationships with fire and ambulance personnel.

Not Interfering With Responder Personnel

One of the biggest complaints I have heard over the years from responders is that "people get in the way." This can be on purposeful interference in some situations. Most localities have some subset of the population where people have an antagonistic relationship with police. Some people could actively oppose emergency response actions. More benign or inadvertent interference is far more prevalent. People can be moving downstairs when the responders are going up, blocking traffic, bothering responders with questions, and engaging in myriad other activities that fall under "generally getting in the way." Emergency response plans need to think of ways in which people might behave and try to foresee situations where this behavior might interfere with response.

Compliance

This is of particular importance. In CBRN/HAZMAT situations, it is likely that response activities will need members of the public to do things. The public may need to evacuate their homes or workplaces. Conversely, they might need to "shelter in place" and do the exact opposite of an evacuation, while their friends half a mile away are evacuating. Mass decontamination processes may require people to remove their clothing in ways that sacrifice privacy and dignity. All of these will require people to comply with instructions. Willingly, cooperative compliance is far easier than coercion. Responders will want people to be compliant with instructions.

HOW DO WE APPLY THIS KNOWLEDGE?

Responders and incident commanders will want a compliant public that is not getting in the way and not making the problems at the incident worse. But how do we move toward such an end-state? To make progress in achieving such a state, we have to look at psychological and behavioral subjects to see why and how people behave the way that they do and understand what responders can do to encourage and enable members of the public to take recommended actions.

Use Design Features to Reduce Likelihood of Harm

This book is about major public events. In many cases, there will be opportunity to design venues from the bottom up. In other instances, there will be opportunity to retrofit or modify facilities. In such situations, physical features of a venue can be used to mitigate the factors that have been associated with incidence of panic. Some factors (presence of a crowd, presence of a real threat) are not easily mitigated. Other factors can be affected. Physical layout of venues can be modified to speed escape in emergency situations.[11] Exits can be reviewed and clearly marked. Electronic signage can be used in an emergency to pass on information or even just large arrows to escape route. Public announcement systems can be built into a venue to allow provision of information. Information can be provided to people before an incident (see below). All of these might serve to be enough of a mitigation to reduce the possibility of mass panic.

Communicate Well

Focus on communicating in a clear manner to the public. Dr. Carter points out that open communication about emergency responses (such as decontamination) that explain why certain measures are necessary is beneficial. Further, people may not always be able to comply, simply for lack of knowledge. Therefore, communication needs to include enough practical communication to allow people to reply.[12] Because responders may be wearing PPE and respiratory protection can interfere with the clarity of verbal communication, consider exactly how you are going to pass information on to the affected people. You may need to have special equipment merely to amplify verbal communication. See also Chapter 18, which discusses communication issues in some detail.

Focus on Health-Based Messages

Dr. Carter's work also points out in her article on crowd psychology and decontamination that communication efforts which are based on health outcomes are useful in achieving compliance. By pointing out the health benefits, not just to the people involved, but to their loved ones, compliance and cooperation can be achieved.[13] While Dr. Carter specifically applied this to mass decontamination scenarios, I believe it will be highly useful in other protective measures, such as evacuation, sheltering in place, and event cancellation.

Provide Information BEFORE an Incident

Part of the reasons for people acting in potentially unhelpful ways during a CBRN/HAZMAT accident is that the broader public generally do not know about CBRN/HAZMAT threats or what the appropriate response in the event of such an incident might be. Carter, Drury, and Amlôt come through once again in a 2020 article on this subject.[14] They talk about the benefit of pre-incident information and stress that the most value would be in a shorter period before an incident. This factor lends itself to application in the major event environment as opposed to general efforts to increase community resilience.

One example cited in that article was the "Report, Remove, Rinse" public information campaign by the National Health Service England to provide information to the public on how to deal with attacks involving acids or other corrosives, which were on the rise in 2017.[15] Another example is the REMOVE campaign, which aims to provide information to the public in advance of a hazardous materials incident.[16]

Even a small amount of knowledge retained by a small amount of a particular group or crowd might be useful in the event of an incident. Having some sort of public information campaign might be useful. A successful mass education campaign is not necessarily the only feasible objective. If some people in the affected group do things like rinse off obvious contamination with bottled water, remove obviously contaminated clothing, or say things like "Hey, that hazard is outside the building, we should not go out into the problem," then such behavior can be copied and spread. With the prevalence of mobile data and smartphones, having some easy-to-read materials rapidly put out on social media in the event of an incident might be helpful.

Give People a Job to Do

If people begin to feel that they are part of the response and not merely an object that needs help from others, this can reduce stress and anxiety. It will help the overall situation in several ways. First, you probably do not have enough labor in the early stages of an incident. Enlisting the help of members of the public for necessary but unsophisticated tasks may free up more highly trained responders for more sophisticated tasks. For example, "Hey, here's a pen and some paper. Would you mind writing down the name and a contact number for everyone in this queue?" and "Can you help me out here? Can you hand these bags out to everyone in the line and use this pen to write their name on the bag? Then people can put phones and wallets in the bags so they don't get lost" are useful tasks. Second, by giving people agency and a purpose that is not just fleeing danger or receiving aid, their anxiety might be reduced. Third, such efforts may make it easier for the whole group to receive instructions and cooperate with the responders. Volunteer management is discussed in detail in a later chapter.

Treat People Well

Being as nice as you can to people sounds like a platitude. People fleeing from harm should not be treated as a hostile element. Taking extra care with people who appear to be vulnerable, such as children, the disabled, and the elderly, is not just a good thing to do in general, it may also be specifically good in a crisis situation. Adversarial and coercive methods have a place in incident response, but this place is a fairly narrow one in the overall spectrum of possible response activities. The other part of this advice is that if you treat people badly, the whole situation gets worse. If you treat people like rioters and criminals, you can turn an orderly situation into one of public disorder, which will make everything worse. If responders turn up expecting a riot, the responders may actually cause one.

Do What You can to Preserve Comfort and Dignity in Decontamination

Specific to CBRN/HAZMAT scenarios, decontamination involves the possibility of discomfort, loss of privacy, and loss of dignity. Discomfort (or even injury) may result from water that is too cold or from misapplication of decontamination chemicals. Sacrifices in personal dignity and privacy will be produced by removal of clothing. These elements of decontamination cannot be totally eliminated, but they can be significantly mitigated. These cannot be mitigated if the first time you thought about decontamination is at the time of the incident and your only tools are two fire engines and some hoses. At the planning stages, think of ways that you can ensure that water is at a comfortable temperature and that privacy (at a minimum, segregation of genders) is afforded. If you show respect and empathy, management of the incident will be easier. Decontamination is discussed in depth in Chapter 17.

Lesson Learned: Consider the Behavioral Aspects of the Responders as Well

All of this discussion has been about how the "public" or "crowds" will behave. Much of this work assumes that emergency responders will behave exactly according to plan or to their training. In the past, however, this has not always been the case. There have been instances of behavior, both individual and collective, by emergency services that have acted to make a situation worse, rather than better. Largely, these have been acts by police agencies. The term "police riot" was coined originally in the aftermath of disorders in Chicago at the Democratic National Convention by one Daniel Walker, to refer to situations where the police or security services are acting in a disorderly manner.[17]

There's no example that I can think of so far in CBRN/HAZMAT incidents, but that is largely because such incidents involving large numbers of the public have been rare. Based on incidents like the Ferguson, Missouri civil disturbances (2014), the numerous "Black Lives Matter" demonstrations across the USA after the murder of George Floyd (2020), as well as the Hillsborough stadium disaster in the UK (1989), it is possible to envisage law enforcement response to a CBRN/HAZMAT situation that range from suboptimum all the way to disastrous. The bottom-line advice from me is, if it is not a riot, do not turn it into one.

The same factors identified in the "public" above are relevant to the emergency services. Responders could face fear and anxiety out of lack of knowledge about how to rationally approach a situation. Leadership, training, and exercises, as well as applying the lessons of this book, are the route to avoiding problems.

REFERENCES

1. Palmer, I. (2004). The psychological dimension of chemical, biological, radiological and nuclear (CBRN) terrorism. *BMJ Military Health* 150 (1): 3–9.
2. Carter, H. et al. (2018). Psychosocial and behavioural aspects of early incident response: outcomes from an international workshop. *Global Security: Health, Science and Policy* 3 (1): 28–36.
3. Caponecchia, C. (2012). Relative risk perception for terrorism: Implications for preparedness and risk communication. *Risk Analysis: An International Journal* 32 (9): 1524–1534.
4. Denman A.R., Parkinson S., Groves-Kirkby C.J. A comparative study of public perception of risks from a variety of radiation and societal risks. *Proceedings of the 11th International Congress of the International Radiation Protection Association, 23–28 May 2004,* Madrid, Spain. 2004.
5. Drury, J. (2018). The role of social identity processes in mass emergency behaviour: an integrative review. *European Review of Social Psychology* 29 (1): 38–81.
6. Carter, H. et al. (2015). Applying crowd psychology to develop recommendations for the management of mass decontamination. *Health Security* 13 (1): 45–53.
7. Halliday J. 'Stay Put' safety advice to come under scrutiny after grenfell tower fire. *The Guardian,* 2017.
8. Heide E.A. Common misconceptions about disasters: Panic, the disaster syndrome, and looting. *The first 72 hours: A community approach to disaster preparedness* 337 2004.
9. Drury J. Interview with the author, Conducted by Dan Kaszeta. 18 January 2021.
10. Carter, H. et al. (2012). Public experiences of mass casualty decontamination. *Biosecurity and Bioterrorism: Biodefense Strategy, Practice, and Science* 10 (3): 280–289.
11. Shiwakoti, N. and Majid, S. (2013). Enhancing the panic escape of crowd through architectural design. *Transportation Research Part C: Emerging Technologies* 37: 260–267.
12. Carter, H. et al. (2015). Applying crowd psychology to develop recommendations for the management of mass decontamination. *Health Security* 13 (1): 48.
13. Ibid, 50.

14. Carter, H., Drury, J., and Amlot, R. (2020). Recommendations for improving public engagement with pre-incident information materials for initial response to a chemical, biological, radiological or nuclear (CBRN) incident: a systematic review. *International Journal of Disaster Risk Reduction* 51: 101796.

15. https://www.england.nhs.uk/2017/08/new-help-for-acid-attack-victims-following-recent-rise-in-demand-for-nhs-help/, accessed 18 February 2022.

16. Carter, H.E., Gauntlett, L., and Amlôt, R. (2021). Public perceptions of the "remove, remove, remove" information campaign before and during a hazardous materials incident: a survey. *Health Security* 19 (1): 100–107.

17. Walker, D. (1968). *Rights in Conflict: The Violent Confrontation of Demonstrators and Police in the Parks and Streets of Chicago During the Week of the Democratic National Convention of 1968*, vol. 3852. New American Library.

PART II

Planning

Interagency Planning and Cooperation

The nature of major events is such that different groups, agencies, professions, and organizations are going to have to work together. For a variety of reasons, it is important for accomplishing the overall objectives that people and organizations find ways to plan, cooperate, and operate together in pursuit of a common overall good.

Human and organizational behavior being what it is, you will have friction, rivalries, and competition for resources as an inevitable part of the planning process. The events that occur in a fractious planning process, including conflicts and friction, cannot help leaving a stamp on the operational efforts during the actual event as well. The ability to craft good CBRN/HAZMAT response plans will require cooperation between many different people and organizations.

Each of the nearly 30 major events that I worked with in my time in Washington suffered from problems of interagency cooperation. There are many reasons for this. The often-lengthy planning periods associated with major events will provide both an opportunity to overcome parochial rivalries and misunderstanding but unfortunately also allow many opportunities for frictions to arise. Old grievances and rivalries can flare up in meetings. The existing bureaucracy, often byzantine in nature, in the safety and security sector in many countries is rendered more complicated by the establishment of additional structures before and during major events.

Interagency cooperation in the emergency services sector has been discussed and studied for a long time. Large events, whether they are large accidents, natural disasters, acts of terrorism, or protracted criminal investigations across boundaries have long required interagency cooperation. The subject of interagency cooperation is well studied, both in general terms and in the context of specific incidents. As one example among many, Major Tim Lannan (Canadian Defence Forces) tells us that his experience is that the nature of rigid government hierarchies is simply not a conducive environment for interagency coordination

CBRN and HAZMAT Incidents at Major Public Events: Planning and Response, Second Edition. Daniel J. Kaszeta.
© 2023 John Wiley & Sons, Inc. Published 2023 by John Wiley & Sons, Inc.

and has difficulty cooperating in response to common threats.[1] Major Lannan's observations certainly have applicability beyond Canada. Indeed, the Canadian bureaucracy looks lean and austere when compared with its US equivalent.

Major terrorist incidents have often resulted in protracted review, occasionally providing some interesting observations in retrospect. As one example of many, the National Institute of Justice (NIJ), a research, development, and evaluation arm of the US Department of Justice, has studied the 2005 London bombings.[2] The authors (K. Strom and J. Eyerman) credited the communication, leadership, cultural differences, legal differences, and structural difference.

Additional factors are important in the CBRN/HAZMAT planning and response sector. Official efforts at interagency cooperation are often undertaken at the middle and upper levels of institutions. Efforts by top management to make agencies and organizations work well together may not always filter down to the operational echelons. The CBRN/HAZMAT expert is usually located several layers down in the hierarchy. The boss of the expert's boss may work well with a counterpart from another agency, but the CBRN person may not be able to count on official top-level efforts to put the person into a good working relationship with CBRN/HAZMAT colleagues elsewhere. Conversely, I have seen the opposite occur. As a Secret Service CBRN specialist, I had very good working relationships with counterparts at the FBI while observing antipathy between the two agencies at middle management levels.

DIFFERENCES IN OPERATIONAL PERSPECTIVES

In the emergency services, and in particular in the CBRN/Hazmat operating environment, there are very large philosophical differences between services and disciplines. How the various response disciplines look at the threat and think about how they react to it differs greatly. My own career has been fortunate in that it has given me roles in multiple sectors and broad exposure to many different aspects of CBRN/Hazmat response, so I have witnessed these philosophical differences in action. It is important to understand these philosophical differences because they provide color and texture to institutional and professional rivalries. The interagency environment is influenced heavily by the cultures, philosophies, and biases built into these professions. The following section is a simplification and generalization of the main points of each of these response disciplines.

Firefighting Services

Firefighting services are often the first to respond to a CBRN/HAZMAT incident and provide a lot of capabilities. Decontamination in a civil setting is usually the responsibility of fire departments due to their abilities to handle large quantities of water. At the risk of great generalization and simplification, their perspectives could be summarized thus:

- Fire services are the traditional "owner" of HAZMAT problems in most places. This can lead to a belief that CBRN incidents are really just a type of HAZMAT incident. (In many ways, this is actually true.)
- Reactive posture: Firefighters don't patrol the streets looking for problems. They wait to be called to a problem.
- Good institutional ethos of protecting life and property.

FIGURE 4.1   Different emergency response disciplines and agencies will need to cooperate and work together.
Source: Photo: United States National Guard, public domain image.

- Works in units: Firefighters don't operate alone. They operate in crews, and larger operations often require multiple crews to work together.

- A good emphasis on containment and mitigation stems from traditional firefighting.

- Good operational culture of operating in PPE. This can sometimes lead to the belief that their PPE is superior or even the only correct PPE for CBRN/HAZMAT threats.

- Often a very good relationship with the community. Fire services tend to have a less adversarial relationship than the police.

- A risk reduction philosophy. Modern fire departments largely take a strict approach on health and safety, thus resulting in processes and procedures that might seem conservative in some emergency situations. But fire departments have liability issues and disability payouts and firefighters want to live to retirement. This isn't to say that there is no bravery or heroism. But lots of firefighters have died and the sector has adjusted to try to reduce or prevent deaths and disability.

Law Enforcement

Law enforcement is heavily focused on identifying and apprehending people who violate the law, with a secondary emphasis on prevention of such acts. Law enforcement is often either an early casualty or a late arrival to the CBRN/HAZMAT arena. Some aspects of the law enforcement perspective could be described as follows:

- Patrolling: Police spend effort roaming around their area of responsibility looking for problems to address. Firefighters and ambulance services do not do this.

- Police can use force: Police had means of coercion, up to and including lethal force, at their disposal.
- Adversarial: In some places, the relationship between the police and the public is strained to the point where some people reflexively view the police as the enemy or, at least, not to be cooperated with.
- Focus on arrests and convictions: Police often have performance metrics based on meting out punishments.
- Operating within a framework of what laws says: Police operations are in an environment more closely bounded by specific laws.
- Less experience with PPE: Police often are unfamiliar with CBRN/HAZMAT PPE.
- Many operations are done singly or in pairs: Many police departments routinely have the majority of their staff working alone or in pairs, not in larger units.

Military CBRN Personnel and Units

Military CBRN doctrine is largely about keeping the military forces active in combat so that offensive and defensive operations can continue. Military CBRN training thus combines survivability with maintaining mobility and firepower. Most often, a broader military support effort is headed by someone from a non-CBRN background, with the CBRN specialists several layers down in the military hierarchy. Within broader military contexts, the CBRN specialist usually faces similar concerns about making his or her issues heard. Specific to CBRN/HAZMAT, this operational perspective manifests itself as follows:

- Decontamination is often of the "good enough to stay in the fight" quality not the "good enough to live and work with for months" quality.
- Rigid hierarchy: Military units have hierarchies and rank, like the police and fire services, but are often more rigid in their adherence to it.
- False alarm philosophy: A higher level of false alarms is acceptable in a military operational environment than in a civil setting.
- Acceptable losses: Military CBRN doctrine and equipment are based on risk assessments that have loss percentages built into them for expedience. Some level of casualties is permissible in order to accomplish a mission. This level of acceptable risk is usually higher than in civil settings.
- PPE: Military CBRN PPE is designed to optimize endurance on the battlefield over actual protection levels.

Medical Perspectives at the Prehospital Level

Emergency medical service providers draw on different experiences than police or firefighters. In many places in the United States, EMS providers are, in fact, part of fire departments. Their operational philosophies are as follows:

- Turning victims into patients: There is a strong ethos of wanting to help people and keep ill or injured people from getting worse.

- Operation under remote medical direction and protocols: EMTs and paramedics operate under a variety of standing medical protocols. If you want to add CBRN/HAZMAT concerns, add them to the protocols. (Many US states have done so.)
- Transporting clean patients: EMS providers, when they do consider CBRN/HAZMAT situations, place a good emphasis on wanting to ensure that ambulances do not get contaminated.
- MCI: EMS services usually have mass casualty incident (MCI) plans, even if actual experience in implementing them varies.

Medical Perspectives/Hospitals and Definitive Care

Providers of definitive medical care, usually hospitals, have their own operational perspectives. Having done many dozens of hospital surveys for the US Secret Service, some of this can be summarized as follows:

- CBRN and HAZMAT are largely an afterthought: Many hospitals have little or no direct experience in this regard. Hospital decontamination is never ideal, but it is often neglected.
- The problem starts at the hospital door: Many hospitals have had blinders on when it comes to understanding the broader picture outside the hospital.
- Subject matter expertise: Larger hospitals have good access to in-house or on-call subject matter experts.
- Internal divisions: Many hospitals are riven by internal rivalries.
- Mass casualty plans: Most hospitals have major incident plans to address mass-casualty situations.

FIXING THE PROBLEMS OF POOR COOPERATION

I am not a group dynamics coach or an expert on organizational theory. Most readers will realize the importance of teamwork and cooperation. The various response disciplines see the CBRN/HAZMAT response problem from different viewpoints, as I discussed in the introduction. It is only through cooperation that we will be able to get something useful done in the planning period and minimize problems. Several major suggestions, such as incident command systems, operations centers, and common language, are so important that they are addressed individually in the next sections of this chapter.

Step Away from the Stereotypes

I learned that one could be very proud of your organization and proud of one's own background while still being able to admit that there are brave, smart, and motivated people in rival agencies. Police referring to firefighters as "hose monkeys," firefighters referring to the police as "blue canaries," and basically everyone referring to the FBI as "the feebs" is interesting banter but not helpful in a broader context. Focus on pragmatism. If you cannot walk away from the rivalry, at least declare a temporary ceasefire.

Liaison Positions

Getting a few people from one organization to work in another is a time-honored way for agencies to have a better working relationship. Rotation of personnel will help improve understanding of rival organizational cultures and language.

You Can't Wait on the Official Effort—Build Your Own Network

Sometimes obstacles exist to prevent formal cooperation between agencies. I often found that I had to make an independent effort to find and reach out to my CBRN/HAZMAT counterparts. Often, like me, they sat several layers down in the bureaucracy and I would have retired and died of old age waiting for my boss's boss to talk to their boss's boss about CBRN and HAZMAT. Building a network of the people who care about CBRN/HAZMAT will serve you well throughout the effort.

Social Networking

The internet brings more potential for communications and cooperation. Particularly in the last few years, the advent of online networks like Facebook and LinkedIn make it easier to network across organizational and bureaucratic boundaries. In the CBRN/HAZMAT world, such networking sites may be useful in identifying who has interest in the subject area in the various agencies involved in the planning effort, even if they are four layers down in the bureaucracy.

INCIDENT MANAGEMENT SYSTEMS

The single most useful tool to promote interagency cooperation is the development and implementation of an incident command system, sometimes referred to as an incident management system. If you are working with others in the emergency response sector to put CBRN/HAZMAT plans into place for a major event in a multiagency environment, then you are working in the environment for which incident command systems were designed.

Why?

Simply put, incident command systems are necessary to integrate different response assets into a combined response, working toward mutually agreed objectives. When many different people from police, fire, medical, civil protection, and military agencies all descend upon the scene of a major incident, there is ample scope for wasted resources, work at cross-purposes, and general chaos. All incident management systems are intended to counteract these risks.

Incident Command Systems in Use Around the World

Although I have worked hard to avoid a US-centric approach, many people will have to admit that incident command systems owe much to the American experience. The emergency services in America are very diffuse and decentralized. This has always meant that it did not take much for an incident to grow large enough to encompass the response efforts of many departments, particularly in suburban and rural areas where every county, town, and village may have their own police and fire departments.

Many cities, regions, and countries around the world have adopted some form of incident management scheme. Some examples include:

- Incident Command System (ICS) [United States and others]
- National Incident Management System (NIMS) [US—Incorporates ICS]
- Firescope [US / California—Historically important predecessor to ICS]
- Standardized Emergency Management System [California]
- Australasian Interservice Incident Management System [Australia]
- British Columbia Emergency Response Management System [British Columbia, Canada]
- Coordinated Incident Management System [New Zealand]
- Gold/Silver/Bronze [United Kingdom]

The basic ICS framework is finding traction in many parts of the world. For example, there has been widespread adoption of ICS in Asia, at least in firefighting operations. The World Health Organization (WHO) has been an advocate for ICS around the world. Generally, incident command systems originated in firefighting and civil protection sectors and have either adopted an ICS framework or grown to resemble ICS. The rare exception is the Gold/Silver/Bronze scheme in the United Kingdom, which originates in the police service. Globally, incident command frameworks are most entrenched in fire service and civil protection services. Medical organizations, the military, and police/security services may have less organizational culture of actually using such schemes, although the situation is slowly changing.

FIGURE 4.2   The Buncefield Fire in 2005 was a demonstration of the UK's Gold/Silver/Bronze incident command system.
Source: Used with permission from R. Mead.

Most of the potential readership of this book is in regions of the world that have adopted some sort of incident command scheme. I am neither advocating a specific system nor saying that you need to change the scheme in use with your organization. Rather, I am insisting that you have some system. More likely than not, you already have one in place. My advice is to actually use your system. My own personal experience is that a very good set of practices may exist in binders on shelves but not get translated into operational reality. Most incident management schemes are useful in routine situations as well; normal operational tempo should provide some scope for exercising these procedures. Incident command procedures that are behind a "break in case of emergency" glass screen are probably not going to get implemented properly.

What We Can Learn from Incident Management Schemes

The many schemes in use around the world consume yards of shelf-space. I suggest that any of the schemes listed above make for weeks really boring reading even if they are very useful fodder for planning efforts for major events. The core ICS documents represent the closest thing the world has to "best practices" for this discipline. We can abstract from them the useful core principles, which really must be put into place for useful multiagency responses to incidents. Broadly speaking, if you digest the hundreds of documents available, the universal principles of incident command systems are widely accepted.

Generic

A good incident management scheme is sufficiently open-ended that a wide variety of response resources can be funneled into a single incident under a single management structure. It cannot be police-only, fire-only, or military-only. Many incident command schemes are firefighting-centric in their conception or origin, but need to be broader in outlook. Getting non-fire assets to interact well with incident command systems has been cited as a potential obstacle. It is inconceivable that a CBRN/HAZMAT situation at a major event will only draw in one type of responder from a single agency.

Span of Control

One person can only be expected to control a limited number of subordinates. Incident command schemes set up temporary operating structures that ensure that a commander is not, for example, in charge of 20 subordinates. Five to seven is often considered the maximum useful span of control.

Common Language

Slang, confusing terminology, and jargon can confuse responders. Using a common lexicon improves incident command. This is also discussed below.

Unity of Command

Every person involved in a response should report to only one leader. People taking actions to prevent safety and health do not need to be hearing from several bosses at once. The concept of unity of command is fundamentally at odds with some of the newer "matrix" management structures that are gaining in number around the world.

Management by Objective

Response actions are directed toward specific objectives, which are prioritized by the incident command structure.

Flexible Organizational Structure

The temporary organization established for a specific incident can expand and contract as needed to reflect the requirements of the situation. Positions in the management structure that are required by the parameters of the incident are filled; positions that are not required are left vacant.

SOME CRITIQUES OF INCIDENT MANAGEMENT SCHEMES

No incident command scheme is perfect. Some criticisms have emerged over the last few decades that highlight some perceived shortcomings that should be addressed as one prepares for a major event.

My ICS vs Your ICS

The unique nature of major events sometimes means that agencies that have seldom or never worked together are forced by the major event into cooperation for the duration of the event. This can mean that rival incident command models meet, with resulting friction. Much of the differences are actually cosmetic. If you peel back the layers on most incident command schemes, you will find that the underlying principles are often the same.

Integrating Medical Providers

Integration of hospitals and doctors into incident command models can pose some obstacles. Hospitals may not need to play a role in the incident command structure for small incidents and accidents. Events with widespread health impact, such as many CBRN terrorism scenarios, will need to incorporate a wider swath of the healthcare system into the incident command system. There are two facets to this problem. One is that doctors sit on top of the medical profession. At a conference a decade ago, an Ivy League surgeon told me that he did not understand the concept of incident command because "I am accustomed to being in charge wherever I go" or words to that effect. The second is that, within their particular strata, doctors are assumed to be equal, thus allowing for significant debate among equals. The healthy debate between medical professionals has been cited as being contrary to the interests of unity of command.

Transfer of Command

Incident command structures are meant to be flexible. Their ability to expand and contract in size is often touted as a key advantage to ICS and related schemes. However, this means that individual roles in the management structure may have to be transferred from person to person as the structure changes. It has been noted that transferring command roles in an effective manner can be problematic. Motivated and well-trained professionals can become attached to their roles in dealing with emergencies. Getting people to relinquish command, often to someone from another agency, can be politically and psychologically damaging from some people's perspectives.

THE OPERATIONS CENTER

A very old and useful technique for managing complex operations is the operations center. Operations centers serve to pool information into one room and serve as a useful tool to achieve "unity of command." They have many names: command posts, command centers, joint operations centers, fusion centers, multiagency centers, and coordinating centers merely to name a few. However, they are all broadly similar in conception. The basic framework is to place representatives from the various departments, agencies, and organizations involved in an operation all in one room so that information can be shared effectively and complex operations can be managed. Because there are so many overlapping concepts and definitions in use around the world, I use the terms "operation center" and "command center" relatively interchangeably.

In principle, operations centers are a useful tool. A well-staffed and well-equipped operations center, supported by good information systems and communications will have many benefits to an operation. Some examples include:

- Situational awareness: Managers can step into one room and get an overall picture of the situation as it is evolving.
- Expertise: An operations center can combine many different types of expertise in one room.
- Coordination of complex responses: Complicated problems may require complicated solutions beyond the scope of a single agency.
- Information sharing: Information from a wide variety of resources can be quickly and easily shared among many organizations.

Problems and Issues Posed by Operations Centers

I have no doubt whatsoever that a major event requires a good operations center. But my own experience has shown me that operations centers are a two-edged sword. I have worked on both ends of the issue. I have sat as a watch-stander in several different operations centers in various civilian and military capacities and I have been a responder on the street sending information to operations centers. While I think that operations centers are an absolute necessity, experience has shown that they can cause problems as well as solve them. Planners need to be aware of the following problems and issues associated with operations centers, so that concrete actions can be taken to prevent them:

Inadequate Resources

Operations centers require a lot of resources. Many operations centers, particularly ad hoc ones set up for a short duration or for a single major event, may not have the resources devoted to them that their permanently established cousins might have.

FIGURE 4.3   Emergency Operations Centers are useful management tools.

Size

Operations centers can be big. The larger the major event, the more departments and agencies there are supporting it. This means that many people may be present. I have been in operations centers with over 100 workstations. Large operations centers tend to develop group dynamics of their own. Beyond a certain size, information dissemination within an operations center becomes a problem. Putting twelve people into a room to coordinate a major operation is useful. I am not sure if putting eighty into a room is particularly helpful.

Interaction with the Incident Management Scheme

If not managed properly, an operations center assumes the role of an incident commander and in doing so may negate many of the positive aspects of incident management frameworks. Unity of command, simple span of control, and other advantageous aspects of incident command systems that were discussed above could be easily rendered useless if an operations center steps into the role of de facto incident commander.

Training

Operations centers only really work well if the staff is well trained. All too often, the members of a particular operations center for a major event only come together on the day of the event. Possibly, if we are lucky, they may come together on the day prior to the event. This does not allow for any useful training. It is unrealistic to expect such an operations center staff to work well together. Plenty of good guidance exists for training emergency operations centers.

Data Overload

Information overload is an identified issue with operations centers. When the concept of an operations center was developed, information technology available was a single telephone line to each position at the operations center, and perhaps a message center scribbling out written messages that could be passed out to the appropriate desk. Now, there is no practical upper limit to the amount of information that can be piped into an operations center. The ability to move the information has increased geometrically, but the ability of a human to process the arriving information is not much better than it was 75 years ago.

At one point in Washington DC in the late 1990s, there was a veritable arms race to see how many computer and television screens could be mounted in a single operations center. The managers that write the specifications for operations centers want to see a room full of television screens and computer terminals. It is very easy to provide more information than an operations center can handle. I have witnessed operations center paralysis due to information overload. This phenomenon was clearly in play on 11 September 2001.

Disconnect Between the Operations Center and Reality

Operations centers rely on a flow of information from the field to maintain their situational awareness. The collective awareness of the operations center staff is only as good as the information inputs available to them. Operations centers can have too much information, too little information, incorrect information, or contradictory information. In all of these situations, the operations center may end up having negative value.

Operations Centers Flatten Command Hierarchies

Traditional hierarchical command and control structures have developed in military and civilian organizations over centuries. A commander really only has an effective span of control over a single digit number of subordinates. While hierarchical command structures have both advantages and disadvantages, operations centers run the risk of cutting intermediate layers of command out. If it is conceived and executed poorly, a centralized operations center runs the risk of abnegating intermediate levels of command.

AD HOC OPERATIONS CENTERS

A large number of operations centers for major events are established on a temporary basis merely for the duration of the event in question. It has been my experience that these operations center often only "find their form" toward the end of the event. If it is a major event lasting only a day, finding your form after a day is pointless.

BEST PRACTICES - HOW TO MAKE OPERATIONS CENTERS WORK

While we have identified many potential troubles associated with operations centers, we need them. We need them to work with us and for us, not against us. While the CBRN/HAZMAT planner will probably not be in the correct position to dictate overall policy, here are some suggestions that can be made during the planning process.

Create One Center for Your Major Event

The multiplicity of command centers, seen above, serves no useful end. Have one operations center, not six or eight. Have a smaller one as a backup, if needed. Multiple operations centers cause "brain drain"—an agency may have to put 8–10 aside just to sit in operations centers rather than managing the incident more directly. This could cripple smaller specialty agencies or units.

Keep the Operations Center Away from the Problem

Locating the operations center in the middle of the target area is a bad idea. There is no need for it. Keep your nerve center well away from the major event itself. All of the value added by an operation center is negated if you have to evacuate it because of a CBRN/HAZMAT incident.

Create Cells

There is no doubt that a large major event will require a large operations center. Take a cue from the larger military operations centers and create cells off of the main body for various specialized elements. The creation of specific intelligence and CBRN/HAZMAT cells, both a common practice in military headquarters, is a good example of this practice.

Develop Operations Center Procedures

The operations center that functions well on a day-to-day basis has books full of protocols and procedures. Yet many temporary centers, such as those often established for major events, have ad hoc procedures done at the last minute or even let the participants themselves come up with something of their own. It may take a support contractor and some months to do it, but it is probably effort well spent to develop at least some rough outlines of procedures.

Make Operations Centers Work for the Incident Management Structure

An operations center should be an asset for incident commanders, not a replacement for them. An incident commander, working at the scene of an incident, should not be forced into conflict with an operations center 10 miles away. In the context of a major event, an operations center should not be a nerve center for a manager to assume control of an incident remotely. As multiple incidents may evolve, the operations center needs to maintain a more global perspective and be free to have its eyes on several incidents simultaneously.

Identify the Participants Early

In many of the events that I have supported, the actually agency representatives to the operations center are only identified at the eleventh hour. If this is the case, then it is difficult to conduct any useful training or exercises.

Train and Exercise

If you have identified at least some of the participants before the actual event, the operations center can conduct useful training and even work through some mock scenarios. If you are very lucky, one of these scenarios can even be a CBRN/HAZMAT situation.

Common Language

There is enormous potential for misunderstanding and conflict in the area of language. As previously identified, common language is a motivator behind the development of many incident command system models. By this I do not necessarily mean French versus English versus Spanish, although this can certainly be important at international events. Enormous potential for confusion exists because the same word or phrase can mean different things to different people. Firefighter, medics, police, and the military all have their lexicon of technical and operational terms. When planners and responders get together to support a major event, opportunity for chaos exists due to differences in terminology. The problem will be exacerbated when agencies and organizations without much history of working together are forced together either by plan or circumstance.

The lack of standardized terminology among response agencies was one of the reasons for the development of FIRESCOPE, the predecessor to the much-used incident command system (ICS) in the United States. I am sure that this is not much of a problem in places like New York, London, or Washington which, have had enough experience with major events so that everyone has at least a casual familiarity with everyone else's slang. But how about Gleneagles, Scotland, site of the 2005 G8 summit? Even simple terms like "operational" and "tactical" mean different things across the US–UK linguistic divide.

Parochial terminology, by which I mean slang or jargon confined to one community, is a large problem in complex operations. One of the most cited examples of terminology problems is the New York Police and New York Fire usage of the word "bus." Since time immemorial, "bus" has meant "ambulance" among police and firefighters in New York City. The phrase "send us a bus" may not mean "send us a bus." A search by a firefighter might be a quick sweep for dead bodies or victims and could take 90 seconds. A search by a law enforcement agency may require a court warrant and be a meticulous examination of a building for contraband or evidence. The internet is rife with examples of police, fire, and emergency medical slang. Abbreviations and acronyms make terminology problems worse. If anything, acronyms and abbreviations are even more parochial than regular words. Most acronyms are organization-specific and are even less likely than slang words to be recognized across organizational boundaries.

Lesson Learned: You Can Have too Many Operations Centers

Operations centers, command centers, fusion centers. . .whatever you call them they serve a useful purpose. But do we need ten of them for one event? It is illustrative to look back at the Republican National Convention in New York in 2004, with the US Secret Service as the lead agency. This event spawned a large number of operations centers. My notes from the time show me that at least 10 major operations centers were set up temporarily during the convention:

- Multi-Agency Coordination Center (In a hotel in mid-town Manhattan)
- Tactical Operation Center (Madison Square Garden, Manhattan)
- Joint Information Center (Brooklyn)
- NYC Emergency Operations Center (Brooklyn)
- Principal Federal Official Cell (Manhattan)

- US Department of Defense NORTHCOM ICP (Manhattan)
- FEMA RRIC (Fort Monmouth, NJ)
- Interagency Intelligence Fusion Center (Manhattan)
- US Secret Service ICP (Manhattan)
- Joint Terrorism Task Force (Newark, NJ)
- Coast Guard Maritime Security Incident Command Post (Staten Island, NY)

All of this was in addition to the normal operations centers run on a daily basis by many of the response entities, such as the fire and police operations centers. Not to mention a "security room" for the Secret Service at every single venue. It is easy to see what went on in this case. It would seem that each of major players on the steering committee felt that they had to have their own operations center. Surely, this defeats the purpose of "unity of command?" I am not sure that the 2004 example is one that should be copied by others, as it certainly does not have the appearance of efficiency.

REFERENCES

1. Lannan, T. (2004). Interagency coordination within the national security community: improving the response to terrorism. *Canadian Military Journal*, Autumn 5: 49–56.
2. Strom K., Eyerman J. Interagency coordination: a case study of the 2005 London train bombings. Federal Emergency Management Agency, National Emergency Training Center; 2008.

General Planning Considerations: Building Capability and Capacity

This chapter highlights various planning actions that must take place early in the process in order to provide value. Large events and long planning horizons will spawn planning efforts, meetings, and bureaucracy. The most successful CBRN/HAZMAT planner learns to "work with the grain" not against it. The earlier the CBRN/HAZMAT planner can get their concerns onto the agenda, the easier it will be to get the proper resources in place. As soon as possible, it will be imperative to identify the planning structures involved. Get representation on all the committees that you think might involve CBRN/HAZMAT. If you don't have enough people to sit in on the right committees, start identifying members of other organizations and entities that help. Do not be afraid to make allies and co-opt people who may have similar goals to look after the CBRN/HAZMAT field in your absence.

ESTABLISH YOUR PLANNING THRESHOLD

Early on, the most important thing that can and should be done in the early stages is to develop "a planning threshold." This concept is known under many different names in other fields of security planning, such as a "Design Threat Basis" or (confusingly) a "Design Basis Threat." Definitions vary, but the basic idea is to figure out a baseline of what kinds of problems you will aim to confront.

CBRN/HAZMAT planners should meet at an early stage and agree on the planning threshold. This planning threshold is the mixed bag of scenarios that serve as the cutoff point for advanced planning preparedness. Training to a specific threat basis provides better overall value than open-ended preparedness measures. You can plan, train, exercise, and procure equipment forever and not reach completion if you let your planning run wild. Having a useful goal to work toward is demonstrably better than open-ended response planning.

CBRN and HAZMAT Incidents at Major Public Events: Planning and Response, Second Edition. Daniel J. Kaszeta.
© 2023 John Wiley & Sons, Inc. Published 2023 by John Wiley & Sons, Inc.

How do we pick our cutoff point? The "worst case" school of thought has us start at the top, at the hard end of the problem, and work our way down. This is not very useful in my experience. Sometimes you end up wasting a lot of time that way. Another way to work out the "cut-off point" is to work from the bottom up. A credible way to do this is to figure out what your response capabilities and capacities are in a normal daily situation move on from there. But if you are reading this book, then you are probably trying to go above and beyond your normal daily operating posture. That means that you either should or want to do more than your current plans or resources permit.

Having a reasonable number of scenarios is a good way to start. If you have more planning scenarios than you have fingers and toes, this is information overload and does not provide useful benefit to the planning and preparedness effort. It is much more useful to have a handful of planning scenarios that serve to focus the effort and cover the spectrum of possibilities adequately. Various efforts have been made in the past to have national-level emergency planning scenarios, both in the USA[1] and in the European Union. Such documents can provide a basic background but are a poor substitute for specific scenarios tailored to your event.

Some scenarios can be guided by existing data. Many countries have useful statistics on existing hazards involving hazardous materials. In the absence of specific statistics for the areas around the major event in question, general figures about what chemicals are most usually involved in accidents and the likely mechanism of accident can form a basis for realistic estimates. The US Department of Transportation publishes this information routinely,[2] and it can be a useful basis for calculating realistic scenarios.

Any scenarios used for planning purposes have to be realistic and grounded in technical fact. If we have 5000 people in a building, is a figure of 5000 casualties a correct planning threshold? In nearly all cases, it is probably not. None of the mechanisms of dispersal for CBRN/HAZMAT threats are terribly efficient. Even a very large dispersal is not going to evenly spread contamination uniformly across a wide area. As planning guidance for incidents in open air, the US Army suggests that in most scenarios, a 5 : 1 ratio between uncontaminated and contaminated casualties can be expected.[3] I have included a sample "threat basis" in Appendix B. There is no need to be terribly elaborate, and a threat basis can be modified as needed. It is just that you need to have something on the paper to start from.

Once you have a cutoff point in your scenarios, you now have realistic objectives to work toward and a rational basis for planning, training, and executing. Having agreed-upon cutoff points allows us to estimate resource requirements. One obvious critique of this approach is that it admits defeat at a certain point and surrenders to the enemy. I think that this is not a really useful way to approach the subject. I think that it is realistic. Every plan ever made in human history has a breaking point at which it can no longer provide a useful basis for operations. No group of planners can imagine and plan for every possible event. No group of responders has infinite resources to plan for every occurrence. Any plan has a breaking point.

If done correctly, this process will give you a good cushion 20% over your cutoff point:

If you prepare and train well, you will have capacity over and above your "cutoff point." Planning toward a specific threat threshold does not mean that anything beyond that threshold is failure or surrender. Emergency responders will have training and resilience. A mass decontamination plan designed to accommodate 400 victims does not disintegrate if 401 victims turn up.

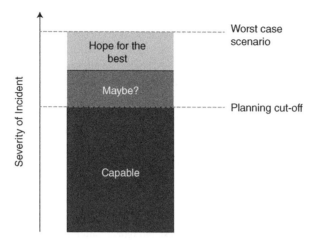

FIGURE 5.1 You will have capacity and capability beyond your cut-off point.

RESILIENCE

Resilience is another important general planning consideration. The CBRN/HAZMAT response effort must plan for resilience. By this, I mean that the planning effort needs to make measures to ensure the survivability and operational continuity of response efforts during a particularly serious and/or lengthy incident. Particularly with short major events, the emergency planning has a timeline directly tied to the timeline of the event. But what happens when an incident occurs and the response efforts are continuing long after the major event would have ended? One person can man a CBRN desk in an operations center for 12 or even 18 hours. But what happens if on hour 17 the big incident occurs and what was supposed to be a one-day major event drags on for days? The following sections highlight issues that need to be considered in advance so that plans can be made to remediate deficiencies.

How Robust is Your Table of Organization?

A CBRN or HAZMAT incident will be demanding. Are there enough people in your organization to handle the essential functions in the event of an incident? This is particularly important in small or medium agencies. I have witnessed planning sessions where a 10-person office was somehow expected to fill two operations center seats around the clock, provide a liaison person to another response element, provide two people to an on-scene incident command structure, and still be able to field a response team on demand. It would have worked for about 8 hours But what about an incident that goes on for 5 days? If you are not manned deep enough, you must either get more people or cut back on what is expected, lest there be a capability deficit on the day of the incident.

Continuity Planning and Succession

Offices, fire stations, operations centers, and laboratories are only real estate, and they can become incident scene in a CBRN/HAZMAT incident. It may become necessary to evacuate or relocate. The New York City emergency operations center, located at seven World Trade Center, discovered this hard fact. Such examples prove the necessity to have identified alternate operating locations for key functions.

CBRN/HAZMAT situations are dangerous. People can die. Supervisors and managers can die or become injured. People can become so exhausted that they are incapacitated. One of the advantages of a military-style chain of command is that it allows a natural line of succession. Many modern organizations use matrix structures that do not delineate an order of succession. Other organizations have a flat organization chart with 5 or 10 subordinates who are nominally equal. Does your plan have a specified succession to various command and supervisory positions? When the head of the unit is incapacitated due to death or injury, response efforts do not need to be distracted by a squabble for leadership.

Managing Continuous Operations

Long operations will require provision for rest and rehabilitation of responders. While this is often recognized in some sectors, it often appears to be less of a priority in others. Some people will work themselves to death if you let them. Others will start to be a safety hazard because they are too tired to operate safely. Obviously, some response tasks are more difficult than others, but every job, including just sitting in an operations center answering the telephone, will start to wear someone down if the assignment goes on long enough. It is important to have a process in place to allow for food, water, rest, and sleep for responders. Some kind of refuge may be necessary because it may not be practical to send people home. Have a rest and rehab plan as part of your emergency planning effort.

EXAMINING AND BUILDING CAPABILITY AND CAPACITY

Preparedness is really about developing capability and delivering capacity. Capability is the list of jobs and tasks that you can do well, or at least adequately. Capacity is how much of the capability you can deliver over what period of time. It is important to look at some general measures that are useful across the board to develop both capability and capacity. Once you have developed a workable set of planning scenarios and have a planning threshold to work against, you can set out to bridge the gap between your present capability and capacity, and your goal. One way to start is with a census of capability and capacity. Since it is easier to demonstrate what I mean than to describe it, Appendices D and E provide useful templates.

Survey the Existing Capability

Capability answers the question "what jobs can we do?" Planners need to be able to assess what kinds of activities existing response organizations can do. Calculating the "delta"—the difference between present capability and desired goal—is not always very easy. A comprehensive self-examination of what an agency, municipality, or government can actually accomplish is a necessary and often painful requirement for preparing for a major event. Sometimes it can be difficult because an organization may not be ready to admit to real or perceived shortcomings. A strategy based on hope and optimism is not a substitute for a rigorous inventory of what is actually possible with the tools and personnel presently on the books. The basic approach is to use the following steps:

Formulate a mission statement: An organization will have one or more missions in a given scenario. Use the planning scenarios and threat basis to derive mission statements. A mission statement should be simple and declarative. For example: "The HAZMAT team will be dispatched to the incident scene, conduct entries to identify any substances involved, assist with decontamination, and take steps to stop or mitigate the spread of hazardous substances."

Develop task lists: Any mission can be broken down into a number of supporting tasks. A task list is literally that a listing of the tasks that may be needed to accomplish a specific mission. Take a blank sheet of paper and start looking at the types of things that need to be done to perform that mission. A detailed planning effort may even make an effort to prioritize the tasks or categorize them into useful categories, such as "mandatory" or "not needed most of the time." Tasks may be collective (the organization does it) or individual (some person or persons do it). It is important to take a broad approach to such task lists. A common failing is to include only CBRN/HAZMAT tasks. It is important to include such things as basic medical tasks, logistics, and communications tasks.

Take Inventory: The only way you can really see if you can accomplish your tasks and achieve your mission is to take an honest inventory of what you have "in stock." By this, I mean much more than a physical count of equipment. Man of the tasks identified in the task list will require equipment, personnel, skills, training, and/or time in order to achieve. The ability to accomplish tasks is based on whether or not people have the requisite skills and knowledge. Effectively, this is a survey of whether your "human capital" (your responders) has the skills, knowledge, expertise, or training required to accomplish the tasks on the collective and individual lists.

The concept of an equipment inventory is easy. Do you have enough stuff? But there's also a "plans inventory." This is an effort whereby you survey the policies, plans, procedures, and other administrative documents and compare it to your mission and task lists. You might have a technician trained to detect radiation and there may be radiation detectors, but there may be no plan or policy or procedure to do so. This is not to say that you need a plan or procedure for everything, but it is extremely helpful to know where the gaps are. If there's no plan or procedure for an essential task, then there's every chance it may get botched in the execution.

Surveying Capacity

Capacity is a measure of how much can be done. By necessity, capacity is more quantitative than capability. For example, a medical provider can have the skills and equipment to treat patients exposed to toxic gases or vapors. This is a capability. But how many people can they treat? What throughput? What time does it take to set up? These are all questions of capacity. In practice, capability and capacity surveys overlap significantly, as there is little point in inventorying skills without also tallying the number of people on the list. The logical approach to follow is to examine each collective task on the task list and apply common sense metrics wherever possible.

Operational trade-offs and overlaps: An issue that is relevant in capacity surveys is the age-old "double-booking" problem. Your planning may indicate that you have a firefighter capable of providing decontamination, a paramedic who has been to a chemical casualty course, and a HAZMAT technician. But what if they are the same person? I once surveyed a regional airport in Wyoming. There were three duty firefighters and three duty police officers.

But two of the duty firefighters were the same guys as the police officers, and they would put on their turnout gear over their police uniform if needed. All too often, I have seen emergency plans evolve that did not take sufficient care to reconcile operational demands on assets.

Measuring throughput: Many of the emergency missions that are necessary in a CBRN/HAZMAT environment require measuring throughput. For example, decontamination capacity and medical transport capacity can only be effectively measured in terms of numbers of victims over a particular period of time. These types of figures can be hard to calculate in the abstract and may require close observation at drills and exercises in order to get an idea of true capacity.

THE SYNCHRONIZATION MATRIX AS A PLANNING TOOL

The best efforts to survey capability and capacity may still leave the planner with a lack of sufficient information to develop a good picture of the whole operational problem. Sometimes another approach is required. One tool that I found useful for finding gaps in resources or capabilities is the synchronization matrix, often called a "synch matrix" or "sync matrix." I have used this tool both for integrating CBRN concerns into wider operations and for planning specific CBRN operations. A sync matrix is a planning document for a specific operation or scenario, and it is often developed by a committee or planning staff as part of a planning process or tabletop exercise. Again, the sync matrix is easier to demonstrate than to describe, so a very simple one is included in Appendix E for illustration purposes. There are many ways to use a sync matrix, but for purposes of brevity, I will describe it at the most basic level. For readers who wish to delve into the details of the military sync matrix issue, there are several military publications in the public domain that describe the process.[4]

Sync matrices start out as discussions and scribbles on large sheets of paper and whiteboards. A sync matrix is a graphic depiction of all the activities that comprise a particular operation over a period of time. In a general military matrix, the vertical axis, starting on the left side of the paper, is a list of all subordinate units, operating elements, and support functions involved or potentially involved in the operation. The horizontal axis, drawn along the top of the chart from left to right, is chronological. The chronology can be linear (i.e. minutes, hours, days), event driven (before attack, during, after, etc.), or a combination of both. A very basic example of a synchronization matrix is included in the appendices.

Getting Started

The way to get started with a sync matrix is to use the tabletop exercise format. Participants from different agencies sit around a table and talk through a basic scenario, preferably one from your threat basis. A facilitator should lead the group through phases of the scenario. The scenario starts with everyone and everything at their normal operational posture. There will be a different line on the vertical axis for each different operating unit and for various important support functions. The police are in their normal assigned locations, firefighters at their station, doctors on duty at the hospital, and whatever else is relevant to the scenario. The first step is to record on the left-most column what this "beginning state" is. The point is to capture what the likely normal posture will be during the scenario. Resist the temptation to do anything different from what your normal plan would be, as this only serves to skew the results of the scenario.

Proceeding Through Episodes

You can think of every vertical column on the sync matrix as an episode in the scenario. The exercise proceeds either on a chronological basis or on an event-driven basis. Sometimes, it makes sense to proceed using a combination of the two. For example, the second column can be a surprise chemical attack, and successive columns can be 5-minute time increments after the attack. Use even smaller time increments if it is necessary. The facilitator will describe the events of the scenario and all of the participants will describe what their actions are at this stage. To do this the right way, there really should be a new column for any significant change from the previous column.

Each participant tells the facilitator what they are going to do at each particular point in the scenario. Actions are recorded in the boxes on the matrix worksheet. For example, a fire department engine company may arrive at "incident + 10 minutes," so that action or event will be annotated in the correct cell.

What the Matrix Shows Us

Incident response, particularly in CBRN/HAZMAT situations, is very interdisciplinary. Response operations require many different and often disparate response elements to work toward the same goal. The synchronization matrix shows us where, when, and how all of these different operating elements are positioned. For example, if a rescue team pulls some victims out of a building and needs decontamination (both for the victims and the responders), but the matrix shows that the decontamination team has not yet arrived or is not yet ready, then we have identified a problem.

Red Ink

Any problem or issue should be noted on the matrix worksheet. Many types of issues can be discovered through this process. The most obvious operational deficit encountered is in time management. Some response elements will be in place faster than others, and this will have an impact on the overall operation. Such synchronization issues are why this entire process was invented. However, other operational deficits are encountered as well. Shortfalls in capability and capacity are easily highlighted through this process.

The End State

The logical end state of the exercise is when you have accomplished the mission or you have come to a point in the exercise where it cannot continue because nobody can progress any farther and gain anything useful. Neither situation represents success or failure. The former is a success because the mission is accomplished and the players in the game have a large matrix that works as a road map for mission success. The "play grinds to a halt" situation is often a success as well, because there will be a lot of red ink that will give you a road map toward improvement.

Analyzing the Red Ink

Every cell on the sync matrix that you have circled in red ink is an "operational deficit". It represents a spot in time and place where some required capability or capacity is not present to accomplish the mission. Every single one of these operational deficits should be captured on a piece of paper as an action for future improvement.

ADDRESSING THE RED INK: OPERATIONAL DEFICITS

You've developed your planning scenarios and threat basis. You've defined a set of useful "cut-off" points to serve as a ceiling for your planning efforts. Organizational task lists have been drawn up and used as a basis for pages of individual tasks. A dozen members of middle management from various agencies drank gallons of coffee during a painful but productive sync matrix drill. A thorough census of capability and capacity has been undertaken. Now there is a long list of shortcomings. You have discovered "operational deficits" in personnel, plans, equipment, and many other areas. Is this a problem? No.

First, remember that nearly everyone who has ever done CBRN/HAZMAT planning for a major event has been in this situation. The glass always seems half empty, and few organizations have enough capacity or capability early in the process. The purpose of the exercise up to this point is not about proving that you have operational deficits. I'm willing to bet that you had them all along. The point is to identify them with sufficient time to do something about them.

What Not To Do

Upon encountering operational deficits, there are some things that you definitely SHOULD NOT do. Do not fall prey to "doom and gloom." In 20 years in this professional field, I have seen more "doom and gloom" and "we're all going to die" pessimism to make the most cheerful optimist into a sardonic cynic. It's not helpful. "Pencil-whipping" the problem (fixing it on paper, but not in reality) does not help, either. Moreover, you think twice before you do any of the following:

- "Administratively adjust" the results of capacity and capability surveys upward.
- Rely on "double-booking"—an asset can't be in two places at once.
- Dilute your planning scenarios to make them "more achievable."
- Reduce training requirements so that more people suddenly become "qualified."
- Compromise safety requirements without serious analysis.
- Routinely write waivers for yourself to exempt your agency from regulations.
- Assume a capability will arrive between now and the event.

None of these tactics are helpful. Some are shady at best, and will not look good an inquiry, should something go wrong. It is best to be honest about shortcomings and work to fix them or find a way to work around them.

Filling the Hole

There are several things that you can do. The obvious answer is to build capacity and capability, by making more plans, conducting training, sending people to school, and by buying more equipment. Above and beyond the obvious, however, the planner needs to get creative. Rarely, if ever, will you get the full amount of personnel, funding, or equipment that you think you will need to overcome all of your operational deficits. The planner, therefore, needs to find capabilities that may already exist (leverage) and make use of "force multipliers." Force multipliers are things that increase the effectiveness of your organization.

Be an advocate for incident command: The strongest tool for using existing capabilities as "leverage" in CBRN/HAZMAT readiness is to use incident command systems in all kinds of incidents on a day-to-day basis. If ICS (or a similar system) becomes used frequently for major conventional incidents, then response entities become used to folding their efforts into a larger, more complex response. This will have value for CBRN/HAZMAT response.

Support all-hazards thinking: A current term of art in emergency planning is the "all hazards" approach. Originally developed to get cold war era civil defense planning adapted to natural disaster work, "all hazards" thinking is used across the response disciplines to get everyone worried about a broad variety of hazards. By definition, an "all hazards" approach will include CBRN/HAZMAT.

Training and Exercises

The single most useful thing that can be done to improve readiness is training. The timeless military maxim "train as you fight, and you will fight as you train" holds true, in my own experience. In stressful situations, well-trained responders fall back on the skills and procedures that were inculcated during strenuous, realistic training. Poorly trained responders are left to figure it out for themselves as they have little applicable "core knowledge" to fall back upon. While the subject of what constitutes effective CBRN/HAZMAT training may yet become the subject of another book, here are a few concepts that I think are useful.

"Tabletop" training is not pointless: Tabletop exercises are basically discussions in which participants talk about what they would do in a particular scenario. I have often seen this sort of exercise rubbished as a time-waster and not a suitable substitute for "real" CBRN training, with lots of running around in suits. There is many a boss who wants to see people running around in suits and respirators, and such people often look down on the "tabletop" exercise as an inferior breed of training. I vehemently disagree, because I have observed, participated, and even led some tabletop exercises that provided great training value. They are cheaper and easier to execute than full-blown field exercises. Consequently, they can be done more often. Tabletop exercises also have the valuable feature of getting lots of thought process of the participants out into an open forum, which can be extremely useful. In a field exercise, the assessment team, a decontamination crew, and the incident commander may not know what each other is thinking. In a well-facilitated tabletop exercise, much of the thought process will be teased out and examined. Gaps in knowledge, training, or procedures can be easily identified without having to have dozens of observers or umpires, as you might need in a field exercise. The tabletop exercise is a necessary component of the synchronization matrix process described earlier in this chapter.

Use the planning scenarios: Part of the rationale for developing planning scenarios and a planning threshold are to give some target for training. Therefore, it makes sense to try to use such scenarios in a training and exercise program. Moreover, observing and evaluating training exercises that are based on the planning scenarios will allow the planner to gain insight on capability and capacity issues.

Use the task list: The point in generating collective and individual task lists is to point out to response organizations what jobs they need to train to do. Prioritize your task list and use it as the basis for training.

Maximize value from persons sent to external courses: All too often, when responders are sent to useful and interesting training courses outside their agency, there is little effort to extract the full benefit of such training. Establish a process whereby personnel who go to specialized courses are required to "back brief" the rest of the unit upon their return. In many situations, I have been to a training course but was unable to put the skills I learned into real use because the framework simply did not exist to apply the skills into practice.

Be wary of mission creep: There is a risk in going too far. Sometimes, procedures get built just because someone has the training for them. Therefore, the "get full value" approach really needs to apply to tasks that support the mission. Never forget that the mission drives the training. The training should not drive the mission.

Engage in realism but not to the point of punishment: Realism has a valuable point in training and exercises. However, CBRN/HAZMAT operations involve equipment, situations, and environments that place the trainee in uncomfortable and unpleasant circumstances. Operational personnel need to be able to do their jobs in unpleasant environments, and there is valid training value in training and exercises that simulate these conditions. If you need to be able to insert an IV line while wearing PPE, you are going to need to practice doing it while wearing PPE. There's no way around the issue.

However, there is a fine line between honing skills in a realistic environment and creating an environment where misery prevails over learning. I am an old Chemical Corps soldier myself, so I know a thing or two about miserable training experiences. But there is not any point in making people train until they get hurt or get so miserable that they just want the exercise to end. I have personally met "old school" officers from former Eastern-Bloc countries like Poland, Romania, and Russia who seemed to believe that CBRN training was not really done until the soldiers in PPE started to fall over unconscious. This attitude is not limited to unreconstructed Soviet types in the East and does not really help. If anything, it makes the preparedness task worse. It alienates responders by making them not want to participate in training. It can even make people afraid of their protective gear. I personally know people who cannot wear a protective mask any more because they were traumatized by harsh training methods and they associate it with passing out or vomiting.

Use large exercises sparingly: I can understand upper management's desire to have all the players involved in a large exercise. Such spectacles can be impressive and often look good for the media. However, such exercises are difficult to execute, cost a lot of money, and take a long time to prepare. They require large venues. While large "capstone" exercises do have a place, smaller exercises conducted more frequently may provide a greater benefit. Because smaller exercises have less of a logistical footprint, it is easier to get management to support them with resources.

DEVELOPING AN ASSESSMENT SCHEME

A very important force multiplier is the development of an incident assessment scheme. Such a scheme will help you make better use of your assets and reduce the amount of time wasted on nuisance situations.

Why Do We Need an Assessment Scheme?

Some type of assessment scheme is required at a major event to assist the overall incident command structure in making timely and wise allocations of personnel and equipment. It seems to the author than 90% of planning in the CBRN/HAZMAT field for major events gets devoted to 1% of the threat spectrum. A diligent planner will spend a lot of time arranging for response to large-scale mass-casualty events, and rightly so. We will be devoting much time and effort to discussions of scenarios that are rare but severe. These are demanding scenarios and advanced preparation is essential. However, we have been fortunate that very few large-scale incidents have ever occurred in the CBRN field. The history of large HAZMAT events in the vicinity of major events is sparse. The number of serious CBRN and HAZMAT incidents that have occurred under the umbrella of major event planning is essentially zero.

The nature of emergency response is that there are typically a lot of general responders such as police, firefighters, and ambulance crews. However, the more specialized the response asset, the scarcer it becomes. While there will be many police, there will be fewer SWAT/tactical teams and hostage negotiators. There may be many firefighters, but only a few hazardous materials teams and urban search and rescue teams. Etcetera. At a large event with many different venues over a large area, it is impractical to dispatch scarce specialized assets for every single scenario and situation that might require their presence. This is wasteful and expensive, because specialized assets may get bogged down at nuisance calls, thus delaying their employment for real emergencies.

Intermediate Scenarios

The security and safety effort at a major event is going to deal with many types of accidents and incidents. Most of these will be routine day-to-day drama. Many large-scale emergencies involving CBRN/HAZMAT substances will often be self-evident in their presentation, such as an explosion or a transportation accident. However, it is just as likely that a serious CBRN/HAZMAT problem may not be obvious in its early stages. The role of an assessment team is to identify the "day-to-day drama" that has potential to be a CBRN/HAZMAT incident.

My definition of "intermediate scenario" is a situation that requires a response asset that might, hypothetically, involve CBRN/HAZMAT threats. There are an enormous type and variety of "intermediate scenarios" that might require a CBRN/HAZMAT response. Here are some examples of intermediate scenarios:

- Unattended suspicious vehicles
- Suspicious packages
- Reports of strange odors
- Letters/parcels containing suspect powders or liquids
- Medical responses to victims with unusual signs/symptoms
- Traffic accidents involving commercial vehicles
- Railway and commercial maritime incidents
- Building fires, particularly where there may be potential for dispersal of dangerous substances.

- Strange sensor readings
- Possible hoaxes and false alarms

Building and Fielding an Assessment Team

One approach to the assessment mission is to field small assessment teams that can rapidly deploy to an incident or accident and provide a skilled assessment of the situation, based on professional expertise, back to the incident commander, command post, or other higher authority. Such an assessment team can serve as an advanced reconnaissance element or scout to determine the need for additional assets, resolve minor issues with their own expertise, and generally serve as a useful additional set of eyes and ears.

JHAT: The most mature example of an assessment team is the US Joint Hazard Assessment Team (JHAT) concept.[5] JHAT has been used in hundreds of medium and large events. I have served as a member of a JHAT, and those experiences are an important component of my motivation for writing this book. Several documents and references speak about the JHAT concept,[6] but none of them are really truly a full explanation of JHAT, as it is a concept that continues to evolve. Within this book, I will refer to an assessment team as a JHAT, although various users of this book will develop and deploy similar teams under many different names.

What is a JHAT? A US-style JHAT is interagency and interdisciplinary. Typically, it is small enough to fit into a vehicle, thus JHATs tend to be no larger than 5–6 people. The JHAT is meant to be representative of the major agencies involved in the CBRN/HAZMAT planning and response effort. For example, one JHAT I sat in was led by an assistant fire chief and contained a Secret Service CBRN specialist, a military CBRN specialist, an FBI WMD specialist, and a Washington DC police sergeant. The basic concept is to have a broad swath of experience in the van, so that when you turn up at a potential incident the JHAT has combined expertise to assess the situation.

JHAT vs. JHERT: It is common practice at US NSSEs to employ a "Joint Hazardous Explosive Response Team" (JHERT) composed of explosives technicians from various agencies. JHERTs and JHATs often work in parallel. The JHERT is intended to assess incidents that may require explosive ordnance disposal ("bomb squad") procedures, such as suspicious packages and abandoned vehicles. A JHERT fills a similar role for explosive responses that the JHAT does for CBRN/HAZMAT response. Many incidents will see the JHERT and JHAT both responding and assessing the situation. If the appropriate expertise exists, the JHERT can act as a "backup JHAT" and viceversa. JHERT and JHAT teams need to work well together and understand each other's mission and procedures.

A joint team? A common critique of the JHAT/JHERT concept is that there it creates confusion by having overlapping assessment teams in operation during an event. There is not any reason that a combined assessment team (a "joint" JHAT/JHERT) would not work, at least in theory. At a minimum, placing a skilled explosive technician in a JHAT would be a useful compromise.

BEST PRACTICES FOR AN ASSESSMENT TEAM

Size and Composition of the Team

Keep the team small enough that it keeps to a single vehicle. If you have to travel around in a minibus or a small convoy of vehicles, then the logistics of supporting the team, moving it around the city, and deploying it to incident sites becomes more complicated. A large team is a drain on resources, both intellectual and financial.

It is necessary to have the appropriate mix of expertise. A two-man team probably does not have sufficient depth of knowledge and is not big enough to do an emergency assessment entry into a hazardous area. The objective is to have expertise across the various disciplines. A minimal objective would be to have police, fire service, military, and medical expertise sitting in the same van. An all-military team, for example, is not a very useful assessment team in a civil setting.

Medical Expertise

Many of the JHATs I served on did not have much medical expertise. I believe that this is an operational shortcoming. There is great benefit to be achieved by having someone with an emergency medicine background on an assessment team. There is a strong medical dimension to most of the CBRN/HAZMAT threat spectrum. Having a JHAT member who can see an incident through the eyes of a medic is very useful in assessment efforts.

Consider Multiple Teams or Backup

Many major events will be too big or too long in duration to be able to be supported by a single assessment team. The frequency of events requiring assessment may vary as well. There might be situations where multiple events requiring assessment accumulate. Even a small major event may need multiple teams in order to operate

Correct Level of Member

You want team members who are not too junior and not too senior. The team members need to be sufficiently knowledgeable of CBRN/HAZMAT response to be able to make useful recommendations. They need to be of sufficient rank and experience to be able to enter the perimeter at an incident scene and gain useful information from the personnel already responding to an event. Very senior members of organizations may have been removed from street-level events for too long. Under many incident command schemes, a high-level official may get dragged into assume on-scene incident command if he outranks the local responders. This would make it difficult for the assessment team to extricate itself from minor incidents.

Equipping and Fielding the Team

Communications: Since the command, control, and communication situation at these events is often complex to the point of being obtuse, each member needs their own communication methods to reach back to their own chain of command. This sounds simple, but believe me when I say that having a Secret Service person, an FBI agent, and a DC police sergeant all in the same van on the streets of Washington DC with access to their own radio networks and getting emails on their smart phones was an unthinkable thought in the recent past.

Vehicle: In a very small event, a dismounted team might be able to do its job effectively. In all likelihood, a vehicle is required. A marked emergency services vehicle of some description is useful. Too large of a vehicle, such as a fire apparatus, begins to restrict mobility and parking, thus reducing the freedom of movement of the team. I was on a small assessment team, and there were times when we were far more worried about where to park the truck than we were about doing our job.

Response Equipment: The team should have basic PPE to allow for survivability. Assessment teams should also have detection and identification equipment. The exact type and variety of sensors will depend on the exact concept of operations that is put into place for the team. The assessment teams that I worked on in the past had some limited detection equipment such as 4-gas meters and hand-held radioisotope identifiers. I do not suggest that a JHAT is going to need to be equipped to perform "level A" entries like a full hazardous materials team, but the ability to wave a detector over a puddle of what looks like water to see if there are any volatile vapors, for example, is highly useful in accomplishing the assessment mission.

Operational Tactics of the Assessment Team

First Mission—Escalation: Assessment teams have two really important missions. The first is to escalate. The team can examine situations and scenarios that may not appear to be major and realize that there is a need for bigger and more sophisticated intervention than was originally thought.

Second Mission—De-Escalation: Historically, assessment teams have fulfilled an important role in de-escalation. Some incidents appear to be more than they really are. Some white powder incidents really are spilled flour on the floor of a bakery. People suffer from food poisoning without it really being a CBRN attack on the food supply. Sometimes radiation detectors alarm for reasons that have nothing to do with dirty bombs or stolen isotopes. The origins of the JHAT concept in the 1990s stem more from the need to "spool down" a big response than to "ratchet up" a small response.

Be an intermediate layer: Improvements in communications technology and the concentration of management into command centers can have interesting effects on the management of incidents. It is now much easier than it was thirty years ago for an official in a command post to micro-manage events in the field. The last thing a hazardous materials technician needs is an assistant fire chief telling him, through a radio earpiece, to do a test again so that he can see it better on the live video link. While I hope that this is an exaggeration, I know for a fact that situations very close to this have occurred. A command post asking every 3 minutes for test results is probably not helpful to the battalion chief or police captain who is trying to manage the incident scene. Sometimes an assessment team can be a useful buffer between the response and the management. The JHAT may be better positioned to get on the mobile phone and explain in some detail what is going on, thus freeing up the on-scene staff to get on with their mission. This helps out both the responders and the command post.

Allow for some level of self-dispatch: While a JHAT is meant to be dispatched by an operations center to be its forward eyes and ears, it is very helpful to let them roam about the operational area. Like any good forward reconnaissance element, a JHAT can sometimes discover things that are not evident to the higher headquarter. A JHAT can provide proactive assessment ahead of dispatch by a centralized operations center. By observing activity in and around the major event, an operationally shrewd JHAT may notice accidents and incidents before they get reported up to a centralized operations center. For example, a JHAT observing a parade may notice a liquid spill in an adjoining street long before it is noticed and reported through other channels. Since a well-composed and well-equipped JHAT is simultaneously listening in on operational communications traffic from several different agencies, a JHAT may "catch wind" of a developing incident minutes before an operations center receives official notification, processes the information, and decides to dispatch the team. These few minutes could be essential to saving lives. Here are some scenarios in which self-dispatch of a JHAT team may come about:

- Fire department radios report a structural fire at an industrial complex. The fire department representative knows from experience that some of the companies in the complex handle toxic industrial chemicals. JHAT decides to visit the incident scene to assess the risk.
- Police radio network reports the arrest of an individual with an amount of powder in a rucksack. While police agencies are working under the assumption that the powders are illegal narcotics, the JHAT decides, due to the proximity of the arrested person to major event activities, to investigate the nature of the unknown powder.
- While travelling on a road, the JHAT team notices an abandoned vehicle leaking an unknown liquid.

In each of these scenarios, it might take significant time for these incidents to be reported up to a central command center, for the command center to decide that they may be of CBRN/HAZMAT interest, and for the JHAT to get dispatched.

Don't let the assessment team delay a necessary response: We must not have a situation where an assessment team become the only official way to pass judgment on whether or not an incident is serious. Emergency planning must still account for the fact that some incidents will require an immediate intervention and immediate dispatch of specialized assets. An assessment team is intended to be a set of eyes and ears for senior management; it is not intended to be the only eyes and ears. Incident commanders need to have the flexibility to send in whatever response that they feel appropriate. Senior managers and command centers need plans and guidelines that give guidance on scenarios in which response assets should not be delayed or impeded pending an assessment from the JHAT. Remember, JHAT is most useful in the "middle ground" of intermediate scenarios. Do not let that get in the way of response to "The Big One."

LESSON LEARNED: DO NOT GET TOO HUNG UP ON WORST-CASE SCENARIOS

It's best to not get too hung up on "worst-case scenarios" and they are not very helpful. My own experience is that CBRN specialists are very good at drawing management's attention to doomsday scenarios. It actually doesn't take much imagination to cook up a scenario where a lot of people die and the responders are totally overwhelmed. I think that such practices are not really all that helpful for planning, though.

It is not useful if planning scenarios implant in the minds of upper management the "too hell with it, we're all going to die if CBRN happens" approach to planning. It is much more useful to develop a useful planning threshold and build your planning toward it. The "worst-case" scenarios make for interesting tabletop exercises and discussions, but the answer tends to always be the same: grab all the gear you have, get all your people, and head into the problem to do the best that you can. Any preparedness measures that we are likely to take have a breaking point.

REFERENCES

1. https://www.dhs.gov/xlibrary/assets/rma-strategic-national-risk-assessment-ppd8.pdf, accessed 9 February 2022.
2. https://www.phmsa.dot.gov/hazmat-program-management-data-and-statistics/data-operations/incident-statistics, accessed 13 October 2021.
3. United States Army (2000). *Army Soldier and Biological Chemical Command. Guidelines for Mass Casualty Decontamination During a Terrorist Chemical Agent Incident.*, 3. Edgewood (MD): US Government.
4. United States Army (1997). *Field Manual 101–5 Staff Organization and Operations.* Washington (DC): United States Government.
5. (a) Gordinier J. FBI informs Airmen of interactions, procedures. Shaw Air Force Base, 25 March 2008. https://www.shaw.af.mil/News/Article-Display/Article/213868/fbi-informs-airmen-of-interactions-procedures/;(b) Graham J. Jeffrey Trail hosts school shooter exercise. Orange County Register, 14 August 2013.
6. Hawley, C., Noll, G., and Hildebrand, M. (2009). The need for joint hazard assessment teams. *Fire Engineering* 162: 9.

Buildings and Venues

Major events happen somewhere. Some are indoors. Some are outdoors. Some are in structures that are open to the air, such as sports stadiums, which combine aspects of both indoor and outdoor events. The location where you hold an event, hereafter referred to as a "venue," will have ramifications on all aspects of prevention, preparedness, and response to CBRN and HAZMAT incidents. This chapter starts with some basic guidance and then focuses on three key areas—ventilation, physical security, and existing HAZMAT threats.

MAJOR TYPES OF VENUE

The types of venues that matter to the event planner are as wide as the built environment itself. Generally, the bulk of venues at major events fall into some broad categories, each of which has its own considerations.

Open Air Stadiums and Large Enclosed Arenas

There are a wide variety of coliseums, stadiums, racetracks, arenas, amphitheaters, and similar venues with overlapping categories that are used for major events. The primary categories for the purposes of this book's analysis are open air and enclosed. The key features of open-air stadiums are that they are exposed to the ambient environment and that they house a large number of people. A handful of stadiums exceed 100,000 in capacity, and stadiums with capacities in excess of 60,000 are commonplace. The National Center of Spectator Sports Safety and Security has a wide variety of resources on general safety and security subjects relevant to such venues.[1]

CBRN and HAZMAT Incidents at Major Public Events: Planning and Response, Second Edition. Daniel J. Kaszeta.
© 2023 John Wiley & Sons, Inc. Published 2023 by John Wiley & Sons, Inc.

FIGURE 6.1 Large stadiums are very common as event venues.
Source: Used with permission from Design Security Ltd.

Arenas are effectively enclosed stadiums. The key distinction is that enclosed arenas are likely to have a very large air handling systems to be able to provide adequate ventilation to the persons inside. As a hybrid category, there are stadiums with retractable roofs that effectively become enclosed arenas when the roof is closed.

Open-air stadiums concentrate a large number of people in a relatively small area, usually with controlled entrances and exits. Of all the potential venues discussed, the open-air stadium is most exposed and vulnerable to off-site releases of CBRN/HAZMAT; this is the so-called *external release*. By the same token, the vulnerability of open-air venues to *internal releases* is constrained by both the amount of material that can be brought through any security screening and the dilution of threat materials by open air. Even enclosed arenas have a very large volume of air that mitigates somewhat against an internal release because of the sheer volume of material needed to pose a widespread threat. A major vulnerability posed by large enclosed arenas is that the large volume of air required for routine ventilation requires a very large air-handling system, which can make it possible for materials in an external release scenario enter the venue. Because there is likely to be recirculation of air within such a venue, an internal release is more effective, in principle, than in a venue exposed to open air. Catering at a stadium provides a vulnerability in the form of threats disseminated in food and drink, as there are limited sources for attendees to buy food and drink.

Most stadiums and arenas will have numerous events at them throughout the year in order to commercially justify their existence. Many of these events will have less-stringent security measures and will be open to anyone with a ticket. This raises the prospect of terrorists conducting extensive reconnaissance and possibly even hiding threat materials at the site well in advance of the major event.

Searching and securing stadiums and arenas will take a large number of personnel and a long time. They are likely to have a fixed system of exits and entrances due to the fact that most events there are ticketed. As an advantage, large arenas and stadiums are very used to operating major events. The incremental changes needed for a "major event" might be a sharp change in operating posture in some types of venue but are likely to be "business as usual" at a professional sporting venue. Most stadiums will have an existing evacuation plan and significant dedicated on-site security personnel. This sort of venue is likely to have a reasonably well-experienced head of security. Many such event venues will have routine medical plans to support attendees at sporting events.

Convention Centers, Exhibition Centers, Auditoriums, and Related Venues

Convention centers and exhibition centers vary widely in size and description to such an extent that it is difficult to draw useful generalizations. Often, this category of property is designed to accommodate a wide variety of events. Many venues are modular and are capable of hosting a number of smaller events simultaneously. Occasionally, the conference

FIGURE 6.2 Conference centers and convention centers of every size and configuration host major events.

facilities at a major hotel may be large enough to qualify for this category as well as the hotel category. The large and flexible nature of convention center venues is one reason major events are drawn to them.

Conventions, trade shows, and similar events are typically the bread-and-butter of such venues. The management of such sites is very used to a wide variety of goods and personnel coming and going. Extensive marketing materials, describing the exact characteristics of the venue, are usually freely available. Temporary construction is often occurring on the event floors. My own personal experience is that, given enough advanced notice, it would not be difficult for an individual or small group to hide devices in a convention center under the guise of legitimate trade show or convention activity, as such a wide scope of activity is normally occurring at such a venue. As with other public venues, it would be relatively easy

It is important to remember that venues of this type are usually commercial enterprises seeking to maximize profit. They typically aim to have minimal "dead time" between events. Many readers will have attended trade shows and exhibitions and are familiar with scenes where one event is being built, while another event is being dismantled across the hallway. This phenomenon can adversely affect the amount of time necessary for security preparations and searches.

Hotels and Dormitories

Attendees at major events are often housed in hotels. At major sporting events, large housing complexes of dormitories may be built to accommodate visiting teams.

Hotels operate by renting rooms to people with money. The entire business model of a hotel relies on people coming and going every day. Security screening of people entering hotels is a rare phenomenon currently in place only in a few "high threat" locations around the world. Therefore, it must be assumed that terrorists can rent rooms at the hotel, either before or during an event. Hotels are vulnerable to internal releases from devices secreted in guest rooms or public area. Catering, both at hotels and dormitories, poses a potential threat vector. Hotels that house people attending an event at another venue that is better protected may provide a "soft target" for terrorist attack.

Security and safety efforts need to be coordinated with management, and in-house security staff and building management. Hotel security can range from nonexistent to world-class. The fragmented and diffuse nature of such properties (i.e. hundreds of rooms) can mean that a comprehensive search will be difficult to manage and very labor intensive. Dealing with the occupants may require special procedures. Many hotels have decentralized air-conditioning systems, with separate intakes and ventilation systems for each guest room. This can make it practically impossible to isolate inside air from outside air in a CBRN incident.

High-rise Buildings

Various places around the world define a "high-rise" building differently. An exact definition here is unimportant. Tall buildings do pose their own safety and security issues, but these issues are largely not specific to CBRN/HAZMAT scenarios. Fire safety issues are paramount, and significant resources are available at the National Fire Protection Association (NFPA) website.[2] Geoff Craighead's book, *High-rise Security and Fire Life Safety*, is an excellent broad all-hazard look at high rises, which does mention CBRN/HAZMAT considerations.[3] It is worth nothing that higher levels of high-rise buildings, depending on their ventilation systems, may provide a significant amount of protection against CBRN/HAZMAT threats at street level due to the heavier-than-air nature of most of the threat materials.

Universities

Universities have occasionally been the hosts of major events. Typically, universities have a wide variety of building and occupants. Many universities are distinct campuses that resemble small cities in their own right. Not only do they have students but also there is generally open access to the grounds, if not the individual buildings, of a university campus. Many universities have science and engineering departments and laboratories that could contain a wide variety of hazards, some of which could be exploited by terrorists or pose an accident hazard to a major event. Many universities operate in a decentralized manner, so that it may not even be possible to coordinate with one central official on any particular matter.

Religious Buildings

Religious buildings, such as cathedrals, churches, synagogues, mosques, and others, pose interesting issues. Religious sites are sometimes part of a major event, such as royal weddings or state funerals. Many religious institutions have relatively unencumbered public access much of the time or even around the clock. This provides the risk of a terrorist depositing a device or secreting material for later use.

Conducting a search in a religious building requires considerable advance planning and liaison. There is usually some sensitivity as to what can be searched and how, and the only way this will work without friction is through advanced planning and coordination. In some circumstances, for example, the use of search dogs may be considered a defilement of the premises, and other search techniques may be required. Many of the major events that occur in religious sites will be weddings and funerals. These have the additional complexity of being family, state, and/or religious affairs simultaneously. The emergency response disciplines must tread carefully in such circumstances.

Sites of Cultural or Political Significance

Major events can occur in and around monuments, palaces, museums, and other significant locations. Most such sites are likely to be significant targets for terrorist attack in their own right, regardless of the occurrence of a major event. The major event probably only serves to temporarily heighten an already existing risk. Many such venues will fall into the "hard target" category due to existing security measures. In theory, such facilities will be the least vulnerable part of a major event. However, the symbolic value of such targets can be high. In addition, the fragile nature of many such sites means that it can be difficult to consider any architectural or structural steps to "harden the target."

Outdoor Venues

Some major events, such as large music festivals or parades, by their very definition happen outdoors. In some cases, hundreds of thousands of people may be in attendance. Relevant examples would be the Glastonbury music festival or Washington DC's presidential inaugurations. By their very nature, it can be difficult to enforce a hard perimeter around a large outdoor event. Likewise, outdoor events are more vulnerable to things affected by weather.

RECONNAISSANCE: SITE SURVEYS AND WALKTHROUGHS

One of the most important tools in the planning stages leading up to a major event is reconnaissance. Just as in military operations, it is vitally important that leaders have an idea of the environment that they are going to operate in. Unlike military operations, where tactical commanders may have to rely on reconnaissance missions, battlefield surveillance, and intelligence collection in order to see where they are going to operate, the major event environment gives us a manageable theater of operations. For the most part, we know where the CBRN/HAZMAT incidents may take place because we know what buildings, areas, and venues are part of the scope of the major event. This gives the emergency planner a very clear advantage over general emergency planning because the sites and venues are available for advanced reconnaissance.

Identifying the Scope of the Event: Making a Master Site List

Early on in the planning process, planners should start a list of the event venues and associated sites and infrastructure that are part of the major event. Sometimes this is called a target list, but not everyone likes that phrase, so I refer to it as a "Master Site List." During a long planning process, this list will probably change in response to changes in the overall event planning. By necessity, the master site list will be a living document.

The master site list should be as comprehensive as possible. The nature of each event is different, so it is only possible to give general guidance on what should go on the list. As a general guideline, the site list should include such things as:

- Venues for scheduled events, such as stadiums, arenas, and conference centers.
- Sites of cultural, political, or historic note, targeted due to their symbolic value, that are in proximity to the major event.
- Support facilities hosting significant segments of event staff, such as press and media centers.
- Hotels/lodging for major segments of the event guests. For example, the housing areas for athletes or the hotels where delegations are staying.
- Major transportation interchanges, such as train stations that will bear the brunt of event attendance or airports that will receive official delegations.
- Critical infrastructure required to support events.
- Industrial and commercial sources of hazardous materials in the proximity to the major event.

A large major event could conceivably have hundreds of venues on its list. My notes from the 2004 convention in New York list about a hundred venues of one size or another. The master site list can be used for purposes of prioritizing planning efforts, as there will be limit to planning resources.

Site Surveys

A useful tool for planning and preparedness is the "site survey." The site survey is conducted by a CBRN/HAZMAT specialist and is a reconnaissance of a site or venue to determine any information that may be useful in the event of a CBRN/HAZMAT incident. The term "site survey" applies to both the process of visiting and analyzing a location and the written report produced. The purpose of the site survey is to collect any information that an incident commander or responders would find useful if they had to respond to an incident. An example site survey is included as an appendix.

How to start: Start with the master site list. Do some basic prioritization of the list, based on length of event, size of venue, and the nature of the attendees. By rank ordering your site list, you have an approximate order of priority for conducting site surveys. The order of priority need not be the order in which you wish to do site surveys. You may wish to get some of the minor ones out of the way early on in the process. In some instances, venues will be built for the event or venues will be subjected to serious levels of renovation. In such circumstances, there is often little point in doing the survey too early as important features of the site may change.

Visiting the site: The major effort in conducting a site survey will be a site visit. Large or complex venues may require multiple visits. Rarely, if ever, can an adequate site survey be conducted by only accessing public spaces. The site survey needs to be conducted in cooperation with the site's managers in order to obtain the correct access.

Talking to the right people: The most important act in the preparation of the site survey is taking a tour of the venue with someone from the facility management. A building engineer is ideal for this. So is a head of security. The CBRN/HAZMAT planner does not need someone from marketing or public affairs to give the publicity tour of the convention center, which is aimed at selling exhibition space, yet that is what I have received at times. The site survey needs to get into the more mundane world of drains, water supply, and ventilation systems.

Including the right information in the report: A sample outline of a CBRN/HAZMAT site survey is included in the appendices It is difficult to prescribe exactly what needs to go into a site survey, as venues and events are all different, but the general template provided in the appendix can be a starting point. The benchmark for inclusion in a survey is the question: "Will this information be useful during an incident?" While different types of buildings and venues will need different types of information to be included in the survey, the guidelines below represent the minimum for a survey. At a minimum, the survey effort needs to address the following questions:

Type/Nature/Size of Event: What event is going to happen at the venue? When? How many people are expected? Are any dignitaries or VIPs expected?

Contact information: Who is responsible for what and how do we find them on the day of the event?

Description of building mechanical systems: How does the HVAC system work? Where are the air intakes? How does one operate the HVAC system? Where are the water and gas supplies? (Ventilation is discussed at length below.)

Known, suspected, or assessed vulnerabilities to CBRN attack: Given your "threat basis," how would you attack the facility if you were the terrorist? (Note: Do not put this sort of information into documents with widespread distribution. Specific vulnerabilities may need to be kept separate from main survey documents for security purposes.)

Location and description of known HAZMAT: Are there hazardous substances routinely used or stored in the venue. What about adjacent or nearby facilities? Take a bit of a tour around the neighborhood and see what type of commercial, industrial, and transportation infrastructure exists around the venue. Take particular note of highways, railroads, and commercial waterways, as these are often routes of travel for commercial HAZMAT shipments. This is discussed in further detail below.

Evacuation routes: What are the existing evacuation routes from the building? Are they adequate or sufficient for CBRN/HAZMAT scenarios?

Location of any potential "safe refuge" or "shelter-in-place" locations: Shelter-in-place will be discussed later in the book.

History of incidents: Have major safety and security incidents occurred at the venue in the past? Has the venue ever had to do an emergency evacuation? How well did it go?

Existing emergency plans: What plans, procedures, and equipment are already in place for fire, medical emergencies, bomb threats, and other types of emergencies?

Routes and locations: Suggestions for emergency access/egress routes, staging of response assets, and locations for mass decontamination.

Photographs and Diagrams: You must assume that the person reading the survey has never set foot in the site and is working under stressful conditions. Even if there was an aggressive program of site familiarization tours, it should be assumed that the person reading the survey missed the tour. A liberal use of photographs and diagrams is encouraged, as it will aid the transmission of useful information. Floor plans and site diagrams, even if they are very crude or copied verbatim from the site's owners, should be included in every survey. Often, basic floor plans can be easily copied from required fire evacuation notices. A floor plan, with a photograph of the door to the mechanical room, combined with a photo of the shut-off switch for the air handling system is far more useful than a written explanation of where the switch is located.

Integrating with other efforts: Other people may be doing a similar survey effort. Police and security providers may be surveying sites to see how many personnel they need and what kind of physical security effort may be needed. If an overall security or antiterrorism search or sweep of the building is going to take place, then a reconnaissance for that effort may be underway. The fire department may be doing fire safety inspections. There may be benefit in conducting some of your survey activities in conjunction with others. I have been in situations where literally nine different people had bothered the facility manager with (mostly) the same questions. The ninth person to ask the questions may not get the full benefit of cooperation. The answer is that nine people shouldn't need to do different independent surveys. But if there are going to be a dozen people asking questions, get in early.

Archiving: I think that it is a useful practice to archive your site surveys, and this was standard policy at US Secret Service when I worked there. Major events seem to recur at some venues, and there is no need to reinvent the wheel each time. However, one cannot just open the file and slap a new date on the top of the survey form. The old surveys are useful but do not replace the requirement for on-the-ground reconnaissance.

Dissemination of the survey report: All too often, site surveys are prepared with diligence and care, only to languish somewhere in a filing cabinet. The written site survey product is meant to be useful to responders and incident commanders should something happen at the site. It can only have usefulness if it is available to people who need it. There is a need to balance operational requirements with security needs, as site survey documents may highlight significant vulnerabilities. Site surveys need to be made available to support responses. This is not to say that everyone is going to be walking around with three or four large binders of surveys. Potential incident commanders and assessment teams can have access to necessary surveys. At a minimum, a set of surveys can be maintained in an operations center for reference.

Site Tours for Incident Commanders and Responders

In an ideal world, every responder would have a full tour of every venue in which they could be called upon to enter. Since we do not work in an ideal world, we must make due with other measures. The planning effort should include a program of site "walk-throughs" for at least some of the major venues. The purpose of these site visits is for key response personnel and incident commanders to have an idea of what they may face in and around the major sites of an event. This is similar to the military concept of a "leaders' recon" and is meant to improve the situational awareness of key tactical leaders. The site survey report should be able to serve as a guidebook for such site visits.

Events with large numbers of venues will mean that it will be impractical for site tours at every venue. In such circumstances, the effort should focus on the high priority venues. Lower priority venues can be the subject of a consolidated "virtual tour" in the form of a briefing, in the event that time, or resources do not permit a full walk-through. A well-prepared and illustrated survey document may be called upon as a substitute for an actual tour of the site and should be written from that perspective.

PHYSICAL CHARACTERISTICS: AIR FLOW AND VENTILATION

The types of threats that we are concerned within this book are at their most dangerous when people inhale them. CBRN materials that get into the human lungs are at a position to do the greatest amount of damage. Materials that are suspended in or carried by the air, such as gases, vapors, aerosols, dust, and particulates, can travel far from their point of origin, far more than liquids. This means that understanding how air flows in and around our venues in critical for protecting them.

References

Three excellent documents are freely available online from the United States government on this subject. They are:

- *Building Retrofits for Increased Protection Against Airborne Chemical and Biological Release.* (2007, for the US EPA)[4]
- *Protecting Buildings from a Biological or Chemical Attack: Actions to take before or during a release.* (2003, Lawrence Berkeley National Laboratory)[5]
- *Guidance for Protecting Building Environments from Airborne Chemical, Biological, or Radiological Attacks.* (2002, US NIOSH)[6]

Buildings are designed to provide shelter to their human occupants. Buildings keep people insulated from the effects of hot or cold weather, block the wind, and form a barrier to keep precipitation out. However, a building perfectly designed to insulate from the outside will eventually suffocate the occupants. As such, buildings also have to allow for outside air to come inside the venue and human exhaust (principally carbon dioxide), either naturally (such as a breeze flowing into a window or doorway) or in some contrived way, such as air-conditioning and ventilation systems. Heat needs to be kept in, or let out, depending on comfort and the differential between outside and inside temperature. Exhaled air, full of carbon dioxide, needs to be let out. Fresh air, full of oxygen, needs to be let into the building.

Protective Features of Ventilation

Because of this diverse mix of desires and requirement, a building can work either in our favor in a CBRN incident or against us. Modern buildings are a mix of both simple and complex measures intended to keep the outside environment out and keep the inside environment in while letting just enough of the inside out and the outside in to main. Buildings are designed around a reasonable expectation of what the outside and inside environments are likely to be. CBRN/hazmat incident are, apart from a few very unique buildings, outside of the normal expected environmental parameters that influence building design.

While buildings can differ greatly from each other, there's a general rule of thumb that we can use to analyze their role in protection from a CBRN incident. This rule broadly states that a building can either defend you from a problem or make the problem worse. The difference depends on where the problem originates and where the people you are trying to protect are located. If hazards are outside, but the people you are protecting are inside, then the building serves as a barrier that can be used for protective purposes. But if the hazards are inside the building AND your people are inside the building as well, the building can work against you, concentrating the hazard rather than letting it disperse.

Ventilation systems exist for the purpose of taking air from outside and putting it into the building and to take the older air from inside the building and put it outside. In CBRN/HAZMAT situations, a building's ventilation system can work either for your or against you, depending on the parameters of the situation.

Where Does the Air Come From?

In the old day, ventilation came from doors and windows. (In many historic properties, it still does.) Modern buildings have systems that cost thousands or even millions of dollars to heat, cool, and ventilate. If you are spending a lot of money to heat or cool the air in a modern building, great energy savings come from ensuring that air does not leak out too much. Air intakes suck air into buildings for ventilation, heating, and cooling. Rarely is a building HVAC system designed or operated without some degree of recirculation, so there is usually a mix of inside recirculated air and outside fresh air being processed at any particular time.

The location of the intakes for building HVAC systems is a critical point that must be understood from a standpoint of CBRN/HAZMAT resilience. First, air intakes could be deliberately targeted by terrorists. A targeted attack directly on some air intakes would require less threat material than a wide-area attack, and the intakes would serve to circulate the material more effectively through a building than many other methods. For example, a chemical device using explosives detonated outside the building would likely shatter some windows, thus increasing ventilation.

The other key issue is that, for CBRN/HAZMAT releases outside of a building, most of the threat materials are heavier than air. An incident or accident releasing dangerous materials at street level (or below street level, such as a subway) would not affect a building as much if the air intakes are significantly higher than street level. Air intakes at street level might need to be immediately shut off in such a scenario. Air intakes ten stories higher up might actually be providing useful benefit by putting clean air into a building.

As indicated in the graphics, knowing where the air intakes are, making them harder to find or access, and having them higher up on the building where practical all provide some degree of mitigation against CBRN/HAZMAT threats.

Physical Security of Venues

Antiterrorism starts with good basic physical security practices. CBRN terrorism is only a sub-category of terrorism. In turn, terrorism is a subcategory of criminal activity. Good security of premises is a basic deterrent to terrorism. Security practices will serve to make every aspect of the cycle of terrorism (discussed elsewhere) more difficult for terrorists. Investing in security invests in antiterrorism. And any investment in antiterrorism provides some degree of protection against CBRN terrorism. As a result of this logic, any rational assessment of any venue's resilience against CBRN threats will have to assess physical security.

References

Readers interested in more thorough treatments of physical security are strongly urged to read these two books:

- Lawrence Fennelly's *Effective Physical Security*. 4th edn, 2016[7]
- Broder and Tucker's *Risk Analysis and the Security Survey*, 2011 edition[8]

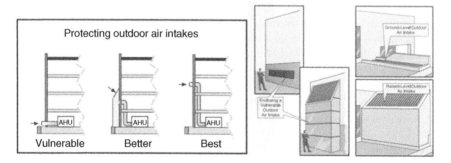

FIGURE 6.3 The location and configuration of air-intakes is important in CBRN/ Hazmat scenarios.
Source: US Centers for Disease Control/Public Domain/U.S. Department of Health & Human Services.

Neither address CBRN issues, but both are excellent primers on physical security. The relationship to the CBRN/HAZMAT threat is best described through the following aspects of physical security.

Specific to the types of event venues commonly used in major public events, the US government has published, free to download, several security and anti-terrorism guidance documents:

- *Best Practices for Anti-Terrorism List for Commercial Office Buildings*[9]
- Field Guide: Conducting BPATS Based Assessments of Commercial Facilities[10]
- Best Practices in Anti-Terrorism Security for Sporting and Entertainment Venues Resource Guide[11]
- Best Practices Matrix[12]

Access Control

Access control measures mean that a venue has some degree of control over who enters the venue and what areas within venues people can enter. Access control includes much of the basic components of physical security such as ID cards and credentials (including ticketing), controlled entry points, card readers, keys, and locks. The more sophisticated access control systems can also provide a high degree of accountability of who may or may not be in a building in an emergency situation.

Situational Awareness

A venue with good physical security has good awareness of what is happening in and around the site. This can be through CCTV coverage, surveillance by guard personnel, monitoring of access control systems, monitoring of building utility systems, and having a robust culture of reporting anomalies to a security operations center. All of these features will serve to deter or detect surveillance or hostile acts or discover unsafe situations, regardless of whether it is CBRN-related or not.

Procedures

Venues with excellent security practices will have developed and exercised a number of processes and procedures for unusual situations. Even basic incident response procedures like responding to a door found unlocked, investigating an unusual package, summoning help for someone with a medical problem, and evacuating the building provide a good basis for additional measures that might be needed in CBRN/HAZMAT incidents. You are not reinventing the wheel; you are adding onto what is already there. If what is already there is good, there is no need to change it.

Accreditations

Within the United States, there is a system of security and safety accreditation. Under the so-called Support Anti-Terrorism by Fostering Effective Technologies (SAFETY) Act, stadiums and arenas can apply for a SAFETY Act designation or a more elaborate certification. This entails a thorough evaluation process including site visits, safety and security evaluations, and whether or not the site adheres to best practices. Overall, this is a good framework. While it is not specific to CBRN/HAZMAT threats, it is inherently useful as an underlying framework. Venues with a SAFETY Act designation or certification have demonstrated their willingness to think about safety and security in a reasonable manner and can be reasonably asked to consider CBRN/HAZMAT threats as part of an all-hazards mindset.

HAZARDOUS MATERIALS AT OR NEAR VENUES

One of the reasons that this book is not just about CBRN events is that hazardous materials used in industry or science can be just as nasty as CBRN weapons. The nature of our modern technological society is that dangerous materials are part of commerce and industry every day. These can pose problems to major events either by accident or by deliberate exploitation by a terrorist. Planning for a major event means that you should understand the hazardous materials that already are in your area of operations. Efforts to improve your situational awareness of hazardous materials should consider several different aspects.

Storage of Materials at the Venue

Many types of buildings routinely use hazardous substances for a variety of purposes such as cleaning, water treatment, or even food and drink uses like carbonation. It is worth surveying what is actually at the venue. Context and quantity are important, as is some degree of judgment. While a single bottle of bleach for sanitation purposes is not really a significant wide area hazard, an entire cargo pallet of bottles of bleach left unattended might provide some target of opportunity.

Routine Deliveries of Gases and Liquids

Many venues will routinely receive deliveries of large quantities of gases or liquids, often for hospitality services. Carbon dioxide gas for carbonation of drinks and large pallets of barrels of beer are only some examples. It is possible that large deliveries could be used to hide other substances, particularly in a build-up to a major event when more deliveries might be expected.

Adjacent Properties

Because some types of hazardous materials could produce effects a long distance away from the point of dissemination, if given sufficient quantity and wind, it is necessary to consider the activities of adjacent properties. Commercial, agriculture, medical, or scientific users of hazardous materials may use, store, or transport such substances in areas close enough to your event to be of significance to your planning processes. The nature of modern technology and commercial occupancy is often such that interesting and dangerous things with chemicals or radiological materials may be occurring in very unassuming premises. Hospitals and university campuses can contain serious hazards down the street from your event. Working with local regulatory authorities or even just the practice of "asking nicely" can yield the information you need. Within the United States, "local emergency planning committees" may be able to provide information on chemicals in a local area.[13] Radiological material will be licensed with state and/or federal regulators.

Proximity to Bulk Transportation of Chemicals

As with premises near an event, transportation routes can carry hazardous substances. Highways, railways, and waterways carry thousands of tons of hazardous materials every day. The type and quantity can vary greatly. It will be fundamentally more difficult to understand this threat as it will vary literally from hour to hour, unlike the more static environment of industrial, medical, and scientific users in fixed premises. This is less about finding specific hazards and more about considering locations where an unexpected hazard may arise.

Lesson Learned: You can Find Surprises in a Venue

A thorough survey of a site can yield interesting surprises. In the course of hundreds of site surveys, I have encountered very odd things. One of the most common lines of banter with my colleagues when I was at the Technical Security Division of the Secret Service falls broadly into "weird stuff we found during a sweep." While a moldy sandwich in the boss's desk is amusing, some things that you encounter in a thorough survey of a building might have either general ramifications on security and safety or specific implications in the CBRN/HAZMAT part of the threat spectrum.

An example of the former is a building where the fire evacuation instructions were so bad that a colleague of mine, in good conscience, had to call the fire inspectors on the spot. Another example was a site with widespread CCTV cameras. The problem was that at least half of the cameras were, in fact, disconnected because an upgrade was botched. In the security room, half of the camera feeds were dead because analog cables had not been replaced when new digital cameras were installed. A cursory walk-through would have assumed that the cameras were functioning. Indeed, many of them were powered up and operating, just not sending a signal anywhere. This is the sort of thing that you want to discover weeks before a major event, not the morning before.

In the CBRN/HAZMAT arena, it is possible to discover some truly interesting things doing surveys. Many of these adventures fall under the category of habit. People who work at a site and have become accustomed to the things they work may have a relaxed attitude to hazards that they work with every day and have become familiar with. I once asked "what's in that room there" and discovered it was several thousand pounds of water treatment chemicals. In another instance, I discovered an interesting but not inherently imminently dangerous amount of ionizing radiation emitting through a wall. The adjacent property was a medical laboratory.

REFERENCES

1. https://ncs4.usm.edu/resources/, accessed 9 February 2022.
2. https://www.nfpa.org/News-and-Research/Data-research-and-tools/Building-and-Life-Safety/High-Rise-Building-Fires, accessed 9 February 2022.
3. Craighead, G. (2009). *High-rise Security and Fire Life Safety*. Butterworth-Heinemann.
4. Persily, A. et al. (2007). *Building Retrofits for Increased protection Against Airborne Chemical and Biological Releases*. Gaithersburg (MD): National Institute of Standards and Technology https://tsapps.nist.gov/publication/get_pdf.cfm?pub_id=861035.
5. Price, P. et al. (2003). *Protecting Buildings from a Biological or Chemical Attack: Actions to Take Before or During a Release*. Honolulu: University Press of the Pacific https://www.osti.gov/biblio/810539.

6. National Institute for Occupational Safety and Health (2002). *Guidance for Protecting Building environments from Airborne, Chemical, Biological or Radiological Attacks.* Washington (DC): National Institute for Occupational Safety and Health; Publication No. NIOSH 2002-139 https://www.cdc.gov/niosh/docs/2002-139/default.html.

7. Fennelly, L.J. (2016). *Effective Physical Security.* Butterworth-Heinemann.

8. Broder, J.F. and Tucker, G. (2011). *Risk Analysis and the Security Survey.* Elsevier.

9. US Department of Homeland Security. *Best Practices for Anti-Terrorism List for Commercial Office Buildings,* 2018. https://bpatsassessmenttool.nibs.org/, accessed 11 February 2022.

10. US Department of Homeland Security. *Field Guide: Conducting BPATS Based Assessments of Commercial Facilities,* 2018. https://bpatsassessmenttool.nibs.org/, accessed 11 February 2022.

11. US Department of Homeland Security. *Best Practices in Anti-Terrorism Security for Sporting and Entertainment Venues Resource Guide,* 2013. https://www.safetyact.gov/lit/hfhtml/BPATS, accessed 11 February 2022.

12. https://www.safetyact.gov/externalRes/refdoc/Matrix.pdf, accessed 11 February 2022.

13. https://www.epa.gov/epcra/local-emergency-planning-committees, accessed 10 February 2022.

CHAPTER 7

Procurement: Buying Goods and Services

If you have months or years to plan for a major public event and you are taking the threat of CBRN and HAZMAT incidents seriously, you are likely to be in the position of requesting and fighting for a budget. You will have to spend much of this money on goods and services. I can assure you that there are people who specialize in trying to part you from your budget. On the other hand, you will actually need things. Earlier in my career, I was spending money, sometimes lavishly, sometimes frugally, and sometimes even unwisely in the run-up to major public events. Later, for several years, I worked for a major industrial player in the CBRN sector. Having been on both sides of the buying and selling equation, I can say that there are plenty of mistakes that can be made. Hopefully, this chapter will help you avoid a few of the major pitfalls.

The world of CBRN/HAZMAT response is heavily laden with equipment and technology. Trade shows and exhibitions have halls full of stuff you can buy. You will need tools to do your job, which means that someone will have to buy them. This situation will put you on the receiving end of salesmen and sales pitches. You may have front-line responders who may see tools and equipment used by other agencies or department. You may have supervisors who go to conferences and trade shows and see all those things on the exhibition stands, and are convinced that they are what you need. Both may be right or wrong, depending greatly on numerous factors. Most of the things that you need to buy are provided by private industry and are sold in commercial transactions. Salesmen are not all bad. Some are, but most are not. They have a job to do, just like you.

CBRN and HAZMAT Incidents at Major Public Events: Planning and Response, Second Edition. Daniel J. Kaszeta.
© 2023 John Wiley & Sons, Inc. Published 2023 by John Wiley & Sons, Inc.

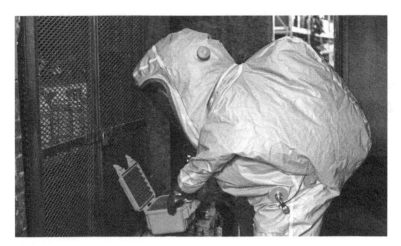

FIGURE 7.1 A wide variety of PPE are available for CBRN and Hazmat environments. Make wise decisions with your PPE budget.
Source: Used with permission from R. Mead.

INTERACTING WITH VENDORS IN AN INTELLIGENT WAY

Any hint of a new budget or a new requirement will get the sales staff to appear as if from nowhere. I say this having actively looked at major event schedules whilst sticking pins on a map of Europe in a sales department. There will be people trying to sell you stuff you don't want and you will be looking to buy stuff that nobody seems to want to sell to you. And even if you aren't in either of those circumstances, you will need to want to get good value for your budget money.

Procure a Capability or More Capacity

At the end of the day, you have a job to do. Success at your job will be defined in a lot of different ways. By the same token, there are a lot of ways to go about accomplishing the mission. The resources you apply to protecting the major event will include many things. When it comes to buying or leasing goods and services, what you really are seeking to do is to improve capabilities and capacities. As discussed in the previous chapters, you should have inventoried your capabilities and capacities. Your procurement activities should be focused on gaining more capability and/or capacity. When you start dealing with procurement issues, do not enter into the process from the standpoint of "I need to buy X" or "I need to get 200 more widgets." It is better to ask yourself the question of "what capability do I need?" The capability may indeed be answered by buying things, but there may be multiple ways of achieving that capability. You should be buying (or leasing or borrowing) a capability. This mental framework will help you define what you really want and helps give a degree of resistance against the sales pitches.

You Will Likely Have to Live with the Limitations of the Products

The product development lifecycle for things in the CBRN and HAZMAT sectors can be years. It is very costly for a company to develop new products, and such product development is often based on the prospect of winning very large defense contracts. You might be able to get some small modification to an existing product, but even the largest major event is not likely to have the budget or the timescale to get a company to develop a whole new detector or decontamination system from scratch.

Customer Service and Maintenance

Some of the best equipment can be made by companies with reputations for below average (or worse) reputations for customer service and maintenance. How quickly can the vendor repair or replace a faulty unit? It will pose serious difficulties if a mission-critical item of equipment will take 12 weeks to return to service, if the event in question is only one week long. Ask for maintenance training, as some repairs may not really need to be done at the factory or by a specialty technician. One tactic is to build a requirement for temporary replacement units to be supplied within a set period of time into purchase agreements and contracts.

Drive a Hard Bargain

Use of equipment in visible manner at a major public event is good publicity for a company in the CBRN/HAZMAT space. If you are a large agency, your purchase might be a considerable sales volume for the vendor. A dirty secret that I discovered when I worked in sales and business development in CBRN detection was that the list price is almost never the end of the story. Price is almost always negotiable. And negotiate on all the extras as well, like service, training, maintenance, and spares.

Consider Leasing

Some categories of equipment are very expensive. Government agencies may not want to be burdened by large inventories of equipment that they may not need after the event is over. Some officials may be wary of substantial outlay of funds for an event of short duration, even if it is high profile. Leasing is a standard practice in X-ray machines and metal detectors, so perhaps this could be a useful model. Industry attitudes to leasing arrangements vary widely. More leasing of CBRN equipment will become possible if more customers, particularly large and prestigious ones, start to demand it.

Consider Borrowing

Many of the items you might need, such as decontamination systems and detection instruments, are expensive. They might not even get used. If you plan ahead and establish good relationships with other departments and agencies, you might be able to borrow some equipment from them. Do not leave this until late in the planning cycle as you might be saddling your responders with equipment that they do not know how to use. This can cause more problems than it solves.

Generics Versus Brand Names

In some areas of CBRN/HAZMAT response, highly priced specialty products compete with lower-cost generic or non-specialty goods. A clear example is that of decontamination chemicals. There are a number of generic equivalents that are nearly as good as specialty decontamination products.

Cost of Ownership is More Than Just the Price

When shopping for equipment, it is important to consider the overall life-cycle cost of the instrument, not just the base price for a unit. Many items require consumables. Some items may require calibration. It is important to calculate the total cost to the user over the expected lifespan of the product. Many items have consumables. Batteries are the obvious consumable, and not every instrument uses off-the-shelf common batteries. Many detection instruments have filters, sieve packs, sensor units (common in 4 gas meters), UV lamps, or other items that will require replacement over time. In some cases, total life cycle cost of the instrument may mean that the consumables will cost more money than the instrument over the long run.

DETECTION AND IDENTIFICATION HARDWARE

I will now use some examples of types of products that are commonly marketed for security and safety at major public event. The first of these is detection and identification equipment. Such equipment is very important for protecting major events, and its proper role is discussed at length in Chapter 15.

Terminology Gets Tricky Here

An item that is a "detector" to one person may be an "analyzer" or a "monitor" to another person. The same item may be called a "detector" by one distributor and a "gas monitor" by another. Arguments over the exact taxonomy are beyond the scope of this book. I use the term "sensor" to include all detectors, identifiers, collectors, monitors, sample collectors, analyzers, litmus paper, colorimetric tubes, spectrographs, dosimeters, and canaries in a cage. I use the term "sensor" to mean any technical means to collect information in the environment and pass it to a potential user in the form of information.

Sensors are Merely Information Tools

Detection is not a means to an end, although it often gets treated as such. A sensor that provides information that is not useful in making decisions is not useful. Sensors are only as good as the brain, whether it be human or software, which interprets the data. It is up to the user to decide what to do based on the information provided by a sensor. Good sensor information can be ruined by application of poor education and training. Information can be misinterpreted. An otherwise useful sensor can be rendered useless by incorrect or poor operating techniques. Conversely, a mediocre sensor can provide great operational value if used creatively by an intelligent user.

What Are You Detecting?

There is also the matter of "what do you want to detect and measure?" But there is no such thing as a "universal hazard detector." Yet I have seen management in response agencies ask for such a thing. There is not any technology available now or in the research and development pipeline that will detect everything of interest in the CBRN/HAZMAT threat spectrum in one instrument. Some "multiple threat" detectors have been truly awful contraptions. Even if a very good example of such a device existed, it would not be as useful as one might think. Chemical, biological, and radiological threats behave differently and the ideal mode of employment for a chemical detector may not be the same as the ideal way of using a radiation detector.

Value of Sensors

The value of a sensor can only really be calculated indirectly. It may not have any correlation to the price tag. Linear relationships do not apply here. A $100,000 detector may or may not produce ten times the practical value produced by a $10,000 detector. And it may not equal ten $10,000 detectors. In some circumstances, the information provided by a sensor may be unclear, misleading, easily misinterpreted, or just wrong. In such cases, the sensor actually has negative value. You can, in fact, spend a lot of money on something that makes your job harder, not easier. Even if no action is taken based on a sensor reading, at a minimum, the time and effort taken to use the sensor is time wasted that would have better spent doing something else. In some instances, responders might make bad decisions based on sensor information. Such bad decisions could cause death, injury, or damage and, in many cases, such decisions result in a situation worse than if nothing had been done.

Information Overload

Sensors are designed and built by scientists and engineers. The scientific community has an understandable desire for the acquisition of data. This can lead to a belief that more information is always better. While this is certainly the case in academic and scientific research, it is not necessarily true for the emergency responder. Any person has a finite capacity for receiving and processing information. Therefore, anyone can reach his or her limit to be able to react effectively. For example, a person in a command post can probably watch three computer monitors and listen to two different radio networks at once and maintain situational awareness. But if you make that person monitor ten computer monitors and listen to six radio networks, his situational awareness tends to be worse, because information will arrive faster than the normal human brain can process it. The same principle applies with other types of sensors.

Speed

Some detectors do not work very fast. This is particularly true with biological detection and identification. In some cases, the detector can take many minutes or a few hours to provide useful information. However, radiation and chemical detectors are often very fast. The detection value of a sensor can be divided into two categories. "Detect to warn" sensors are those that operate fast enough to provide operators with information that allows immediate protection decisions to be made. For example, gas and vapor detectors often need to operate fast enough to allow people to put on respiratory protection, such as a filter respirator or breathing apparatus.

Not every sensor works that fast. Many biological sensors operate better in a "detect to treat" mode. This means that they cannot provide information fast enough to allow immediate protective action. But such sensors can provide information in a manner timely enough to allow for effective treatment of the people exposed to the material. For example, detection of anthrax within a few hours of exposure allows for effective intervention by administering appropriate antibiotics.

PERSONAL PROTECTIVE EQUIPMENT (PPE)

The other equipment area that seems to get the salesmen out in force is personal protective equipment (PPE). PPE, such as masks, suits, boots, and gloves, takes up a lot of the table space at trade shows in the CBRN/HAZMAT business. There is a bewildering range of products available to purchase and a bewildering array of vendors pushing them.

References

There is a good reference book from this publisher called *Personal Protective Equipment for Chemical, Biological, and Radiological Hazards* by Dr. Eva Gudgin Dickson.[1] This book goes into great detail about how PPE works and how to select the right PPE for CBRN environments. The US National Institute of Occupational Safety and Health (NIOSH) also has resources that are useful.[2] The US Department of Homeland Security's 2007 *Guide for the Selection of Personal Protective Equipment for Emergency First Responders* is now quite dated but comprises a very useful discussion of the subject.[3]

Why Do You Need New Stuff?

The first thing you should ask yourself is "why do I need to buy new PPE?" You should review whether your current equipment is adequate to the task. A seriously outlay of funds on protective clothing could use up your resources for other things if your current stockpile is adequate. This is an area of technology where radical improvements come along occasionally, but much of the time improvements in products are incremental. On the other hand, if you are using clapped-out worn-out gear, a major event might be the opportunity to get funds from management for useful protective capabilities.

Different Roles and Different Scenarios Will Need Different PPE

A paramedic will likely require different PPE from a security guard. If you have paid attention to the other chapters of this book, you have developed a baseline of threat scenarios and, with luck, a synchronization matrix showing who does what, and where and when they do it. By examining the various threat scenarios, you can see which personnel will be in a particular threat environment. These notional situations will form the use cases that form the basis of your decisions as to the type of PPE you will select.

You Will Need More Suits Than People

If you have 100 staff and you need to provide them with protective clothing, 100 sets of "Large" will not work. People have different sizes. If you have a large and variable workforce, you will need rather a lot more sets of PPE than your average headcount to account for all of the size variation.

Consider the Regulatory Issues

Equipment designed to protect people from hazards generally comes under the purview of health and safety regulations and legislation. The regulation, testing, and certification of PPE is an entire professional discipline unto itself. There are laws and rules saying what item of equipment can be used for what role against what threat. There can be strict penalties for breaking these rules. What is legal in one setting may be illegal in another. Make sure that equipment you buy is legal for the use that you intend to put it to.

No PPE Is Perfect

There is no PPE that can reasonably protect the wearer against every hazard indefinitely. In designing such equipment, scientists and engineers must make design compromises. The classic design compromise is between level of protection on the one hand and a combination of endurance and comfort on the other hand. Really good protection comes at a sacrifice to comfort, endurance, and (inevitably) performance. Make sure the equipment you are buying fits the operational scenarios that you have in mind.

Firefighting "Turnout Gear" Is Not Useless

In the 1990s, it was widely viewed by most of the military CBRN specialists that I was working with that normal structural firefighting clothing (so-called turnout gear) was useless in CBRN incidents. In turn, the civilian HAZMAT responders I worked with somewhat later laughed at the charcoal "MOPP Suits" of the US Army and rightly pointed out that these would have serious deficiencies in many HAZMAT situations. But is normal firefighting gear worthless? It turns out that, when you study the phenomenon, civilian firefighting turnout gear is not useless in CBRN situations.[4] Not only that, further advances in firefighting clothing are likely to incorporate improved protective features.

Military PPE

In most countries, the civil health and safety authorities do not regulate military equipment for battlefield use. Such items are built around military requirements, which can be significantly different from civil protection specifications or missions. Some equipment available on the market is a nearly identical civil version of military equipment. This may be what you need for your mission. Or not. It may not even be legal for your missions.

DECONTAMINATION

One of the other great areas of the CBRN business is decontamination. There are numerous manufacturers out there who will sell you sprayers, foams, solutions, lotions, and every sort of accessory. Much of this market is driven by military requirements for equipment decontamination. Most of this stuff is very good and meets its specifications. But does that mean you need to spend a lot of money on it?

At the risk of offending the very good manufacturers in this sector, many of whom I know and respect, the fundamental problem in the CBRN decontamination segment of the defense industry is that they are competing with free or very low-cost alternatives in a way that the other segments are not. Sensor manufacturers are competing with the naked eye and the sense of smell. These are not your best options. The PPE manufacturers are competing with normal clothing. With sensors and PPE, you get comparatively less value out of the free or generic alternative and you need to spend some money to get value.

With decontamination, the competitor is soap and water, or even just water. Water, even dirty water or seawater, is not bad at removing contaminants. Many types of contaminants are degraded by water, whereas for others it is more a matter of removal. Adding soap helps, because soap helps the physical removal process. Having the water warm is helpful for decontaminating people, both as a matter of comfort and safety. Moreover, every city that is going to host a major event has firefighting equipment designed to put water onto things. If you have some hoses and fire engines, and maybe some soap, you've already got decontamination capability. (Much of military CBRN decontamination equipment is based on ways either to save on water or to deal with the fact that battles happen away from fire departments.)

When dealing with possible purchases in the decontamination segment, my advice is to look at the areas where you can't necessarily always fix the problem with lots of soap and water. Will you have to, say, fumigate the inside of some ambulances to return them to service? Will you need to decontaminate sensitive equipment, such as medical electronics or the interior of a helicopter? In this case, a fire hose, a bucket of soapy water, and some sponges are not your best option. If you follow the processes that I explain elsewhere in this book, you can perform an analysis of your possible missions in the CBRN/HAZMAT environment. In doing so you can determine what, where, and when you may need to conduct decontamination. Then, you start asking yourself how you are going to do this decontamination. Some of it may not be easily done by fire department personnel with water. Save your decon budget for specialty products and equipment in the spots on your matrix where water is not suitable.

Finding a Home for It All After the Event

It will be much easier to justify budget spent on equipment if you can find a rational use for it after the event. The nature of CBRN/HAZMAT response equipment is that, unless you have a big disaster on your hands, much of the equipment will be unused. New detection equipment, even if heavily used in a short-duration major event, will be low on hours and within warranty.

In an ideal world the best course of action might be that you keep the new equipment. Now that you have developed a robust CBRN/HAZMAT response capability, you have resources to use in the future. But you may also consider passing some of this stuff on to other departments and agencies who have a critical need for it. If you are, for example, a temporary entity like an Olympic organizing committee or some sort of ad hoc multi-jurisdictional task force set up for a specific event, you will go out of existence. If, at the beginning of the planning process, you spend some time thinking about the eventual disposal of assets, it could be a great windfall for some of the departments and agencies involved in the effort. Having residual capability to spread around your municipal or metropolitan area might help justify some of the budget needed.

Lessons Learned: Some Problems You Cannot Overcome by Buying Stuff

Technology, hardware, and software cannot solve all of the problems you might be having. Splashing out money on equipment does not necessarily buy you all of the capability and capacity that you might need to deal with a crisis situation. If your plans and procedures are fundamentally flawed, but are executed by people who are very well equipped, then you are still setting yourself up for failure. If you do not have enough trained personnel to accomplish essential tasks, a warehouse full of masks and suits is not going to make much difference. No amount of equipment purchasing is going to make up for fundamental structural failings in your planning and procedures.

Splashing out money might make a difference in some kinds of personnel shortfalls. You could hire contractor personnel for many essential functions. You could, for example, bring in contractor EMTs and paramedics. You could use security contractors for some tasks to free up police staff. However, these things are expensive if you do them as a last-minute panic. If you are going to use contractors to make up for deficiencies in staffing, plan this in as far upstream as possible.

There are many things that you can do that basically are free. But the thing about stuff that doesn't cost anything is that there is no salesman advocating them. Even I am not totally free, in that this book costs some money to buy. But my point is that there are a lot of things that are procedural and operational that do not incur any financial costs. "Hey guys, maybe you should lock that set of doors today" is an example, as is "hey, maybe park the fire engine over there today, it blocks that line of sight, just for this event."

One problem I've seen is middle and upper management falling prey to sales pitches and buying equipment and then trying to shoe-horn it into operational procedures. Equipment is at the end of the equation. Figure out your mission first. Your mission dictates the number and type of personnel you need to have. Then you can calculate what type of training and equipment is needed to do the job. Equipment supports the mission. It does not drive the mission.

REFERENCES

1. Dickson, E.F.G. (2012). *Personal Protective Equipment for Chemical, Biological, and Radiological Hazards: Design, Evaluation, and Selection.* John Wiley & Sons.
2. https://www.cdc.gov/niosh/ppe/default.html.
3. US Department of Homeland Security (2007). *Guide for the Selection of Personal Protective Equipment for Emergency First Responders Guide 102–06,* 2nde. United States government.
4. US Army Soldier and Biological Chemical Command (2003). *Risk Assessment of Using Firefighter Protective Ensemble with Self-Contained Breathing Apparatus for Rescue Operations During a Terrorist Chemical Agent Incident.* United States Government.

CHAPTER 8

Preparedness in the Medical Sector

Medical preparedness is a critical component of major event CBRN/HAZMAT planning and readiness. Broadly, medical support to major events is provided at various levels. At the field care level, this consists of first aid rendered by people with relatively little or even no training, volunteer and salaried emergency medical personnel, dispatched ambulance assets, and other emergency responders such as fire, police, and security personnel who have various degrees of training above the average layperson. In addition, you may set up significant temporary additional medical provision at event venues that may consist of additional personnel, higher levels of training (e.g., doctors or nurses present), and more equipment. At the hospital level, definitive care is rendered. And above hospital-level care is the overarching sphere of public health.

REFERENCES

It should be noted that this is not a guide on how to treat casualties, nor is it an attempt to tell physicians, nurses, paramedics, EMTs, or other medical providers how to provide medical care. There are excellent resources for this purpose, and many of them are cited below.

The Textbook of Military Medicine

This is really a set of diverse volumes rather than a single book. The US Army Medical Department's Borden Institute has published a number of thoroughly researched volumes on many subjects in military medicine, including several of particular value in CBRN/HAZMAT response. I consider these works to be canonical. All of the following are freely available for download at https://medcoe.army.mil/borden-3-textbooks-of-military-medicine.

CBRN and HAZMAT Incidents at Major Public Events: Planning and Response, Second Edition. Daniel J. Kaszeta.
© 2023 John Wiley & Sons, Inc. Published 2023 by John Wiley & Sons, Inc.

Medical Aspects of Chemical Warfare (2008)
Medical Aspects of Biological Warfare (2018)
Medical Consequences of Radiological and Nuclear Weapons (2013)

Some of the other volumes available in that series will be useful for guidance on conventional injuries from explosive devices and psychological/behavioral issues. The following two books provide encyclopedic knowledge in the areas relevant to this chapter and I refer medical specialists to them for more detailed information relevant to this chapter. They are relatively focused on the US operational environment, but their lessons will be adaptable elsewhere:

American Academy of Orthopedic Surgeons. *Special events medical services.* Jones & Bartlett Publishers, 2010.
McIsaac Joseph H. *Hospital Preparation for Bioterror: A Medical and Biomedical Systems Approach.* Elsevier; 2010.

SIZING UP THE PROBLEM

Every possible scenario or incident will create different types and numbers of casualties. In order to get very far with the medical aspects of a planning process, it is necessary to ask a few questions. How many casualties you are likely to be able to deal with? What do CBRN/HAZMAT casualties look like (e.g., signs, symptoms, condition, degree of care that they will need) when they appear to the medical provider? What is the timescale over which you they will appear?

Number of Casualties

The short answer is that there is no way to know ahead of time how many people will be injured or sickened by an incident. The longer and more useful answer is that you should use the threat basis for your major event to guess some numbers of casualties based on your planning scenarios. You must do so knowing that you will never be right. The idea is to work out a basis for evaluating your capabilities capacity and for planning to, if needed, temporarily increase your capabilities and capacities. Accuracy is less important than overall magnitude. By this, I mean that if you have a plan to deal with 500 casualties who need hospital-level care, and you are confronting with an incident that produced 507 casualties, you will probably be able to cope. But you will not cope if you planned for only 100 casualties and received 300.

What Do CBRN/HAZMAT Casualties Look Like?

One of the key components to effective planning and preparedness for the emergency medical sector is an understanding of mechanism of injury. It is important to understand the means by which CBRN/HAZMAT scenarios affect people and create victims who will need help from the medical system. Different types of causative agents will produce different types of victims, which will have varying effects on the medical system that is trying to care for them. Therefore, it is useful to look at the types of injuries and illness we are likely to see. The various mechanisms of injury will drive the required logistics and plans. This book is not the place to seek diagnosis or treatment advice for exposure or injury from CBRN/HAZMAT causative agents in definitive detail. For technical details, please see the references listed above.

Chemical and HAZMAT incidents will have the highest potential for casualties requiring acute care. Chemical warfare agents (CWAs) and a wide variety of commercial and industrial HAZMAT have the greatest potential for immediate casualties. In a majority of scenarios involving CWAs and HAZMAT, it is respiratory distress and breathing difficulty that will provide the most acute immediate problems. In situations involving nerve agents, blood agents, some blister agents, and many categories of industrial HAZMAT, it is likely that respiratory problems will be the mechanism of injury that requires the most acute care. Other symptoms may cause pain, discomfort, and disability, but it is the respiratory symptoms that will require the most urgent and intensive field interventions. Oxygen administration, airways, assisted ventilation, and suction may be required, all of which entail manpower and logistics. A few of the threats (nerve agents and cyanides) have specific antidotes, which we will discuss in Chapter 16. Decontamination (also discussed in later chapters) will be important in cases involving actual deposit of material on skin and clothing, but it is less important than airway and breathing. Many types of chemicals will have latency, and signs and symptoms will not be apparent at the field level.

Nerve agents call for specific measures. Nerve agents have very specific signs and symptoms (see text box), but also have specific medical countermeasures. Incidents involving

---

Nerve Agent Signs and Symptoms

Nerve agents can enter the body by inhalation or absorption through the skin.

Inhalation—Seconds to minutes after exposure

Mild: pinpoint pupils, dimness of vision, headache, runny nose, salivation, tightness in chest

Serious: Mild symptoms, plus difficulty breathing, generalized muscle twitching, weakness, paralysis, convulsions, loss of bladder and bowel control

Liquid exposure to skin—Minutes to hours after exposure

Mild/Moderate: Muscle twitching at site of exposure, sweating, nausea, vomiting, weakness

Serious: Mild symptoms, plus difficulty breathing, generalized muscle twitching, weakness, paralysis, convulsions, loss of bladder and bowel control

A useful acronym that I learned in training is SLUDGE:
Salivation
Lachrymation (tears)
Urination
Defecation
Gastrointestinal distress
Emesis (vomiting)

Sources: Field Management of Chemical Casualties Handbook, USAMRICD/Public Domian/U. S. Army Medical Center of Excellence and *https://www.medcoe.army.mil/borden-field-mgt-of-cb-casualities.*

nerve agents are a specific case in CBRN/HAZMAT incident management because of the potential for rapid fatalities and the specific countermeasures needed to treat moderate or severe nerve agent injuries.

Combined Trauma

It is also important to note that use of explosives to disperse CBRN agents may cause conventional injuries. Victims may present with a combination of chemical and conventional injuries. Wound contamination is possible. Fragmentation and debris from a chemical device may cause both conventional injury and percutaneous exposure to a chemical agent. Plans should accommodate the possibility that conventional injuries may be more immediately life-threatening than CBRN injuries.

Time Scale

It is very important to note that different CBRN/HAZMAT injuries and illnesses appear with differing rates of onset as shown in Figure 8.1 below. A common misperception is that CBRN and HAZMAT materials cause immediate casualties. Some do and some do not. A large percentage of the types of injuries described above will present themselves an hour, hours, or even days after the actual incident. The rule of thumb is that chemical injuries are generally faster than biological or radiological ones. But there is some significant overlap between categories. Route of exposure plays into the timing as well. For example, victims of nerve agent attacks in Syria became very ill very quickly after inhaling aerosolized Sarin. But Sergei and Yulia Skripal became acutely ill only some hours after dermal exposure to a liquid nerve agent. The graphic below summarizes the timing of major categories of threats.

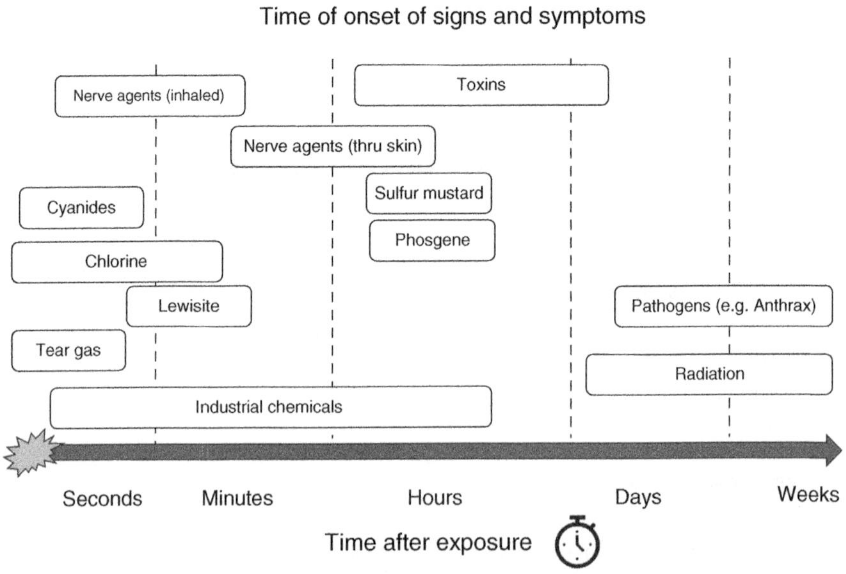

Time of onset of signs and symptoms

FIGURE 8.1   The time between exposure and effects varies greatly from material to material. Author's own work.

The timescale is significant because some incidents can cause an acute spike in need for medical care, while other incidents and materials will spread the requirements for acute care over a longer period of time. For example, an incident involving nerve agent in vapor or aerosol form will have casualties presenting very quickly, with the amount likely tailing off as the threat has dispersed. An incident involving sulfur mustard would end up with casualties showing blisters only some hours after dispersal. An incident involving anthrax spores may see no acutely ill patients for several days.

PREPAREDNESS AT THE FIELD LEVEL: THE CONCEPT OF "SPECIAL EVENTS MEDICAL SERVICES"

Major events include large numbers of people. The statistical truth is that when enough members of the public are assembled, there will be people who need medical assistance, regardless of the environment. People will sprain their ankles, suffer from heat/cold injuries, and have heart attacks. Fans will drink too much alcohol. As a result, venue management and public safety authorities have long known that they have a requirement for first aid and medical support at major events. The provision of front-line medical care at major events, or "special events medical services" (SEMS), can range from a handful of first aid volunteers at a small event to very large and comprehensive services at major sporting events, with hundreds of medical providers ranging from volunteers to trauma surgeons. This book urges events to set up SEMS provision, in one form or another, at major events. Setting up and executing SEMS is a field unto itself; I will concentrate on integrating CBRN/HAZMAT considerations into existing SEMS structures.

First Line of Defense

Field medical personnel that are stationed at major event venues serve as an early warning network in many circumstances. Persons seeking medical attention from venue first aid staff may be one of the only early indicators that a CBRN/HAZMAT release has taken place. Effectively, such teams are chemical detectors, in their own way, because they can observe signs and symptoms early in an incident and they have communications directly back to an operations center.

Preparedness for SEMS Providers

The planning effort cannot assume that SEMS providers will have CBRN/HAZMAT scenarios embedded in their planning effort, and preparedness efforts need to cover such personnel.

Awareness and training: At a minimum, SEMS providers need to be able to recognize early signs and symptoms of the various CBRN threat materials. Because SEMS teams are an early warning network, training should focus on recognition. Medical training should probably emphasize the general patient care interventions (ABCs) that will be needed in the event of mass casualty events. It is airway and respiratory issues that will be the most immediate producer of lethality in many situations.

Communications: The advantages of SEMS providers as the first line of defense are largely negated if they cannot communicate with the overall event security and safety effort. Therefore, there needs to be a communications pathway for SEMS teams to interact with incident management structures.

Equipment: CBRN/HAZMAT scenarios will pose a threat to responders. Consideration should be given to survival PPE equipment so that responders do not become victims in chemical scenarios.

Languages: The public information problem at many major events is complicated by the multinational nature of the attendees at some events. Particularly at major sporting events, there may be people present, who speak dozens of non-native languages. While it is clearly impossible to cover every contingency, public health authorities can have specific critical messages translated into a handful of the major languages in order to be ready for this eventuality. Various telephonic interpretation services could be used. This applies both in the field and in the hospital.

Integration with Response Planning

The providers of SEMS, whether they are governmental or contracted, need to be included in the overall CBRN/HAZMAT planning effort. They should be integrated into the incident management structure for events at their venues. One common mistake that I have noted is that in many places, the planning process for major events is run by people from the fire and police services, often to the exclusion of medical experts. It is useful to have some medical perspective, both from the pre-hospital and clinical levels incorporated into the planning process.

Get Some Medical Knowledge Onto Assessment Teams

The usefulness of assessment teams is well described elsewhere in this book. It is highly useful to place experienced medical personnel onto assessment teams. In some circumstances, a skilled pair of eyes looking at a victim can provide quicker characterization of the nature of the incident than sophisticated equipment.

PREPAREDNESS AT THE HOSPITAL LEVEL: GETTING READY FOR MASS CASUALTY INCIDENTS (MCI) IN THE CBRN/HAZMAT ARENA

Mass casualty planning for hospitals is well understood as a component of disaster medicine and normal operations. A bus crash, mass shooting incident, or hotel fire can cause mass casualties. But the entire premise of this book is that CBRN and HAZMAT incidents can cause mass casualties. There are numerous books and articles[1] in general medical preparedness that are of use, but I recommend the McIsaac book mentioned above as a specific tool useful for the CBRN environment.

Include CBRN/HAZMAT into Triage Plans

Triage is a well-understood concept and has long been an important component of mass casualty incident (MCI) plans. Some triage will happen at the field level, and some will happen at the hospital level. MCI scenarios place a burden on medical resources and triage serves to utilize medical resources in a way that maximizes the overall survival rate. Heroic measures may save some victims who are in a serious state, but in doing so scarce resources may be denied to others who have a very good chance of surviving. As triage is a critical part to MCI plans, CBRN/HAZMAT injuries need to be incorporated into existing triage guidelines. Fortunately, military references, such as the *Textbook of Military Medicine* volumes mentioned above, contain excellent CBRN triage guidance that can be adopted into the civilian sector.

Understanding Capability and Capacity

Not every hospital has a full understanding of its true capacity in every scenario. In many situations, the capacity to treat patients may be limited by more narrow factors than simple bed capacity and the numbers of trained medical personnel available. Availability of pharmaceuticals, specialty clinicians, and specialized medical devices (such as ventilators) will be the determinants of overall capacity, not bed space. Merely treating the Skripals in 2018 depleted a British hospital of its ready supply of atropine.[2] Again, the best method for roughly determining capacity may be through exercises.

Security

Mass casualty situations, particularly ones that require decontamination and segregation of areas into "clean" and "unclean," will require support from security staff. The hospital may get barraged with self-referring patients, so there may be a need to defend the hospital itself and go into a form of lockdown. CBRN/HAZMAT incidents will place a strain on hospital security arrangements. This may be an area where mutual aid (i.e., personnel from police departments outside the affected area) or the military might be useful.

Staff Protection and Training

As with field EMS staff, PPE and training will be needed for medical providers. The extent of PPE and training will vary greatly on hospital configuration, decontamination strategy, and the overall CBRN/HAZMAT mass casualty plan. Let the plan dictate the requirement for training and equipment and do not let it work out the other way around.

Hospital Internal Communications

McIsaac particularly stresses the need for good communications within the hospital in such environments as CBRN incidents.[3] Mass casualty incidents (MCIs) and similar disasters place a lot of strain on hospitals and will greatly increase the need for information and messages to flow within the hospital. Mass casualty plans may involve using areas where existing communications are nonexistent, such as loading bays as decontamination areas. Departments and staff that don't normally talk to each other much may have to be in close coordination. Internal telephones might get contaminated. In your planning process, work out who needs to talk to whom, and test this with drills and exercises. You may need to develop additional methods for use in emergencies, such as additional handheld radios or even runners. In some places, amateur radio operators have been used to augment hospital communications in emergencies.[4]

Be Prepared for a Wide Variety of Syndromes

Not every casualty from a CBRN/HAZMAT incident will necessarily be from exposure to such agents. You may have combined trauma, such as blast and fragmentation injuries associated with a "dirty bomb." You might encounter a variety of conventional trauma injuries if people have fled a disaster and have been trampled. Stress and anxiety may cause acute problems or exacerbate people who are vulnerable. Heat stress may afflict responders. People having been decontaminated in cold weather may suffer hypothermia. It is important to look at the full spectrum of possibilities.

Stockpiling

It is necessary to consider having stockpiles of necessary consumables, such as PPE, medicines (e.g., atropine for nerve agents), and decontamination supplies. Experience in India during COVID-19 showed that oxygen supplies can be a bottleneck.[5]

Hospital Decontamination Plans

Decontamination will be discussed in detail in Chapter 16, but it has important implications for medical preparedness. In the scenarios where decon is a necessity to save lives, it needs to be done as soon as possible. Almost always, this means doing it in the field. But there are also other very good reasons for performing decontamination in the field, rather than at the door of a hospital. It is far easier to perform medical interventions in an environment where the provider does not have to wear protective equipment. Contaminated victims are likely to spread contamination to any surface that they are laid on or sit on, so ambulances will become contaminated and will need to be taken out of service lest they cause injury to others. For these reasons, emergency planners at the clinical level need to be a strong advocate for decon in the field.

Do not take any of this as advice against having a decent decontamination plan at hospitals. In a perfect world, no decontamination would ever be needed at the hospital door. However, many scenarios may see potentially contaminated patients turning up at hospitals and clinics. Scene control at a large incident may be hard to establish, and the emergency medical system may be hard pressed to corral those with chemical injuries and handle them at the incident scene. Many scenarios may not involve agents with prompt symptoms, thus providing a situation where people may simply go home only to get sick later in the day. The "worried well" (see below) may turn up in force, and they may have legitimately contaminated victims interspersed among them. Therefore, a hospital needs a decontamination plan, even though we hope that it will not be necessary.

PREPAREDNESS AT THE PUBLIC HEALTH LEVEL: BIOMEDICAL SURVEILLANCE

Various types of surveillance can be a useful tool for the detection of biological warfare agents, albeit in a consequence management mode after the fact. Medical surveillance is the systematic collection and reporting of data on medical patients, usually collected and reported by doctors, clinics, and hospitals. Because biological warfare agents have latent periods (i.e., a delay between exposure and the onset of signs and symptoms), it is possible that the first indication of a biological attack may be the appearance of dispersed and varied victims with a discrete set of signs and symptoms. An unexpected surge in particular signs and symptoms or a sudden surge in internet search for information on particular search terms may be indicative of an attack. Scenario J in the latter part of this book addresses aspects of public health surveillance in its various episodes.

Syndromic Surveillance

The spectrum of medical surveillance can range from informal efforts to very elaborate reporting schemes. At one end of the spectrum, there are simple schemes whereby medical providers are given a list of suspicious conditions and a phone number to call if they see something odd. Elaborate medical surveillance schemes may include such efforts as data mining of internet searches for signs and symptoms, sales of medicines, data reporting from insurance forms, analysis of patient at hospitals and clinics, and reporting schemes from all of the GP practices in a particular area. Many excellent articles on medical syndromic surveillance are available.[6]

Medical surveillance schemes should make a serious effort to gather information from across the health sector, not just from physicians. Pharmacists, nurses, EMS staff, alternative health providers, and even veterinarians can all have a part to play in detecting suspicious circumstances. Suspicious deaths of birds was an early indicator of West Nile virus in New York City.[7] Medical surveillance can pose issues with regards to data protection and health privacy laws. It may be necessary to ensure that personal identification information is stripped from any reporting done under a medical surveillance regime.

Pathogen Surveillance in the Air

One possible way to do surveillance for pathogens is air sampling. A special event could install air samplers in areas of concern in and around major event venues. Samplers could use dry filter units to concentrate particles out of the air, and such samples could be periodically collected and analyzed by a competent public health laboratory. This would, for example, be an excellent way to look for anthrax spores. Such surveillance has been done for years by the US federal government in the form of something called the "Biowatch" program.[8] Whether or not the Biowatch program is effective is up for debate,[9] but such techniques are certainly an option.

Wastewater Surveillance

Yet another form of surveillance is the detection of hazards in the effluent in a municipal wastewater system. Pathogens, and possibly other hazards, may be detected in the sewage in a city by analysis of the contents of wastewater. Such studies have long been used to estimate the prevalence of illegal drug use in some cities and regions.[10] Various analytical techniques to detect the SARS-Cov-2 virus during the COVID-19 pandemic[11] give us the hope that some types of biological hazard to major events could be monitored through wastewater surveillance.

DEALING WITH THE "WORRIED WELL"

CBRN/HAZMAT incidents are likely to have a psychological component. Medical plans must address the likely appearance of "worried well" in many scenarios. These are people who are physically unaffected but who believe that they might be affected by the incident. It is in the medical sphere where the worried well may become the most operationally important. The employment, real, threatened, or imagined of CBRN materials will cause many people to suffer from worry and anxiety. Many people will think that they are ill from exposure to such materials, even if they are not. A few useful studies have been done. Lt Col Fred Stone, a PhD clinical social worker and US Air Force officer, provided an interesting overview in a study that he did for the USAF's Air University.[12]

The problem is exacerbated by the fact that many CBRN/HAZMAT substances do not provide immediate signs or symptoms of exposure. Illness may take a long time to emerge. Other substances may have signs and symptoms that are vague, nonspecific, or commonly produced by mundane and/or routine conditions. Headaches can be a mild symptom of many chemical warfare agents. The so-called flu-like symptoms of the early stages of anthrax can easily be mistaken for more benign conditions, and vice versa. Many of the recognized symptoms of an acute anxiety attack are similar to some of the symptoms of mild exposure to nerve agents. If a causative agent is not firmly identified, it may be difficult or impossible to discriminate between types of casualties, and it is also possible that anxiety and stress may exacerbate mild signs and symptoms of actual exposure.

Nothing I write should be construed to suggest that "worried well" never need medical attention. It is altogether possible that many people in the "worried well" category will require some help. There is an acknowledged medical condition known as "crowd syndrome" characterized by anxiety, sweatiness, shortness of breath, and related symptoms.[13] A person in acute distress may need medical attention, but we must be cognizant of whether or not they require more than merely general patient care.

We do not have much history to work from in the area of CBRN incidents, but the examples that we have demonstrate distinct potential for generating large numbers of "worried well." Hypochondriacs clearly exist in most populations, and nearly every doctor will have experience with them. It is clear that stressful situations may increase the incidence rate of "worried well" seeking assistance. There is specific case history of worried well flooding the response system in the handful of serious CBRN incidents in the past. In Goiana, Brazil, in 1997, an incident involving an improperly disposed-of cesium medical radiation source caused radiation sickness and contamination; 120,000 people presented themselves for treatment, of which only 249 were actually contaminated with the cesium[14]; 5000 of the first 60,000 who sought treatment were displaying psychosomatic symptoms of radiation poisoning.[15]

The Aum Shinrikyo nerve agent attack on the Tokyo subway system in 1995 resulted in a similar phenomenon. Estimates vary, but authorities have estimated that between 73% and 85% of the people seen by medical providers after the subway incident were suffering primarily from psychogenic symptoms.[16] The anthrax terrorism in 2001 in the United States resulted in a similar situation. While 22 people developed anthrax and five died, many thousands were offered treatment. It is estimated that between 10,000 and 32,000 people were offered antibiotic treatment following the attacks.[17]

Preventing and/or Handling the "Worried Well?"

Generic readiness measures are difficult to suggest for the phenomenon of "worried well" as the nature of an incident will greatly affect the incidence rate. Because the case history of actual incidents is a bit sparse, beyond making planners aware of the phenomenon, I can only make the most general of suggestions.

While I do not necessarily know what can make the problem better, I do know what can make it worse. A response plan can do worse than to merely minimize various aggravating factors. Arbitrary, highly visible heavy-handed government actions, even if intended for the public good, can serve to provoke panic and anxiety rather than reducing it. Some examples may include:

- Responders wearing more aggressive PPE in areas where it is not strictly needed. Responders wearing PPE can be depersonalized in the eyes of the public and may provoke fear. This may have been mitigated somewhat by the COVID-19 pandemic, but we do not know how long this will last.
- Aggressive actions by police and security services.
- Impersonal and harsh quarantine measures. Authorities may react, out of fear or ignorance, by imposing quarantines as a reaction to conditions that are not contagious from person to person.
- Closing roads or buildings or modes of transport unnecessarily.

Public information has a role to play as well. Although we address public information and media relations elsewhere, I must stress that direct, clear, and factually correct statements to the public and media are necessary to combat the "worried well" phenomenon. Public statements will be scrutinized very closely and the major media organizations will quickly find fault if there are factual errors. This will only serve to sabotage credibility. Public statements regarding "the anthrax virus" in 2001 (anthrax is a bacteria, not a virus) did not help build credibility, as anyone with five minutes time on the internet discovers that anthrax is a bacteria, not a virus.

Few things are a worse gaffe than having some middle-grade public health civil servant known only to his colleagues and neighbors read out a mumbled and highly technical public health notice on television. Get someone the public knows and trusts to do it if you can, not a doctor who has never addressed a television camera in his life or a politician who mangles the science. Bad news is not improved by botching the delivery.

Mental health professionals, to include social workers, counselors, and mental health nurses, and related professionals need to be incorporated into emergency plans. They have skills and experience to deal with some of the aspects of the "worried well."

LESSON LEARNED: INTENSIVE CARE CAN BE THE CRITICAL SHORTFALL

If we take a systematic look at the ways in which people might get sick or die from CBRN or industrial HAZMAT, we come to a few interesting conclusions. The chemical hazards, whether they are irritants, vesicants, nerve agents, riot control agents used in excess, cyanides, or corrosives, point to signs and symptoms that involve the respiratory system. If chemical exposures get to the point of requiring hospitalization or threatening imminent death, many or even most of these scenarios require serious airway, ventilation, and/or other respiratory support. Such measures may not be the only medical interventions required, of course, but they may very well be needed in addition to other medical interventions. If you start looking at biological warfare threats, such as anthrax or botulinum toxin, one also gets into many scenarios where much support to the respiratory system may be needed in order for a patient to survive.

All of this leads to the question that was asked in various stages of the COVID-19 pandemic. Do you have enough equipment, personnel, and hospital space to provide the intensive care that victims may need? Depending on the scenario and the material involved, you might need a lot of intensive care support all at once (nerve agents, for example) or spread out over a number of days (biological agents and blister agents, for example).

The lesson for the major events operating environment is the same as that drawn from COVID-19 response. You should understand your intensive care capabilities and capacities and your abilities to expand such capabilities and capacities and re-allocate resources if needed in an emergency situation. Merely looking at hospital beds and ambulance capacity is the underinformed layman's way of looking at medical capacity. A patient who needs a ventilator can die in a hospital bed because a ventilator is not available. A patient who needs atropine can die for lack of atropine. You need to drill down to a deeper level of detail to understand the capacity of your systems.

REFERENCES

1. (a) Moran, C.G. and Webb, C. (2008). Lessons in planning from mass casualty events in UK. *World Health* 359: 3–8. (b) (2018). *Disasters and Mass Casualty Incidents: The Nuts and Bolts of Preparedness and Response to Protracted and Sudden Onset Emergencies.* Germany: Springer International Publishing. (c) Barbera, J.A. and Macintyre, A.G. (2003). *Jane's Mass Casualty Handbook: Hospital: Emergency Preparedness and Response.* Surrey, UK: Jane's Information Group.
2. Haslam, J.D. et al. (2021). Chemical, biological, radiological, and nuclear mass casualty medicine: a review of lessons from the Salisbury and Amesbury Novichok nerve agent incidents. *British Journal of Anaesthesia* 128: e200–e205.
3. McIsaac, J.H. (2010). *Hospital Preparation for Bioterror: A Medical and Biomedical Systems Approach,* 7–9. Elsevier.
4. Ibid, 219–227.
5. Bikkina, S., Manda, V.K., and UV Adinarayana Rao (2021). Medical oxygen supply during COVID-19: a study with specific reference to State of Andhra Pradesh, India. *Materials Today: Proceedings* https://doi.org/10.1016/j.matpr.2021.01.196.
6. Pavlin, J. (2007). Medical surveillance for biological terrorism agents. *Human and Ecological Risk Management* 11: 3.
7. Eidson, M. et al. (2001). Crow deaths as a sentinel surveillance system for West Nile virus in the northeastern United States, 1999. *Emerging Infectious Diseases* 7 (4): 615.
8. National Research Council (2011). *BioWatch and Public Health Surveillance: Evaluating Systems for the Early Detection of Biological Threats: Abbreviated Version.* National Academies Press.
9. United States. Government Accountability Office, Timothy M. Persons, and Chris Currie. *Biosurveillance: DHS Should Not Pursue BioWatch Upgrades Or Enhancements Until System Capabilities are Established: Report to Congressional Requesters.* United States Government Accountability Office, 2015.
10. Lai, F.Y. et al. (2016). Cocaine, MDMA and methamphetamine residues in wastewater: consumption trends (2009–2015) in South East Queensland, Australia. *Science of the Total Environment* 568: 803–809.
11. (a) Venugopal, A. et al. (2020). Novel wastewater surveillance strategy for early detection of COVID–19 hotspots. *Current Opinion in Environmental Science & Health* 17: 8–13.(b) Hata, A. et al. (2021). Detection of SARS-CoV-2 in wastewater in Japan during a COVID-19 outbreak. *Science of The Total Environment* 758: 143578.
12. Stone F. The "worried well" response to CBRN events: analysis and solutions, Counterproliferation Paper 40, Air University, US Air Force, 2007. www.dtic.mil.

13. Richmond, C. (2010). *Special Events Medical Services*, 88. Sudbury (MA): American Academy of Orthopedic Surgeons and Jones and Bartlett.
14. Ibid, pp. 4–5.
15. Petterson, J. (1988). *Perception vs. Reality of Radiological Impact: The Goiania Model*, 84. Nuclear News.
16. Stone F. The "worried well" response to CBRN events: analysis and solutions, Counterproliferation Paper 40, Air University, US Air Force, 2007. www.dtic.mil. pp. 6–7.
17. Ibid, p. 9.

CHAPTER 9

Preparedness in the Law Enforcement, Security, and Intelligence Sectors

Books on CBRN and HAZMAT threats often focus on what makes such incidents and accidents unique. Likewise, terrorism studies often try to draw a distinction between acts of terror and more conventional crime. But terrorism and accidents in this CBRN/HAZMAT space also share many features with more "normal" hazards. The risk of both terrorism and accidents are seriously mitigated by sound application of basic security and safety principles. Even very mundane and routine security measures can provide a safer environment, notice hazards before they get serious, and hinder possible terrorist acts by deterrence and detection.

Effective policing and law enforcement, even in the broadest of general terms, also serves to reduce risk in this area. Even things as routine as traffic enforcement and neighborhood policing can discover, deter, or disrupt hostile acts such as surveillance or even an attack. Although CBRN threats are different, routine policing plays a role in prevention. Further, response to CBRN/HAZMAT situations will require police support. Private security also plays a role in this segment. Issues specific to private security will be addressed in Chapter 11.

Intelligence collection and analysis is important as well. Intelligence and security services have a job to do, and their routine duties will involve collecting information about threats to major events. The CBRN angle poses interesting challenges, but does not undermine or change the basic principle that intelligence and security services have a fundamental duty to detect terrorist plots. Finally, the legal system must be prepared to deal with the aftermath of CBRN terrorism. This may involve forensic evidence collection, laboratory analysis, and effective prosecution. The weighty issues of forensic response to such incidents rediscussed in Chapter 19.

CBRN and HAZMAT Incidents at Major Public Events: Planning and Response, Second Edition. Daniel J. Kaszeta.
© 2023 John Wiley & Sons, Inc. Published 2023 by John Wiley & Sons, Inc.

UNDERSTANDING THE PROCESS OF CBRN TERRORISM

Government policy often makes a distinction between antiterrorism and counter-terrorism. Antiterrorism is generally composed of actions designed to prevent terrorism. Counter-terrorism is composed of actions responding to acts of terrorism while they are happening. Law enforcement, security, and intelligence services have a key role in antiterrorism.

Preventing a terrorist incident is always preferred to reacting to an incident once it has occurred. Prevention requires proactive thinking, as well as traditional passive security measures. The most important thing that security services, police, and law enforcement can do to support the safety and security of the major event is to conduct actions that detect, deter, and/or disrupt the "Process of CBRN Terrorism."

The process of CBRN terrorism is a simplification that represents the basic steps and procedures by which terrorists select a target, acquire CBRN materials, and make an attack. There have been many thousands of acts of terrorism over the last 150 years. Few have been truly random acts of opportunism. Most major terrorist acts have been the result of some planning process, whether improvised or deliberate. A mentally ill lone actor may simply get up one morning and make his mind up to shoot into the crowd standing on the street corner, just because the voices in his head tell him to do so. This sort of incident is hard to detect or deter. But most adversarial events have some sort of activity ahead of time that could be detected, deterred, or disrupted. Terrorist groups seeking to perpetrate violence in the name of some cause will generally follow some kind of process to select a target and execute an attack.

Steps in the process or cycle of terrorism: There are logical steps that terrorist groups or a well-organized perpetrator may take in order to execute a terrorist attack on a major event. Many experts have come up with similar procedures,[1] but this list really just represents a broad consensus of the principles in use for many decades. Each step can be detected, deterred, and/or disrupted by police activity, security processes, or intelligence activities.

Target selection: Terrorists will select a target, using criteria that make sense to them. Target selection criteria may not make any sense at all to the security services. Ideology, resources, security features at possible targets, and many other factors will be part of this process. Experts have considered a wide array of factors that may go into the target selection process.[2] In some groups, ideas of a charismatic leader or concepts from a perverse ideology may heavily influence the process, as may have been the case in the Tokyo Sarin attacks. Additionally, target selection may be modified or changed due to the results of reconnaissance.

Assigning personnel: A group may need to decide how many people are needed for an attack and which people to assign to the attack. Sometimes additional personnel may be recruited for part of the process. For CBRN attacks, technical specialists may need to be recruited, and this will take more time than recruitment of normal group members. Group leadership, group dynamics, and other factors (ideological purity, special talents, etc.) may play a role in this.

Selection and acquisition of materials: The group may make a deliberate choice to use a CBRN weapon, as opposed to more conventional means. Materials will need to be manufactured or acquired. Group members may need training on technical tasks and may seek information on how to properly accomplish this task.

Construction of a device: Except for very rudimentary attacks, some sort of device or mechanism of dispersal will be needed. The technical expertise needed to manufacture the threat materials or the operational expertise to steal materials may not be the same as the engineering knowledge required to construct an effective device. Some types of device will require significant effort and resources, or may require multiple attempts.

Testing of a device: In many situations in the past, terrorist groups have tested the improvised weapons that they have constructed. It may be necessary to conduct multiple tests to get a device to work as designed. If a group feels the need to get their attack done in an optimum manner, the group leadership may insist that various tests be performed to ensure that their device(s) function in the intended manner. This has often been the case with conventional devices and weapons. In particular with CBRN weapons, there is less experience and knowledge available in their construction, so the probability of a weapon failing is higher unless some testing is done.

Reconnaissance and Surveillance: Terrorist groups usually conduct extensive reconnaissance of potential targets. In some examples, several iterations of reconnaissance have been performed, first for target selection and then for development of specific plans for execution of the incident. For example, several of the September 11 hijackers conducted several reconnaissance trips prior to the infamous attacks.[3] Surveillance efforts may be simple observation or could take on aspects of rehearsals, such as abandoning a package or attempting entry into a secure area to see what the normal security response might be. In many historical situations, the terrorist group members conducting reconnaissance were not the same individuals who conducted the attack.

Briefing: Many terrorist attacks have been perpetrated by small groups specifically assembled for the purpose. Historically, many terrorist incidents were perpetrated by people who do not normally work together. A group leader may need to provide instruction to such a group.

Rehearsal: Simple schemes may need no rehearsal. However, complicated plans are more likely to succeed if rehearsals are undertaken. Many terrorist groups have rehearsed attack plans before implementing them. Even the poorly executed Tokyo Sarin attacks had at least a limited rehearsal.

Movement to contact: The perpetrators will need to move themselves and/or the CBRN device to the place where the attack will be executed.

Execution: Execution is the actual perpetration of the act. Some plans have multiple phases. Terrorists may need to use conventional force to gain access to a target area or may need to employ ruses and distractions. The attack execution itself may involve multiple phases. One example is the planting of "secondary devices" designed to detonate an interval after the initial attack. They are generally intended to injure or kill responders. Eric Rudolph, perpetrator of the bombing of the Olympic Park in Atlanta during the summer 1996 Olympics, as well as a number of other bombings, frequently planted secondary devices.

Exploitation: In many situations, terrorist groups conduct activities designed to exploit the situation caused by the incident itself. The simplest may be sending messages to media claiming credit for the event. The release, either actively or passively (by letting them be discovered by security services) of "suicide videos" of the last statements of suicide bombers is another example.

Escape: In the post-9/11 environment, it is easy to forget that not all terrorism is suicide terrorism. The majority of terrorist plots have at least a nominal plan for the participants to escape. In many circumstances, terrorists have blended in with escaping members of the public in order to evade detection or capture. The Tokyo Sarin attackers all escaped from the subway system.

Logistics: Logistics is not a step in the process. It consists of the ongoing support required to conduct all of the stages of the process. In many cases, establishing a logistical support structure must come first. Some planned terrorist attacks may require significant resources. Transportation, operational funds, means of communication, accommodation, food, and other necessities will require support. Because logistics underpins all of the steps above, it represents a vulnerability for security services to exploit. All aspects of the logistics chain behind a terrorist event represent opportunities for the plot to be discovered.

ANTI-TERRORISM: PREVENTING OR DETERRING AN ATTACK

Every step in the "process of terrorism" represents opportunities for the authorities to disrupt the activities of terrorists and prevent, delay, hinder, or otherwise diminish a terrorist attack. The authorities can break this process by both active and passive measures. Police agencies and the criminal justice system are generally focused on apprehending wrongdoers, prosecuting them, and obtaining legally sanctioned punishment for them. However, this is not the only useful mission of police and security institutions. Often, the "success" of the police services is defined by how many people were arrested and how many successful convictions were obtained. An arrest means an offense was committed. The fact that an offense was committed means that prevention and deterrence measures failed. If an offense occurred and someone is arrested, in some ways this is a failure of security, because the offense was not detected, deterred, or disrupted. Sometimes, authorities neglect to consider that preventing bad things from happening is more important than seeking convictions in court. It is the role public security and safety agencies of any country to act firmly to prevent terrorism, and not just lock up the perpetrators after the fact.

Legal environments differ in every country, and sometimes within different regions of a country. Proactive measures to disrupt, confuse, or otherwise hinder terrorist activity can cover a wide variety of behavior. Not every measure is legal, available, or even advisable in every circumstance. Some prevention and disruption measures may backfire if they break the law. When such measures exceed the expectations of the society they are meant to protect or are performed poorly, the political leadership or citizenry may react. Harsh, heavy-handed measures by police and security services can actually serve to make the problem worse by increasing disaffection with the state, thus leading to increased extremism.

Some examples of measures to detect, deter, and disrupt, while by no means exhaustive, are listed here.

Intelligence Collection and Surveillance

Security services can undertake a wide variety of measures to collect information on groups or individuals that may be willing to undertake acts of violence. The whole panoply of intelligence disciplines may be used to prevent terrorism. Intelligence and surveillance activities can be overt or covert in nature.

Placing suspected individual members of terrorist groups under obvious or easily detected surveillance is an overt method, often used for purposes of disruption as well as intelligence collection. By forcing group members to go to great lengths to avoid surveillance, security services can make it harder for groups to meet or communicate, as well as making operations more time-consuming. Heavy-handed surveillance efforts are a powerful tool and must be used with discretion. Indiscriminate surveillance efforts can have unintended consequences. Surveillance efforts could backfire and push people along a radicalization pathway.

Police officers who live and work in the community can be an incredible surveillance asset. They could easily be in a position to observe changes in activity, notice suspicious actions, or collect information from the public that could be useful intelligence.

Public Engagement

Governments have undertaken a wide variety of general and targeted information campaigns to counter-terrorist activity, with a similarly wide level of success. Further, there have been efforts to get people to report suspicious online activity. Efforts to focus such campaigns on collection of information relevant to CBRN terrorism may bear fruit. This also works on a more grass-roots level with community policing. If police have a good relationship with the people that they are meant to serve, there may be more members of the public willing to report suspicious activity.

Active CBRN Detection Measures

Some CBRN threat materials, principally but not exclusively radioactive materials, have signatures that are detectable from a distance, particularly if efforts to hide them are substandard. Some security services have engaged in proactive detection efforts to try to detect such materials. Controversially, it appears that the US FBI used radiation detection technology around mosques in the years after the 9/11 attacks, a project that caused criticism.[4] Done in a general manner, this technique is unlikely to be fruitful, as widespread detection efforts will require large amounts of personnel and equipment that might be better utilized. However, in conjunction with intelligence information to focus the search effort, active detection techniques may be useful.

Covert Operations and Infiltration

Intelligence services sometimes seek to undertake actions deliberately designed to disrupt the operation of terrorist groups. Groups may be infiltrated, not just for the purposes of intelligence collection but also for the purpose of disruption. Even the rumor of infiltration can cause an atmosphere of suspicion within groups that hinders the effective planning and execution of operations. While many people have criticized the FBI's COINTELPRO program for its excesses in the 1960s, there is no doubt that it heavily affected the ability of white extremist groups (such as the Ku Klux Klan) to operate.[5]

Conventional Policing

Efforts to provide increased conventional police presence, both uniformed and plain clothes in and around major event sites during the weeks and months up to a major event have both direct and indirect values to antiterrorism. Suspicious activity may be detected or deterred. Reconnaissance activities may be more difficult to conduct. There are clear instances in recent history of terrorist acts being disrupted because of conventional police activity. Development and maintenance of good relationships between the police and the civilians in and around the major event site will mean that police personnel will be more likely to gain useful information from the populace. Even routine traffic enforcement can gain intelligence, improve safety (thus reducing the risk of HAZMAT accidents), or even get lucky and disrupt a terrorist attack.

Physical Security

A safe and secure event rests on a bedrock of sound physical security practices. Preventing access to restricted areas, safeguarding property, and detection of intrusion are all important activities. Physical security measures can serve to make terrorist reconnaissance difficult, deter attacks on well-protected targets, or disrupt the execution of an attack by forcing unforeseen delays. Surveillance and intrusion detection technology can aid in prevention and countersurveillance by identifying possible terrorist group members and aid response to incidents by providing high quality surveillance data to responders. Physical security measures at facilities that handle or house commercial and industrial chemical can help to prevent their theft or dispersal through sabotage.

Placing too much emphasis on venue and site security will create a hard target, but will not do much to prevent attacks on soft targets outside the perimeter of the hard targets. At some point, there is a point of diminishing returns for hardening major event venues, particularly when thousands of members of the public are traveling to and from the events.

Information Security and Data Protection

Information security is the protection of data. In this modern digital age, much of the surveillance and reconnaissance done by terrorists can be done in the digital realm. A surprisingly large amount of information useful to terrorists is available online, so we must consider information security measures to be a part of the countermeasures available to security services. Even seemingly harmless information about security measures, when compiled and aggregated with other information, can provide useful intelligence to an adversary. Much of the physical security infrastructure now relies heavily on the cyber domain, such as CCTV and access control. There has been much concern about "cyberterrorism" as an end unto itself, but we must not forget that cyberterrorism may be used as a multiplier to boost a CBRN attack.

Intelligence Analysis, Data Fusion, and Intelligence Dissemination

Collection of information is only one part of the intelligence process. Effective analysis of the information and the dissemination of useful intelligence products to organizations that can act on the information are just as important as the collection process. In many instances in recent history, intelligence agencies had information that could have prevented attacks, but the information was not analyzed effectively, quickly, or in the correct context with other information. Nor was it always shared in a timely and efficient manner. Many of the recommendations of the 9/11 Commission involved the collection, handling, and sharing of intelligence.[6]

Effective Interagency Cooperation

It is very rare for a single security service to be responsible for all aspects of a major event. The size and scope of major events means that many different agencies will need to work together; this is a theme that I repeat several times in this book. Security and antiterrorism operations are most effective when intelligence is shared effectively across organizational boundaries. Security efforts for major events are likely to bring dozens, if not hundreds, of agencies into a common effort. Interagency cooperation is mandatory in such an environment.

Effective International Cooperation

The ease of communication and transportation across international boundaries means that even purely domestic terrorist groups often operate in international contexts. Information sharing across international boundaries is needed to combat threats to major events.

Border Control

Effective border control procedures can serve both as deterrents to the travel of terrorists and as an operational disruption to their activities. In recent years, border control measures have disrupted some terrorist operations by preventing some group members from joining an operation. Even if border security does not always result in arrest, detention, deportation, or some other overt result, information of value may still be collected. Passport information, visa information, entry and exit details, and other information can be collected and used in intelligence analysis.

Making Access to Materials Difficult

CBRN terrorism requires special materials. Measures to make it difficult to obtain or manufacture threat materials will make execution of CBRN terrorism more difficult. A wide variety of nonproliferation and security measures, from the strategic to the most basic tactical level can be useful in interdicting terrorist operations. These measures range from simple security provisions to safeguard materials all the way to international measures to detect and deter trafficking.

PREPARING POLICE TO OPERATE IN CBRN ENVIRONMENTS

Police and security agencies will be needed not just to prevent hostile acts. They will be needed during and after CBRN/HAZMAT incidents as well. However, the nature of these tasks is such that advanced planning and preparedness is needed. Although the situation, globally, is less dire than when the first edition of this book was written, many agencies are insufficiently trained or equipped for police and security support to major events in a CBRN/HAZMAT environment. The procedures discussed in Chapters 4 and 5 are useful to help identify situations where police personnel may be needed, as well as where and when police personnel might need PPE.

Understand Resource Requirements

One area that will be particularly demanding of police personnel is the handling and exploitation of crime scenes. Thorough investigation of crime scenes requires resources. The burden in personnel and equipment will be increased once the complications of operating in a CBRN/HAZMAT environment are taken into consideration. We must assume the presence of hazards at the crime scene(s), which will necessitate that many of the personnel involved in forensics or cordon duty will be required to operate in PPE. This will require the use of two-person teams for entry, with the likelihood of fixed entry times, based on the endurance of both the team members and the PPE.

In many locations, the protocol for responding to contaminated crime scenes will rely heavily on specialized national-level teams. Specialty teams exist, but they represent a pool of finite depth. It is important to critically question how quickly such assets will be able to arrive on the scene of an incident or how quickly their capacity will be expended during the management of medium or large incidents. The need for large quantities of people to process hazardous and/or contaminated crime scenes could exhaust many specialized units. Reliance solely on highly specialized assets may not be the soundest course of action. Planners should consider training and equipping other investigators to be able to work in contaminated crime scenes. At a minimum, trained technicians with several days of CBRN training could serve as a useful auxiliary to a more veteran unit. Teams who have environmental crime or clandestine laboratory experience may have most or all of the requisite skills and training already, as such environments are not patently different from a CBRN scene. Crime scene technicians with little or no experience in hazardous crime scenes could be trained to safely operate in such environments a week or two of training. Consideration could be made to pairing less-experienced investigators with senior personnel to stretch the number of entries that can be made into contaminated crime scenes.

Implement Incident Management Systems

In many countries, including the United States, incident command systems (such as ICS) first arose in the firefighting services, not in police operations. Many police and security agencies are "late adopters" of such systems. Because of regulatory or legal requirements, many police agencies have, over time, formally adopted or have been forced to adopt incident command procedures. However, many have, anecdotally, done a relatively poor job of incorporating them into practice. Getting incident management systems out of the book and into use in the field is a key part of law enforcement and security preparedness.

Equipment Considerations—Fielding CBRN Survival Kits

A key obstacle to performing police missions in a CBRN environment is the lack of equipment to allow police officers to survive in dangerous environments. In many places, few police personnel have ready access to PPE as shown in Figure 9.1. This leads to firefighters referring to police as "Blue Canaries."[7] One approach to fight the "Blue Canary" is to field a basic CBRN survival kit. Some departments and agencies have done this. Such kits often contain a filter-based respirator, a suit, gloves, and boots or boot covers. I was part of such an effort in the US Secret Service.

A tremendous amount of tactical capability can be gained by giving field police personnel respiratory protection and a basic capability to protect against liquid exposure. Of course, fielding such equipment places a training burden on the organization, but in my experience, basic familiarity with such a kit only takes a small effort from a competent instructor. Some law enforcement agencies take a cue from the military and incorporate some basic survival skill training into initial training for recruits.

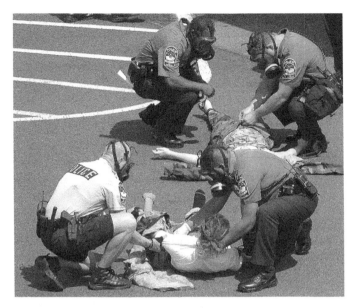

FIGURE 9.1   Specialized PPE for police is now more widely available.
Source: Photo: U.S. Air Force, public domain image.

Integrate Basic Skills and Awareness into Police Training

Police can play a vital role in early recognition of CBRN/HAZMAT incidents, but only if they are given the skills and knowledge to do so. Firefighters and HAZMAT technicians are usually dispatched assets; they do not patrol the city looking for problems. Police, however, are on the street looking for problems. Therefore, it is merely common sense to give the police some basic instruction on what to look for. A training program should include signs and symptoms of chemical injuries and basic indicators of CBRN terrorism, all of which are widely available in many reference documents.[8]

VIP Protection Considerations

The nature of major public events is such that very important persons (VIPs) are likely to be in attendance. These may be of such prominence that they bring their own security details with them. Presidents and Prime Ministers go to many major events. Indeed, it was the experience of managing major event security that earned the Secret Service the role of leading "national special security events." The presence of VIP protection teams will complicate response to major incidents. The larger the event, the more likely it is that you will have multiple VIPs with their security teams. A major international event, like the Olympics, could conceivably have dozens of presidents and prime ministers.

While this book is not about VIP protection, there are three main points to be made. First, what makes sense for public protection may not make sense from the viewpoint of a VIP protection team. The way they react to an event may run contrary to what would make the most sense for protecting the public in a given scenario. Second, training and procedures will vary between different agencies providing the protection. Third, and most importantly, advanced planning and liaison can work well to mitigate any problems from the first two. For example, the US Secret Service routinely manages a seemingly chaotic confluence of security personnel from around the world arriving in New York City every year. Significant effort goes into coordinating the entire event to prevent undue incidents.

LESSON LEARNED: COUNTERSURVEILLANCE—PRETEND LIKE YOU ARE THE BAD GUYS

There are methods that are relatively useful for detecting terrorist reconnaissance and disrupting possible terrorist attacks. Countersurveillance is the detection of surveillance. Police and security services should not rely solely on luck or coincidence for their routine operations to encounter or disrupt hostile reconnaissance. Many measures may be taken to detect the target selection and reconnaissance phases of the terrorism process.

While exact tactics and techniques vary, one general approach is to critically examine high value targets, events, and venue within your major event environment from the perspective of the possible attacker. Ask yourself "if I were going to conduct an attack, where and when would I need to be to conduct the necessary reconnaissance." Such an analysis would yield the places and times where hostile reconnaissance would be most useful in aiding an attack. You can then deploy both overt and covert measures to detect and deter such reconnaissance. For example, you could install CCTV coverage and increase both uniformed and plain-clothes police presence in the identified areas.

The same principle could be extended to disruption of an attack. If there are times and places, which would potentially be a good window of opportunity or excellent vantage point to stage an attack, you could position both obvious and unobtrusive response assets in the areas rated as most problematic.

In both cases, reconnaissance and actual attacks, it's likely that you will not have enough resources to cover every conceivable vulnerable spot all of the time. You can develop a matrix or scoring system to help you discern the most troublesome combinations of time and place and place other ones in some sort of rank order. This methodology works not just for CBRN incidents but for all kinds of conventional threats as well.

LESSON LEARNED: USE THE 1-2-3 RULE

A very useful practice that has been adopted in the United Kingdom and elsewhere is the "1-2-3 rule" for the assessment of possible CBRN incidents. Because their role places them on the streets, on patrols, and in community-based policing roles, police may be the first to encounter a potential CBRN incident. In recognition of this fact, police forces in the United Kingdom have adopted something called "Step 1-2-3,"[9][10] which I believe is a useful codification of common sense.

1-2-3 is guidance to police officers that recognizes that incidents with multiple casualties can be CBRN incidents. One person who is ill and has collapsed for no apparent reason is simply treated as a normal medical response in accordance with existing protocols. Two people collapsing at the same time, however, is a very suspicious circumstance and the situation needs to be addressed with caution. The presence of three or more unconscious people for no apparent reason is quite possibly a CBRN/HAZMAT situation, and this should trigger response plans. When used properly, "1-2-3" is both an early warning tool and a valuable piece of safety advice.

REFERENCES

1. United States Army Training and Doctrine Command (2007). *A Military Guide to Terrorism in the Twenty First Century*. Washington (DC): US Government.
2. Drake, C. (1998). *Terrorists' Target Selection*. New York: St. Martins Press.
3. National Commission on Terrorist Attack upon the United States. The 9/11 Commission Report. Washington, (DC): 2004. p. 242.
4. Kaplan D. Nuclear monitoring of Muslims done without search warrants. US News and World Report. 22 December 2005.
5. http://vault.fbi.gov/cointel-pro/kkk
6. 9/11 Commission report, 399–419.
7. Manto, S.E. (1999). *Weapons of Mass Destruction and Domestic Force Protection: Basic Response Capability for Military, Police & Security Forces*. Carlisle (PA): United States Army War College.
8. US National Fire Academy, FEMA (1999). *Emergency Response to Terrorism Self Study: FEMA/USFA/NFA-ERT:SS*. Emmitsburg (MD): National Fire Academy.
9. Doel S. Terror tactics. Police Magazine (UK). The Police Federation of England and Wales; 2006, pp. 16–17.
10. Chilcott, R.P. and Wyke, S.M. (2016). *"CBRN incidents." Health Emergency Preparedness and Response*, 166–180. Oxford, UK: CABI Publishing.

CHAPTER 10

Preparedness in the Firefighting, Rescue, and Hazardous Materials Disciplines

Most CBRN and HAZMAT incidents do not involve a lot of traditional firefighting. But in most of North America, hazardous materials response and rescue of victims in urban environments are tasks largely associated with fire departments. Other agencies may provide capabilities in this area as well. Inevitably, even in countries where fire departments are not the lead agency in CBRN/HAZMAT response, fire services tend to become involved because they have the ability to handle large quantities of water, which is useful for decontamination as shown in Figure 10.1. In most places where I have worked or served, fire departments have reasonable experience in normal hazardous materials response, which is a good basis for preparedness. This chapter is written largely from the perspective of adapting fire service responses to major CBRN/HAZMAT incidents, but the information here is still relevant if other agencies are responsible for dealing with this class of threat.

ADAPTING THE FIRE SERVICE RESPONSE TO CBRN/HAZMAT INCIDENTS AT MAJOR EVENTS

Fire departments may have training, procedures, and equipment for adequately addressing HAZMAT incidents of typical commercial and industrial scope normal found within their area. Adapting the existing knowledge and experience to major incidents is often problem is one of scale. The scope of routine incidents is normally much less than the potential size of CBRN/HAZMAT incidents at major events. Principally, the number of potential victims is far higher at major event than in most industrial and transportation accidents, which comprise the vast bulk of HAZMAT callouts. Provisions for decontamination and rescue will need to be larger. A secondary factor may be the size of venues. Sizes and distances may be longer at major events, given the size of venues. Consequently, HAZMAT teams may need to make more entries in the hot zone than in smaller events, because the combination of longer distances and

CBRN and HAZMAT Incidents at Major Public Events: Planning and Response, Second Edition. Daniel J. Kaszeta.
© 2023 John Wiley & Sons, Inc. Published 2023 by John Wiley & Sons, Inc.

FIGURE 10.1 Early in an incident, fire engines can provide emergency decontamination.
Source: Photo: US Air Force, public domain image.

limited air supply mean less "time on target" is available for essential tasks. At the planning and preparedness stages, it is important to have plans that accommodate these factors.

Such planning must heavily concentrate on logistics. The scale of major events means that the fire service components of the potential response need to consider their logistical requirements carefully. The planning stage is when logistical shortfalls need to be identified, not twelve hours into a response operation. The synchronization matrix process described in Part II can help to highlight where there may be any shortfalls.

Supply of Consumables

Water, fuel, and power are important logistical considerations. Decontamination operations will need lots of water. Provision of adequate water is important to advanced logistical planning. Similarly, provision of electrical power may be important in sustained operations. Provision for refueling of response vehicles may also be needed.

Supply of bottled air is important as well. Firefighters rely heavily on SCBA for respiratory protection, supply of air can become a critical failure point in operations. Planners need to make sure sufficient replacement of air cylinders are available and that refills occur. It is important to note that, generally, the decontamination team outside the hot zone can operate in a lower level of PPE than the entry teams. In many circumstances, this means that the decontamination crew can probably operate using filter respirators, thus reducing the air cylinder replacement burden.

Personnel Sustainment

Personnel sustainment is a considerable planning factor as well. There is a limit to human endurance. Operations in PPE will increase the burden on response personnel. Sustained operations must include plans for relief of personnel. It will be necessary to work out how many people will be needed in a given scenario.

Protracted operations in hazardous environments will increase the likelihood that responder personnel will need medical assistance. It is the standard practice among US HAZMAT teams to do pre-entry and post-exit medical monitoring of team members entering a hazardous environment. However, this practice is not universal around the world. I highly recommend it as a useful health and safety measure that will reduce injury, illness, and disability. Even very simple and quick checks for blood pressure, body weight (I've seen a number of teams with a bathroom scale in their kit), body temperature, and pulse rate, will help to reduce the incidence of injuries. I have personally lost three pounds in water after a Level A entry. In fire departments where many or most personnel have medical training, there should be little obstacle to implementing medical monitoring. But in many parts of the world, fire brigades do not serve as the local EMS provider and may have to rely on the local EMS provider for such support.

RESPONDER SAFETY

Finally, planning for responder safety is a huge consideration. Terrorist incidents add various safety and security concerns to an already dangerous occupation. There are several areas for concern. Protracted operations in PPE pose health hazards due to dehydration and heat stress. In fact, there are circumstances where the stress of long stretches in Level A or Level B may pose a higher risk of death than the CBRN/HAZMAT material itself.

Because of the World Trade Center collapse, structural collapse has accounted for more firefighter fatalities at terrorist incident scenes than any other mechanism. While structural collapse due to building fires is a known risk and one that is relatively understood by firefighters, terrorist use of explosives can cause a wide variety of damage that is different in nature from normal fires. The mechanisms that cause structural collapse in terrorist bombing situation can be different from those in a normal fire.

A large safety concern is secondary devices. Terrorists have been known to plant additional hazardous devices at an incident scene for the express purpose of targeting responders. Many terrorist groups will view the emergency services as their true enemy. In some instances, "secondary" devices may be more destructive than the "primary" device, whose real purpose was to draw responders into a trap. Normal firefighting and HAZMAT response does not concern itself with secondary devices. Training and education are required to improve awareness among fire/HAZMAT responders.

PLANNING FOR DECONTAMINATION

Almost invariably, fire departments will be involved in the planning efforts for mass decontamination. This is, naturally, due to the ability to access and move large quantities of water. Firefighting equipment remains the most feasible means of putting lots of water onto a problem in most civil settings. The scope of the decontamination problem will be larger at major events simply because the number of victims may be much higher. This will be a significant factor in planning and preparedness, because decontamination is often one of the most resource-intensive aspects of CBRN/HAZMAT response.

The concepts of the threat basis and the planning threshold, discussed earlier in the book, are particularly important here. The planning threshold should include a requirement for decontamination. Because of the different types and categories of decontamination (see Chapter 16), a planning threshold that breaks down the various types of people and things likely to need decontamination will be a useful basis for planning.

Decontamination operations are likely to be one of the most demanding parts of any CBRN/HAZMAT response. Decontamination is hard work and it will be necessary to account for sufficient numbers of staff to be able to do it correctly. Hard physical exertion while wearing PPE means that personnel will have to be rotated. A lengthy decontamination operation may need four people for every position on the chart. With enough advanced notice, it may be possible to augment firefighting personnel with other types of personnel for various aspects of decontamination.

The success or failure of decontamination operations may depend on having enough water in the right place at the right time. In arid climates or areas beyond municipal water supplies, this may be a critical factor. The site survey process will be helpful in identifying any shortfalls in this area. The planning process needs to identify the areas where water will be in short supply. It is also important to remember that the same water supplies may be needed for traditional firefighting, so your logistical plan should account for both decontamination and firefighting needs contemporaneously.

RESCUE IN THE HOT ZONE: WHO AND HOW?

Major CBRN or HAZMAT incidents at a public event may differ from the usual transportation or industrial HAZMAT accident in that they may create situations where a large number of people need rescue. Traditional hazardous materials response in transportation accidents and industrial problems concentrate heavily on aspects of response, such as detection, mitigation, and decontamination. What they often overlook is rescue as illustrated in Figure 10.2.

Analysis of the Problem

Some CBRN/HAZMAT situations will cause people to be incapacitated and also harm them to the point that they need help immediately, lest they die. Within the CBRN realm, this would be largely cyanides, nerve agents, and conventional injuries that are associated with some sort of CBRN device, such as blast or fragmentation injuries from a bomb. In a situation like this, there is a narrow window in which to save lives. But you also do not want to add to the problem by getting responders injured or killed.

References in this area are few. One that I can recommend is by US firefighter Robert Wagner. He has discussed the issues of CBRN rescue at length in a 2021 thesis at the Naval Postgraduate School entitled *Saving our own: Maximizing CBRN Urban Search and Rescue Capabilities to Support Civil Authorities*.[1] He focuses on a nuclear detonation environment, but the overall analysis is credible and broadly applicable to many other CBRN situations, and it is available online.[2]

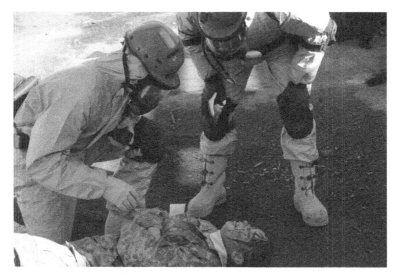

FIGURE 10.2   Rescue in a CBRN/Hazmat environment is possible, but needs to be thought through thoroughly.
Source: Photo: US Army, public domain image.

Victims in the Hot Zone

Not everyone will be able to get out of the hot zone under their own power. People could be injured (both from conventional and CBRN/HAZMAT sources of harm), disoriented, and/or entrapped. In many scenarios, victims may need decontamination. Decontamination will take time, because a decontamination process needs to be established and then the process itself consumes time. People will need to get to the decontamination area. But if people can get to the decontamination area under their own power, how bad off are they? The people who need the most help are those who are injured and cannot walk or crawl or run to the decontamination line. There have been numerous training exercises where the fire department set up a textbook decontamination line while victims who otherwise would have survived were dying inside the building.

Use Firefighting as an Example

If we look at a generic rescue situation from a structural firefighting perspective, firefighters arrive at the scene of a house fire. They have PPE designed to protect them in many firefighting environments, but of course not all of them. The on-scene commander has to make a very quick evaluation of the hazards. Is someone alive inside? Can firefighters get to them? Is the building going to collapse? Are there other hazards present that are going to make the situation even more hazardous than it already is? Firefighters understand house fires, so there is a good knowledge base for making this type of risk assessment. No fire chief wants to visit the widow of three firefighters who died searching for someone in an empty building that collapsed. But if there is someone trapped and there's a way to get the person out, there are processes to do this.

Deficiencies of Traditional Approaches

CBRN and HAZMAT situations present unknowns. The methodical approach by HAZMAT responders to an incident is not one conducive to rescue. The typical HAZMAT team response is a follow-on response to an initial callout involving an industrial or transportation accident. Rescue of victims did not figure much in the methodical drills I learned in 1999 at HAZMAT technician training. Conduct a scene assessment. Establish scene control and perimeters for hot, warm, and cold zones. Initiate incident command and appoint a safety officer. Establish a responder decontamination corridor. Do an initial reconnaissance entry. Take some readings and samples. Come out through the decontamination line. Work out a detailed plan. Perhaps do another reconnaissance entry. By the time some victims are discovered, they are either dead (and thus it's a body recovery exercise and not time critical) or have managed to escape on their own.

Compromising in the Middle: Developing a Rescue Protocol

Is there a way to compromise? Do you have to wait for a slow and methodical technical analysis of the hazards in order to try to save people? My answer is a qualified no. The 2003 US Army publication[3] on firefighting protective ensemble mentioned in Chapter 7 has devoted some considerable thought to the problem. The report is, as of the time of writing, available in several locations online and I encourage readers interested in rescue in CBRN incidents to read it.[4] Based on the 2003 report, it is possible to come up with a basic rescue protocol to rescue people in a CBRN incident while wearing firefighting protective clothing and well-fitted self-contained breathing apparatus (SCBA).

Putting together my own expertise and blending it in with the 2003 report, here is a possible guideline for rescue in such environments:

Conduct a rapid risk assessment: Are there live victims present? The presence of people actually alive in the hot zone means that the situation is not so acute as to totally preclude a rescue. The presence of at least some living victims works as a de facto chemical detector. If there is no sign of live victims, it may not be worth the risk to conduct emergency entries merely for reconnaissance purposes. This might be a situation where the risk exceeds benefits. CCTV or unmanned means (drones or robots) might be useful for such reconnaissance in order to seek the presence of living victims. Consider the presence of secondary devices that may be intended to harm responders

Has the situation peaked or is it getting worse? If concentrations of vapor hazards are increasing rather than decreasing, it is too dangerous to conduct an emergency entry because the potential exposure cannot be estimated. If the method of dissemination was a single event, like an explosion, the Army document suggests that 10 minutes after an explosion is a reasonable rule of thumb to follow. If something like a spray device or leaking tank is involved, and it is continuing to emit liquid, gas, vapor, or aerosol, then the concentration of hazard is likely to be increasing, not decreasing.

Conduct a rescue: A team of properly clothed firefighters, with SCBA and gloves, can enter and retrieve victims. They should avoid liquid contamination as much as possible. Victims with grossly contaminated clothing should have as much of that clothing cutoff as possible as quickly as possible. Do not mess around too much with lifesaving measures in the hot zone. If they are still alive by the time you get there, the handful of extra minutes in getting them away from the immediate hazard may be better spent in moving them. Keep the total entry to less than 30 minutes. The Army study found that in a nerve agent environment (the most acute hazard they could envisage), 30 minutes in turnout gear and SCBA minimized but did not totally eliminate the risk of mild nerve agent exposure to responders.

Turnback guidance: Rescuers in structural firefighting clothing and SCBA should be briefed to immediately turn around and seek decontamination if they get any liquid contamination anything other than impermeable boots. If the rescuers cannot find live victims where they thought they would be, they should turn back and go through decontamination. Again, there is the issue of secondary devices. Suspicious objects that might be secondary advices should trigger a turnback.

Detection instrumentation: Rescuers could carry detection sensors with them if they are available. However, rescue efforts to save known live victims should not wait for arrival of specialty instrumentation. A photoionization detector, commonly in use either on its own or as part of an industrial gas monitor, can provide reasonable detection of many types of gases and vapors and should be carried if available.

Decontaminate both the rescuers and the victims: Conduct an emergency decontamination of both rescuers and victims with low pressure water. If it is cold, be very careful about the hazards of hypothermia. If there is a piece of clothing contaminated with the suspected agent, bag a sample of it and leave it in the warm zone. It may be useful for agent identification, particularly if the threat material evaporates quickly.

Conduct medical surveillance of the rescue team: Because you might not know what the identity or concentration of the threat is at an early stage of incident response, the team that did the emergency extraction needs to be medically evaluated at the earliest opportunity. Remember, many of the threat materials have delayed onset of medical effects.

Technical extrication is difficult: If victims are entrapped and there is a need for more advanced technical rescue, this complicates the emergency rescue mission. This will have to depend greatly on training, equipment, and the judgment of personnel at the incident scene. It is not easy to assess this in the abstract in advance of an incident. Using a tool to force open a locked door is well within normal firefighter tasks and a rescue guideline should not necessarily prohibit such basic tasks. But if victims are trapped within a collapsed building and need proper urban search and rescue training and equipment to extricate, such a scenario is likely to put an emergency rescue mission out of bounds.

CBRN/HAZMAT MITIGATION

Fire department and HAZMAT team responders are best positioned to conduct various emergency mitigations of CBRN/HAZMAT incidents. This section discusses a few tactics and techniques that might be considered for mitigating an incident. By "mitigating" I mean any measures taken with an incipient CBRN/HAZMAT situation to either prevent an imminent release of materials or make an existing release less hazardous.

The basic premise behind most HAZMAT responses is to stabilize the situation so that no further damage to life, health, property, or the environment occurs. Most civil HAZMAT responders are versed in a variety of mitigation techniques designed to contain the problem. This brief exploration should not replace a detailed study of product control and mitigation tactics and techniques commonly taught in HAZMAT technician courses. The major categories of mitigation techniques are as follows, in no particular order:

Diking, Damming, and Diverting

This is the use of dirt, sandbags, or other material to build a temporary channels, barriers, or similar measure to direct a release of liquid into a holding area, divert a release away from an area where it will cause more harm, or to divide water from liquid contaminants.

Dilution or Dispersal of Gas or Vapor

Use of air, water, or water fog to lower the concentration of harmful vapors or dilute a liquid. For example, positive pressure ventilation fans could be used to disperse harmful gases, vapors, or aerosols. It should be noted that this tactic does not usually do much to neutralize hazards, but moves them or spreads them to a large area but with lower concentration.

Firefighting

In some HAZMAT situations, the underlying cause of the dispersal of hazardous materials is a fire. In such cases, firefighting is a valid mitigation technique. Similarly, cooling hot containers to prevent an explosion is broadly in this category.

Moving, Plugging, and Patching

If a tank, container, or device is leaking material, then effecting emergency repairs might be a plausible tactic. Normally performed by hand, there are numerous kits and accessories designed for these tasks, as well as a long history of field expedient improvisation.

Explosive Ordnance Render-Safe Procedures

EOD technicians have a wide variety of tactics and techniques for rendering ordnance or improvised devices safe.

Movement of Material

Moving containers or materials to a place where they are less of a hazard is a valid technique. Even the basic technique of shifting or moving a container or vessel so that the hole is higher up might stop or slow release of material. Pushing a rucksack that is emitting vapor into a broom closet is not entirely unhelpful in some scenarios.

Manipulation of Valves and Fittings/Product Transfer

These are commonplace techniques which are often useful in more traditional industrial and transportation HAZMAT accident. Things being released through a valve may be stopped by manipulation of the valve. Transferring materials into another vessel is also a valid tactic.

Absorption, Adsorption, or Chemical Treatment

Things can be added to liquids to absorb or adsorb them. Chemicals can be used to treat solids or liquids.

Foaming

Various types of foam, such as firefighting foam or soaps and detergents, can be used to cover liquids to prevent or at least slow evaporation.

Overpacking, Manual Removal, or Covering

This tactic involves placing a threat into a container or the act of covering of a liquid or solid (i.e., a powder) with sheeting or other material to prevent evaporation or further spread. It could be as simple as putting a bucket over a small device or putting plastic sheeting over a puddle of dangerous liquid.

Venting, Flaring, and Controlled Release

These are more industrial-type processes. Release of a relatively non-dangerous gas, vapor, or liquid to reduce pressure in a container. This could be used as a tactic to avoid an explosion of containers like tanks. Flaring is the deliberate burning-off of combustible material to reduce overall danger in a situation.

The extent to which civil HAZMAT product control techniques are applicable during a CBRN terrorist incident will depend entirely on the situation, as well as the initiative and ingenuity of the responder. Certainly, the conventional product control strategies described above have some application.

Forensic Considerations

Because of forensic considerations, it is worth assessing just how much containment, mitigation, and product control is satisfactory to the accomplishment of the strategic goal. By all means, stabilize the situation and keep the bad stuff from spreading. But remember, the responder's role is not to remediate the scene for immediate reoccupation. The responder is saving lives, protecting property, and keeping the environmental damage to a minimum. A serious CBRN or HAZMAT incident might take weeks, months, or even years of specialty work to remediate.

LESSON LEARNED: FOLLOW THE HART TEAM

The problem of rescuing people from contaminated environments is still not one that is readily embraced across the emergency response disciplines. Such situations combine the worst aspects of urban search and rescue with chemical environments. Sharp corners and debris can easily tear up "Level A" and "Level B" gear. The whole issue of who does rescue, and how, in a CBRN environment, is still not fully addressed.

It is important to look for positive examples in this area, and one of them is the Hazardous Area Rescue Team (HART) concept started in the United Kingdom.[5] This is an initiative that came out of the ambulance services in the United Kingdom, not the fire services. (In the United Kingdom, emergency medical services are primarily performed by ambulance services that are part of the National Health Service, not the fire departments.) The HART teams are a useful template to follow. They provide useful rescue capability in a variety of hazardous environments, including CBRN and HAZMAT situations. They are also capable of a number of basic and advanced life support procedures in a contaminated environment. The overall concept is not radically dissimilar to the work I did in VIP protection for the US Secret Service HAMMER team, only with the public as the beneficiary, not just a single VIP.

REFERENCES

1. Wagner, R. (2021). *Saving Our Own: Maximizing CBRN Urban Search and Rescue Capabilities to Support Civil Authorities.* Diss. Monterey (CA): Naval Postgraduate School.
2. https://calhoun.nps.edu/handle/10945/67190, accessed 1 March 2022.
3. US Army Soldier and Biological Chemical Command (2003). *Risk Assessment of Using Firefighter Protective Ensemble with Self-Contained Breathing Apparatus for Rescue Operations During a Terrorist Chemical Agent Incident.* US Government.
4. https://apps.dtic.mil/sti/pdfs/ADA440863.pdf, accessed 2 February 2022.
5. Price, J. (2016). Hazardous area response teams: celebrating 10 years in the making and counting. *Journal of Paramedic Practice* 8 (8): 390–393.

CHAPTER 11

Preparedness and Response in the Private Sector and "Third Sector"

All too often, emergency planning and emergency response training focus on public sector provision of emergency services to the neglect of the private sector and "third sector." By "third sector" I mean charities, nongovernmental organizations, nonprofits, religious groups, and individual volunteers. This sector has had significant roles in disaster response and recovery for many decades and cannot be ignored or dismissed. In particular, this chapter discusses how to integrate private sector and third sector personnel and entities into prevention and response.

PRIVATE VERSUS PUBLIC SPACES

When we start talking about major public events and urban areas, the terms "private" and "public" get very intermingled very quickly. The demarcation line between private and public will vary greatly across the world and even within countries or cities. A high percentage of major "public" events occur at venues that are, in fact, privately owned. In London, where I live, there's often no visible demarcation between something that you might think is public (such as a street) and a plaza or square, which may actually be private.

One can also make a number of distinctions between property owners, property managers, tenants, and contractors. The legal relationships and where responsibilities lay can be quite complicated, disputed, or nuanced. It's not unreasonable to have a privately owned building, managed by a different company who is the property manager, occupied by dozens of private companies or individuals as tenants, and serviced by security and maintenance staff from outside contractors. This building could then be surrounded on all sides by public property in the form of sidewalks and streets owned by the local municipality.

CBRN and HAZMAT Incidents at Major Public Events: Planning and Response, Second Edition. Daniel J. Kaszeta.
© 2023 John Wiley & Sons, Inc. Published 2023 by John Wiley & Sons, Inc.

FIGURE 11.1 In many cities, many areas that the public routinely use are technically private property.

Just because something is a major incident or terrorism does not negate the fact that all of these private entities have responsibilities. Historically, terrorism has sometimes fallen into the "too hard to do" category for property owners and property managers. Regardless of the situation, there is a duty of care to the public who enter properties and a duty of care to their employees who work there. This duty of care does not begin and end with any one person or entity. It won't always be easy to make the case that private entities can or should do much to help the emergency planner. But this is the environment in which we must work.

PRIVATE SECURITY PERSONNEL

Private security staff play a large role in keeping society safe. In many modern cities and at many types of major public events, there will be more private security personnel than personnel from the public sector emergency services. Most event venues will have a significant private security footprint. Some private security functions are provided by in-house staff, but more often than not, the security staff are provided on contract by manned guarding firms.

The private security contractor can be the most neglected aspect of major event security preparedness. As a practical reality, private security contractors will be present in some form at most major events. The presence of private security personnel can range from the incidental, such as small numbers private security guards at facilities near event locations all the way up to

FIGURE 11.2 Private security staff are essential in most, if not all, major events. *Source: Photo: US Air Force, public domain image.*

thousands of security guards that are integral to event security efforts. As a general practice, the rank-and-file security personnel at most sporting venues around the world are private security guards. Private security has yet to evolve much of a CBRN or HAZMAT perspective and very little has been written on the subject. However, certain aspects of the private security industry's role in major event security are applicable to this discussion.

Profit Motive

The private security industry is typically driven by loss prevention. The customers and employers of private security personnel do not want to have loss to property, injury to employees or visitors, or have situations that expose them to legal liability. The employment of security staff is often considered as one of many forms of risk management. In many contexts, employment of security staff is one of the most expensive forms of mitigating risk, so many other measures are employed to reduce the number of security personnel required. If it were ultimately cheaper to have fewer or no security personnel, the commercial motivations of businesses would make such a thing happen. If there are security personnel at a private site, mostly it is because there are direct and indirect financial motives for it to happen. This is significant when it comes to major events, because major events may require additional security personnel. However, the additional expenditure may have to be justified in terms that a private company's financial team understands.

Limitations of Private Security

The slang term "rent-a-cop" is commonly used to describe private security staff. Rarely it is accurate. The tasks and missions that private security providers can perform are typically restricted by laws, regulations, and the specifications provided by the customer. These tasks and missions often take the form of internal policy or clauses in a contract. The managers and supervisors responsible for major event security may find their ability to direct private security personnel prescribed by law, contractual vehicle, or policy.

When it comes to major event planning, it is incumbent that you understand what laws and regulations say that a private security provider can do, as well as what the existing contractual relationships provide for. More than once in my career I have accidentally uncovered situations that were problematic, either legally or contractually.

Rivalry with "Official" Security and Law Enforcement

Police often look down on security personnel as "rent-a-cops" that are in an inferior position in the security and law enforcement hierarchy. In some area, relationships between police and private security are strained. In a countervailing manner, in many places private security staff may be former or off-duty police personnel.

Reservoir of Knowledge

Security staff may have worked at a venue for many years and have developed a unique perspective on their niche in the security business that may not be equaled by police or government security service personnel. A guard who knows every room of the floor of the building that he works in is of great value in antiterrorism. Further, the security industry is replete by former or retired military or police personnel, some of whom may have relevant expertise.

You Get What You Pay For

It is very easy to assume a general lack of knowledge on the part of private security providers. This viewpoint is usually based on the lowest common denominator, a 19-year-old drop-out or semi-literate who somehow managed to get a security guard license. There is a wide variety in the security industry, and private security is like any other field in the service sector. All sorts of people end up there for a variety of reasons. Many security personnel are former or retired military or police personnel themselves, often with decades of experience. There may be a significant knowledge and experience available in the contracted security staff, which may not be apparent based purely on job description or pay grade.

Physical Security, Crime Prevention, and Antiterrorism

As has been repeatedly stated in this book, security against CBRN and HAZMAT threats rests firmly on a bed of good physical security practices. Competent use of private security staff boosts overall physical security and situational awareness. If CCTV is used for security purposes, this can be a useful detection asset. (See Chapter 15.)

Response Protocols

If there are security staff at a local venue who are widely knowledgeable about the facility, build them into your response procedures. They can guide you through a facility, unlock doors, and general expedite assessment of possible incidents.

FACILITY MANAGEMENT STAFF

The personnel who actually manage the physical aspects of a particular site, such as facility managers, facility engineers, and maintenance staff, are the unsung heroes who make day-to-day life in urban environments possible. This book has already discussed some of the major aspects of venues in Chapter 6. The site surveys and analysis of a venue's vulnerabilities will only be truly effective if you do them in conjunction with the appropriate staff at a particular venue. There are a few things that you will definitely need to work out with the facility management staff.

Situational Awareness

Facility management staff are actually a very good source of situational awareness. If you want to know if something is seriously out of place or different, it is often the cleaner who cleans a particular floor every day that can notice it. The person who receives incoming shipments every single work day is probably best suited to notice new or unusual parcels or deliveries. It will not be the police or the ambulance services that will be the first to notice an unusual number of staff illnesses or absences.

Establish and Rehearse Occupant Emergency Plans

Part of the duty of care to building occupants and visits is the establishment of "occupant emergency plans." Nearly everyone in the modern world understands fire drills in buildings, and occupant emergency plans evolved out them. Fire is a known hazard and everyone understands the need to get out of the building in a rapid but orderly manner. Similarly, bomb threats, gas leaks, tornadoes, and other hazards have caused the development of other procedures. Large buildings have "occupant emergency plans," and these may be required by policy, by law, or by insurers. The development and exercising of such plans is a valuable component of emergency planning.

Emergency evacuation: Everyone understands fire evacuations and fire drills. If there is a CBRN/HAZMAT problem inside a building, getting people to leave the building is often a sane approach and uses the building's own structure to keep the problem bottled up.

Include shelter-in-place scenarios: Sometimes, occupant emergency plans have not given adequate consideration to CBRN/HAZMAT scenarios. Often, the general reflexive approach to occupant safety is to evacuate the building. But, as will be discussed elsewhere, evacuating a building is not always the best way of protecting people. Sometimes, shelter in place is a more useful tactic to provide protection to people. In some parts of the world that need to deal with sudden adverse weather, such as tornados, the concept of seeking refuge inside a building is not a foreign concept.

Vertical evacuation: "Vertical evacuation" is the concept of moving up inside a building instead of emptying the building out. Bear in mind that storm shelter areas are often in low-lying areas and may or may not be the best place to seek refuge in CBRN/HAZMAT scenario, because most of the CBRN/HAZMAT threats are heavier than air. Make sure your planning process is flexible. If there is a CBRN/HAZMAT threat outside the building, and your air intakes are 20 stories up at the top of a high-rise, then it makes much more sense to move people upstairs inside the building than to evacuate everyone out onto the street, where the problem is most acute.

De-conflict evacuation routes: Evacuation plans that have the net effect of dumping hundreds or thousands of people into an area that has been identified for other purposes, such as decontamination or staging of response vehicles, can play havoc with your overall management of an incident. Use the site survey process and the synchronization matrix process discussed in previous chapters to identify potential problem points.

CHARITIES, VOLUNTEERS, AND "UNAFFILIATED RESPONDERS"

If a large incident occurs, there is a distinct possibility that many people will turn up as volunteers, both skilled and unskilled, to assist with the response effort. There are also charities and NGOs, which provided extensive support during and after disaster situations. These range from well-established and well-respected organizations like the Red Cross, who have formal roles and established procedures to very ad hoc local efforts from neighboring communities. NGOs with established track records in disaster response like the Red Cross are not likely to pose problems and issues for you. Problems, issues, and difficulties result from more ad hoc or amateur efforts, even ones that are very well intended. For further guidance, I suggest reading Chapter 15 of the American Bar Association's excellent book *Homeland Security and Emergency Management: A legal guide for state and local governments*[1] for additional guidance on volunteer management.

Volunteers represent a potential pool of labor and resources. It can often be assumed that some number of people will turn up wanting to help and the incident management scheme should be prepared to handle such an eventuality. Several authorities have studied this problem, particularly in response to disaster events.[2]

Problems

The problems with unaffiliated responders are numerous. People may deliberately or inadvertently misrepresent their skills and training, or their skills and training may be inadvertently over-estimated by responders. Incident managers and on-scene commanders cannot be expected to know what to expect from volunteers. Not every doctor is a trauma surgeon. Volunteers may be hard to incorporate into a command system and may not follow orders, operate in an unsafe manner, or do more harm than good. Unaffiliated responders are more likely to be unfamiliar with hazards and could hurt themselves or others. Having loose or no accountability of extra people at an incident scene is a breach of scene control. There have been many tales, both documented and apocryphal, of expensive equipment going missing from incident scenes and looting taking place inside police cordons. It is not impossible that criminals or terrorists would attempt to cause further damage or achieve financial gain by infiltrating the response effort in the guise of volunteer responders.

Opportunities

It is quite possible that a pool of volunteers may include a wide variety of desperately needed skills. Many professionals who are near an emergency situation have an ethical and professional duty to help out if they can. Doctors, nurses, other medical providers, and off-duty emergency response personnel of every description may present themselves for duty. For example, members of the British Medical Association were confronted with one of the 7 July 2005 bombings at their door step and provided critical support to casualties. There may be large scope for putting useful skills to work in a response effort, even in relatively unskilled roles. For example, credentialed off-duty police officers from other jurisdictions could guard abandoned property, and speakers of other languages could work as interpreters in a setting where many people may not speak the host language, such as international sporting events. Some skills, such as language ability, may be self-evident, but most skills will require documentation and verification.

Managing the Issues

The best course of action is to assume that volunteers of some description will turn up. You can establish some guidelines for how to manage them.

Establish a policy: Like other aspects of incident response, it is always better to have a policy ahead of time. I think that a blanket "no volunteer" policy is unrealistic, possibly antagonistic to some professions, and likely to be broken in a severe incident anyway. Therefore, it is incumbent upon planners and responders to establish who will be responsible for various aspects of volunteer management. Establish both rules and guidelines for what volunteer responders can and cannot do. There is no harm in having a policy that says "the following forms of identification are allowable as identification as a doctor" and "a physician who is board certified in X can do the following XYZ tasks under the direction of the on-scene medical commander." It is superior to making it up as you go along.

Evaluate the volunteers: You should find a way to evaluate arriving volunteers. First, you want to verify who they are and what actual credentials and licensing situation is. Second, evaluate them for utility. What is it that they can do for your situation? You might have too many of one thing and not enough of another. If there is no formal role for them in the particular disaster, consider nonspecialist ways in which they can:

Engage the professional organizations, registries, and NGOs: Since major events often have a significant planning timeline and are usually well known, there may be opportunities to engage with professional associations and NGOs, such as medical associations and the Red Cross as examples, to establish scope for volunteering in advance of a serious incident. Professional associations can be very valuable resources and could provide advice on how to check the credentials of volunteers and how to establish the level of competence of a volunteer responder. There are also existing schemes like FEMA's Community Emergency Response Team (CERT) that may be useful as a basis. In medical professions, the US government is operating a scheme called the Emergency System for Advance Registration of Volunteer Health Professionals (ESAR-VHP),[3] which is intended to alleviate some difficulties in this area.

Volunteer Coordinator as a Management Role: A volunteer coordinator should be established in the incident management system if a significant number of volunteers start to turn up. This serves to delegate the problem down one or two levels in the incident management hierarchy but also serves to provide a central point of contact. The volunteer coordinator is responsible for documenting everyone turning up and checking identification. Check their identification and verify any operational credentials. Basic checklist could be made available to screen volunteers. A central reporting point for volunteers should be established for the purpose of managing any "walk-ons." Pre-planning will help this task considerably.

Keep in control of the situation: Keep a record of volunteer responders along with the rest of responders at the scene of an incident. Once you put someone to work, regardless of the task, you assume a degree of responsibility for their actions. It could be viewed in legal cases in a way similar to as if they were a paid employee. Therefore, incident managers need to exercise control over the who, what, where, and when of volunteer response. Crime scenes and hot zones are probably out-of-bounds for all but the most tightly defined and regulated volunteer responders. An off-duty HAZMAT technician from a neighboring city with a mutual aid agreement in place, with clearly understood and verifiable credentials, is one thing, but a "walk-on" who says "I'm a volunteer firefighter back in Country X" is something completely different.

Logistics and sustainment: Some volunteers will turn up with nothing. Eventually they will need to be fed and watered. They will get tired. You will have to work out a scheme for dealing with the logistical needs of volunteer responders who turn up without logistical support. One role for NGOs may be the sustainment of volunteer responders.

PPE FOR THE PRIVATE SECTOR AND THIRD SECTOR RESPONDERS

As a practical matter, CBRN/HAZMAT incidents may provide situations where personal protective equipment is needed. In emergency situations, private sector personnel may not have the level of PPE or training to be able to remain in position or in their roles. They may have to evacuate to a place of safety. Likewise, arriving volunteers or NGO staff may have to be kept well away from hazards because they will not be able to protect themselves. In the past "well, the private sector has no training or PPE so we don't even bother to talk about them in CBRN/HAZMAT planning" has been an excuse. There are several ways to approach this issue.

First of all, remember that much of this book is about prevention, deterrence, and advanced planning. It is all of the stuff that is "left of the bang"—before an incident—on a chronological timeline on a PowerPoint slide. The private sector has a lot of work that they can and should do to aid antiterrorism by good basic security principles and situation awareness. It is important to remember that, just because people do not have PPE or the relevant training, that they do not all instantly become inanimate casualties or victims waiting for a response. You will want the building managers to do things with ventilation and drainage, or open or close doors, or other such things. You will want private security guards to evacuate floors or buildings, or manage a shelter-in-place plan. The people you need to deal with do not just disappear.

Second, the stuff that is quite far to the "right of the bang"—longer term response and recovery, will need private sector assets. This is also the point at which the third sector, those charities, NGOs, and volunteers, may come into the peak of their usefulness. This will be at a point in incident response where you have, most likely, divided up the hazard into zones, and you can safely put people to work in cold zones, even if that's a mile up the road.

Third, you actually can look at the provision of PPE and training. What PPE you are allowed to issue to people, and the training that is needed for it varies widely. This is a subject that is quite heavily regulated. It may be possible to train and equip private security staff and key employees with relevant PPE. But you will need to work it out procedurally, and that may not always be easy.

Being heavily regulated is not actually an excuse to say "ah well, let's not do it." There are various things you can do. The respiratory protection industry has been working on "escape masks" and similar devices for use to help people escape dangerous environments. These may have a role. There are other things that can be done as well. Having many boxes of FFP2/ FFP3 masks for emergency use in the event of, say, biological aerosols or radiological incidents is not actually such a bad idea. Is it the ideal PPE for every threat? No. Is it a valuable mitigation measure that will reduce exposure to several classes of hazards for a lot of people? Yes.

Lesson Learned: Do Not Assume an Anfinite Supply of Labor

The London Olympics in 2012 provides an excellent lesson for management of private security provision.[4] There was a panic as the event drew near. It turned out that there simply were not enough security staff to secure the Olympic event venues. The primary contractor had promised a certain amount of personnel, and despite significant efforts, the contractor was simply unable to hire enough staff. The armed forces were obliged to step in to fill the gap. Certainly, an unnamed industry leader, like the one that was involved here, should have understood the market dynamics of their own sector better than they did.

In most places, there are training and licensing requirements for security guards, even if these are minimal standards in some places. Security guard staff cannot be conjured out of nowhere. The ability of the security industry to disgorge additional staff on short notice is limited. If you are going to need lots of additional staff, you need to spend money on it many months upstream of the event.

REFERENCES

1. Abbott, E.B. and Hetzel, O.J. (2018). *Homeland Security and Emergency Management: A Legal Guide for State and Local Governments*, 3rde. American Bar Association.
2. Barsky, L.E., Trainor, J.E., Torres, M.R., and Aguirre, B.E. (2007). Managing volunteers: FEMA's Urban Search and Rescue programme and interactions with unaffiliated responders in disaster response. *Disasters* 31 (4): 495–507.
3. https://www.phe.gov/esarvhp/pages/about.aspx.
4. Booth R. London 2012: concern mounts over potential shortage of security guards. *The Guardian*, 8 July 2012.

CHAPTER 12

The Military—Preparing for Military Support to the Civil Authorities

Military support is often needed or requested to provide CBRN/HAZMAT. The majority of nations have some CBRN defense capability in their armed forces, and in some countries the military is a deep reservoir of capacity and expertise in this domain. In most places, this capability resides in the army, not so much in navies and air forces, for historical reasons. There is a belief, strongly held in some quarters, that CBRN is a "weapon of mass destruction problem" and is only really addressed by the military. While I think that statement is highly simplistic for many reasons, there is not any question that a "send in the army" mindset does exist.

It is my own first-hand experience that the military has no monopoly of knowledge in CBRN/HAZMAT response. I say this having served as a military CBRN officer. It is very important to realize that military CBRN defense cannot be copied, cut-and-paste style, into the major event arena. A lot of mistrust has been spread around when military units attempted to lecture local emergency responders on the CBRN threat. In reality, the militaries have a lot they can learn from the civil sector responders. I know this because I was a military CBRN specialist and I learned a lot from civil responders.

UNDERSTANDING MILITARY CBRN PHILOSOPHY

The military, by both design and necessity, views CBRN defense in ways that are quite different from their civilian emergency response disciplines. Because of these differences in philosophy, technologies, procedures, and equipment designed for military use may not have the same applicability in the civil emergency response situations that this book covers. Regardless of whether or not a military approach is the right one, it is a fact that the military CBRN philosophy figures large in this field, due in part to the presence of large numbers of former military CBRN specialists (my own self included) providing advice to the civilian emergency response sectors.

CBRN and HAZMAT Incidents at Major Public Events: Planning and Response, Second Edition. Daniel J. Kaszeta.
© 2023 John Wiley & Sons, Inc. Published 2023 by John Wiley & Sons, Inc.

FIGURE 12.1 Some military CBRN defense capabilities, like the M1135 reconnaissance vehicle, may be too specialized for many civilian missions. *Source: Cpl. Kwon Yong-joon/U.S. Army.*

Military CBRN defense philosophy dates from the First World War. Chemical warfare is, at the end of the day, just a different set of operating conditions and environment that can occur on the battlefield. Although exact philosophies of military CBRN defense vary from nation to nation, the basic underlying principles are similar. The basic premise of CBRN defense is that CBRN weapons may be used on the modern battlefield and that measures must be taken to ensure that the war can be continued despite the presence of lethal or incapacitating agents in the war-fighting environment.

The general military philosophy toward the CBRN threat can be summarized in the following operational imperatives:

Keep the Army Fighting: Military equipment and CBRN defense doctrine focuses not just on keeping soldiers alive, but keeping them doing their jobs. A soldier's job is to fight a war, not wear a gas mask. The wearing of a gas mask is a tool to the soldier to continue to do his job. A military protective mask may not be the most perfect protection available. But it has been designed to be wearable for a long time, to fit as many soldiers as possible, and to not require specialist training.

Acceptable risk/Acceptable Losses: There is no military planner who can reasonably plan for a zero-casualty environment. The politically and socially acceptable levels of casualties in military operations have changed over the course of the last century, but zero-risk is not the operating principle. CBRN defense equipment could be designed around a zero-risk principle, but such equipment is usually not useful on the battlefield, as it would be too heavy and too expensive. Military CBRN defense seeks to minimize risk in a reasonable manner, but cannot eliminate risk.

Detection philosophy: Military operation can generally tolerate a higher level of false positives than civilian operations. Also, soldiers need quick warning to be able to effectively don protective masks. This means that military detection and identification equipment have been designed to optimize speed at the risk of accuracy. Nobody likes false alarms, but the penalty for a false detection event in an artillery battalion is a degradation of performance, while a false detection event in a crowded stadium can be, potentially, a worse disaster than an actual detection.

Adapting equipment to the military environment: Military equipment must be reasonably sustainable within the established logistics system. For example, military PPE is often designed to be worn for days or weeks. As another example, equipment manufacturers have long ago learned that taking scientific equipment and painting it green is not the best course of action, as the rigors of the modern battlefield are well out of the comfort zone of many scientists and engineers.

Categorization of the Threat: Military thought on CBRN threats also provides us with the classical classification of chemical warfare agents. Agents are classified by their mode of physiological action (nerve, blister, blood, choking, etc.) and their persistency (persistent vs. nonpersistent). For the purposes of major event planning, this classification is not that useful.

RELATIONSHIP BETWEEN MILITARY AND CIVIL AUTHORITIES

Civil–military relations are often complex. The political and practical issues surrounding the use of military resources in civil settings are complicated and are very different from country to country. In some places, use of soldiers for peacetime domestic missions is rare. Aspects of it may even be illegal. In other places, it is routine and expected. It is not within the scope of this book to comment upon or analyze where the acceptable boundary between civil and military operations is placed, because the answer will always be different. In any case, real or perceived crisis situations tend to shift these boundaries.

The Nature of "Military Support"

Even the question of what actually constitutes, "The Military" differs from country to country. In many nations, there are quasi-military police services, such as Italy's Carabinieri, Spain's Guardia Civil, and the various Gendarmeries, all of which are considered part of the armed forces and typically come under the defense budget and/or the defense ministry. Much CBRN/ HAZMAT response capability for use in civil settings resides in such organizations all over Europe. There are many other examples. The United States has its Army and Air National Guards, which serve as both reserve components of the US Armed Forces and as state militias for use by state-level authorities for use in civil emergencies. Indeed, National Guard capabilities are instrumental in CBRN/HAZMAT support to major events in the USA, but are legally and administratively fundamentally different from active duty military support from the Federal government.

FIGURE 12.2 The 94th Civil Support Team is an example of a National Guard CST. *Source: Photo: US Army, public domain image.*

International Support and Mutual Assistance

It is important to think beyond national frontiers. Many countries are members of bilateral or multilateral treaties or organizations. Other countries can lend support during a crisis or in support of a major event. A recent example is the various US military support lent to Japan during the 2011 earthquake and concomitant nuclear incident. In many instances, it is military capacity in CBRN/HAZMAT that is more readily deployed across national boundaries than civil responders, who may not have the transport and logistics at hand to support an international deployment.

What Is It Reasonable to Expect from Military Support?

The military can often provide specialist support. This could mean personnel, equipment, expertise, and capabilities directly related to CBRN/HAZMAT threats. Such support could take the form of pre-packaged CBRN defense units, such as "Civil Support Teams," NBC defense battalions, companies, or platoons. Military support could conceivably include any of the following categories of assistance.

Detection and Identification

Military units can provide CBRN detection capabilities. The detection capability may be very similar to that in use in the civil sector, or it may be completely different. Military detectors often give generic responses or are optimized for speed rather than accuracy. It is important that you understand the capabilities, limitations, and differences from the equipment you are used to.

Laboratory Support

In many countries, the leading national-level laboratory or laboratories for analysis of CBRN substances actually belong under military control, so laboratory support may be a key capability. Many militaries have mobile laboratories that can push some intermediate capability, between field detection and definitive laboratory analysis, closer to the problem in the field.

Decontamination

Military CBRN units are trained and equipped to perform decontamination of large items of equipment, e.g., tanks and armored personnel carriers. In addition, personnel decontamination has long been a staple task for such units.

Medical Support

Modern militaries have doctors, field medics, ambulances, and deployable field hospitals. Many militaries have more training in identifying and treating casualties from CBRN agents. However, in peacetime, much military medical capability is in the reserve components. Mobilizing a trauma surgeon away from a hospital supporting a major event only to deploy him back three days later is a waste of everyone's time and effort.

EOD

Strictly speaking, explosive ordnance disposal (EOD) may or may not be considered a CBRN capability. However, in many militaries, EOD personnel have an extensive knowledge of CBRN weapons. EOD support will be critical to major public events.

General Labor and General Security Duties

Military units represent a potentially useful reserve of physically fit people who can perform general tasks in PPE. Some scenarios may require two hundred men with shovels. If you factor the threat environment into account, the two hundred men with shovels may need to be fit, young, and have PPE. Military personnel have a long history of general support duties during and after disasters.

Military personnel may be able to augment security at major events. Military personnel may perform venue searches. Also, in an indirect capacity, in an emergency environment, civil police may be much more needed for certain critical tasks and some security tasks, such as ones not requiring direct interaction with the public or conducting direct law enforcement roles, could be devolved to the military. For example, guard staff at public buildings could be augmented or supplemented by military personnel, so that police and other security personnel could be used in other capacities.

Aviation

Military aviation is a commonly requested asset after major disasters. Aircraft support, both fixed wing and helicopter, may be vital to some response efforts. In addition, some militaries have aircrew equipped and trained to fly in PPE. This expertise is almost nonexistent outside the military and may prove useful in some scenarios.

Transportation and Logistics

All modern militaries have a substantial logistical footprint. This means that significant numbers of personnel and equipment are in support units designed to transport goods and support ongoing operations.

Mortuary Support

Militaries may be better organized for emergency mortuary operations than their civilian counterparts. Some militaries will have deployable teams that are specialists in this area. In addition, some military mortuary teams may have equipment and/or contingency plans for handling contaminated human remains. It should not be automatically assumed that this is the case, however.

ISSUES AND PROBLEMS WITH MILITARY SUPPORT

It is easy to declare: "We need the army!" But "the army" can mean anything from an experienced senior Colonel with a PhD and 30 years CBRN experience to an 18-year-old lad who can barely put on a protective mask. Some experiences demonstrate that those who shout the loudest for the army often have the least amount of knowledge about what military support can and cannot be expected to provide. There are numerous issues involved with military support, some of which may degrade into problem situations if not effectively managed.

Terminology Differences

Terminology differs between different agencies and career fields. The gap between military and civil terminology can be severe. "Clear a building" means one thing to a firefighter and another thing to an infantry soldier.

Military Equipment Is Not Necessarily Superior

There is a commonplace assumption that military equipment, particularly in the CBRN sphere, is somehow always inherently superior to civilian equipment. While this may often be true on an individual case-by-case basis, it is not uniformly true across the board. Military equipment is designed for military missions, and these missions may not match the exact uses that a civilian responder may have in mind. For example, firefighting SCBA provides a much higher level of respiratory protection than military protective masks, but for a much shorter duration of time.

Difference Between Military and Civil Decontamination

Decontamination in a military setting is not necessarily the same thing as decontamination for civil responders. For example, one type of personnel decontamination in the US military is the "MOPP Gear Exchange"—essentially a procedure for switching dirty PPE for clean PPE. Such a process is very specific to US soldiers and US equipment, and not very applicable in a civil setting.

Military Hazard Predictions

Military CBRN specialists may turn up at an incident with tools to predict the area affected by a CBRN hazard. This capability may range from very old paper charts and tables up to sophisticated computer models on a laptop. I would advise a high degree of caution with military hazard prediction models. It is important to understand that military prediction models are largely based on notional battlefield situations outside of dense urban areas. Many forms of military CBRN hazard prediction uses a number of risk management assumptions that are designed to err on the side of safety and are deliberately designed to produce a predicted hazard area bigger than warranted by scientific calculations. If you ask a military CBRN specialist, or use some of the more primitive military hazard prediction tools for a hazard area prediction, you might get given a useless output.

Military Medical Support

Because medical care is heavily regulated, the military follow emergency medicine protocols that may not match what is expected or permissible in a civil environment. This can be managed if everyone knows about it ahead of time. For example, as a military reservist, I learned to administer intravenous (IV) fluids as an advanced first-aid skill as part of a two-day "Combat Lifesaver" course. But in my regular job in the US Secret Service, I was not allowed to administer an IV, which is considered an advanced life support skill, i.e., a paramedic-level skill. In the Secret Service, my ability to practice emergency medicine was governed by the state of Maryland medical protocols under which I was trained and licensed. As long as the legal aspects are managed and that civil and military providers know what the other can provide, this issue can be managed. Where issues will occur is in the unplanned scenarios, if military medics are dropped into an incident without much advanced warning or planning.

Integration into Civil Incident Command Structures

Military units are accustomed to a defined chain of command. Military officers are typically familiar with a staff structure that is not dissimilar to the various functions of an incident command system. This should be advantageous. However, be cognizant of situations where a military officer is ordering civilians around. If it is going to be the case, make sure you plan for it ahead of time. Some organizations, such as US National Guard units, are well accustomed to integrating into incident command structures. In other places, there is little experience with such integration, so friction can be expected.

Reservists—Robbing Peter to Pay Paul

Many of the assets that may be called in to support a major event will be reserve, territorial, or militia-type units. In some countries, such as Switzerland, the vast majority of the military is a citizen militia. One phenomenon that I noticed as a member of such an organization (in my case, the Maryland Army National Guard) is that membership of reserve components can be quite common among members of the emergency services. A rather large number of military medics may very well be medical staff in their civilian careers. At one point, I was assigned to a military reserve unit where approximately 40 percent of the noncommissioned officers were police officers of one variety or another.

Planners need to be cognizant of the fact that military reservists come out of the community and that their mobilization may just move a response asset from one box to another. Planners should take stock of how much this might affect their plans. Consideration might be made of using reserve forces from well outside the area affected by the major event. Also, it is important to note that this issue can run in both directions. Relatively minor staff in response agencies may turn out to be a Sergeant Major or a Colonel upon mobilization.

SOME IDEAS TO HELP INTEGRATE MILITARY SUPPORT

Ask for a Capability not a Particular Unit

Do not expect to be an expert in what capabilities a particularly military unit can bring to bear in a crisis. There have been situations where civil authorities request a particular battalion or regiment based on a dated or even false assumption about the sort of equipment and capability that a unit may have. It is also best to request military support in terms of explicit capability and capacity. For example: "we need to move cargo by helicopter" is more useful than "send us the 113th Aviation Battalion," which has only reconnaissance aircraft.

Include the Military into Planning Efforts

In many places, military support is considered a last resort or a heavy-handed tool to be called in when things get bad. This often means that military planners are left out of the planning stages for major events because the arrival of military support is, in some people's minds, an admission of failure. In reality, failure to properly plan is a more unconscionable sin than admitting that you need help. The capabilities that military support may or may not provide to your preparedness and response effort will not be made easier by having them descend upon your event without planning and coordination.

Military staffs understand planning and liaison. Often, a competent staff officer in a military headquarters, even a reservist, will have more formal training on contingency planning and preparedness than his civil counterpart. If there is any chance, however remote, that military units are going to be drafted in for some mission during your major event, you should seek out a representative of these units to participate in at least some of the planning effort. Within the United States, every state's National Guard has a military support to civil authority staff that will be happy to work with you.

Identify Potential Conflict Areas Early

It is likely that fire, police, and emergency medical responders already interact to a great extent in the area of the major event. That is not to say that there is not conflict or tension. We know that there often is. However, we generally know where these conflicts are. Because the police, fire, and medical responders may not have had much experience working with the military, the areas of conflict, the "pinch points"—will not be so well known. Early discussions and participation in the planning process and exercises will help to identify these potential conflict areas.

Cross-pollinate

If military support is part of your major event plan or even if it is a strong possibility, then consider having a military representative on the assessment team(s) that are being assembled for the event. In addition, it will be easier to integrate military support into a civil response if there are some representatives from the supporting units in the relevant operations centers.

LESSON LEARNED: CIVIL SUPPORT TEAMS

As mentioned above, the various state National Guards in the United States are often useful demonstrators of military–civil cooperation. In the 1990s, the US government explored many methods of helping state and local governments prepare for terrorist use of CBRN materials. Many of these involved the use of the military and defense department. While these various projects and concepts varied greatly, one that I consider to be a reasonable success is the National Guard Civil Support Team (CST).

The CSTs, which had a few other provisional names before CST, were suggested in a "Tiger Team" report in 1998.[1] This report resulted in the formation of specialty 22-person National Guard teams across the country. The first ten active-duty teams were Guardsmen (both Army and Air Guard) on active-duty status, spread regionally around the country. A number of other states developed reserve-status teams. I was the founding member of the 32nd CST in the Maryland National Guard and served as its operations officer for 4 years. Every state now has a CST, as do the major overseas territories. A few of the larger states have 2 CSTs, and there is even an overseas CST (Army Reserve, not National Guard) in Germany.

The CSTs are specifically intended to be the interface between civil emergency responders and the military in CBRN-type scenarios. A CST brings a mobile laboratory with trained technicians, a survey team, significant secure communications capability, and a variety of skills. The training that CST members receive is quite impressive and it spans both military CBRN training and civilian HAZMAT training.

In particular, the CSTs have an excellent track record in supporting major events around the United States. Numerous sporting events and various other large public events have received specialty CBRN support from CSTs. My advice to American practitioners is to pull in a Civil Support Team early on in the process and use them to help out.

REFERENCE

1. United States Army. Department of Defense Plan for Integrating National Guard and Reserve Component Support for Response to Attacks Using Weapons of Mass Destruction. United States Department of Defense, 1998.

CHAPTER 13

Other Preparedness Issues

Not every aspect of preparedness for CBRN/HAZMAT incidents at major public events falls neatly into the categories delineated in the previous chapters. By necessity, this chapter is a bit of a mixed bag and describes the various additional preparations that are likely to be needed to support the preparedness for a major event, both in terms of specific CBRN/HAZMAT measures and broader safety and health aspects.

TRANSPORTATION AND LOGISTICS

If you have followed the general processes described earlier in the book, you will have developed several different operational scenarios for responding to major incidents. Regardless of what plans you have developed, there will be logistical and transportation requirements. You will need stuff and you will need to move things (and people) around.

When you work through various planning scenarios and develop your synchronization matrices, as I suggested in the earlier chapters of the book, basically every square on a synch matrix incurs logistical requirements. People need food and water. Batteries get used up. Firefighters, being who they are, put water on things. The fire engines use fuel to do this. And so forth. A good staff officer could calculate all of these things. Use your skills and experience to work out the logistical requirements for various scenarios, well ahead of the day that you might need them. This means that, for any given operational plan, you can develop a logistics annex spelling out your requirements.

Just because you require something at a specific time and place does not mean that it will magically appear. See Figure 13.1. You will need to work out the pathways and mechanism by which goods, services, people, and transportation assets can be made available in your time of

CBRN and HAZMAT Incidents at Major Public Events: Planning and Response, Second Edition. Daniel J. Kaszeta.
© 2023 John Wiley & Sons, Inc. Published 2023 by John Wiley & Sons, Inc.

FIGURE 13.1 Having a large stockpile of emergency supplies is not helpful if you cannot move it.
Source: Photo: US Dept of Defense, public domain image.

need. My own experience is that phoning around during a crisis looking for a diesel generator is probably not as good a course of action as knowing that you thought about this ahead of time and have one ready to go in a warehouse.

COMMUNICATIONS TECHNOLOGY

The response to numerous terrorist incidents and major disasters has been needlessly delayed or complicated by lack of effective communications. Ensuring adequate communications capability and capacity is key to developing preparedness. Historical events like September 11 and Hurricane Katrina are often used as examples, but problems occur in much smaller circumstances. Response to CBRN/HAZMAT at major events will be made more difficult if communications are ineffective or if the existing capability is degraded by the circumstances of the incident.

Build Communications Infrastructure into Your Planning Process

Emergency services communication is not conducive to eleventh-hour quick fixes. A box of special radios that turns up on the night before the event is probably not going to cure any systemic illness in the communications sector. Planning for interagency operability and communication between disparate elements of an incident management structure is essential for success.

Use Technology But Do Not Let It Become a Sinkhole for Funds

The communications business is a big component of the defense industry and communications systems are heavily marketed to the emergency services sectors. Lots of heavyweight industrial vendors will be competing for money and contracts and will have "just the right thing" for your event. Very useful technology exists that can fix problems and make your life easier, but there is also plenty of scope for spending the entire equipment budget on communications equipment. I remember a time in the US Army when CBRN recon teams had state-of-the-art modern radios and chemical detection equipment from thirty years in the past.

Changes to Communication Plans Require Training

I have literally sat in an army vehicle with a radio that nobody could use. The latest and greatest, a state-of-the-art frequency-hopping FM US Army "SINCGARS" radio was thrown at my Army company the night before an exercise and nobody had been trained on it. We could not use them. Changes to equipment or procedures require training.

Reduce Reliance on Public Infrastructure

9/11 taught us that public communications infrastructure could be jammed to the point of hindering response. Far more traffic is carried on wireless services now than in 2001. An enormous amount of day-to-day communications in the emergency services relies on mobile phones, smart phones, wireless broadband, and related technologies. The conventional wisdom is that such public bandwidth is already stretched during major events,[1] so it is not hard to imagine scenarios where the public network begins to fail to meet the needs of the responder.

Decentralized Operations

In a good army, soldiers do not stop fighting merely because their company commander cannot talk to the battalion commander. An abiding principle in military organization for centuries has been that commanders need to be trained well enough to be able to exercise some independent judgment in the absence of direction from higher headquarters. Before the invention of the telegraph and the radio, large organizations were forced to operate in an environment where information took weeks or months to convey. Yet wars were won, continents conquered, and major engineering efforts completed without daily or hourly contact from headquarters. The rise of instantaneous voice and data communication and large centralized operations centers competes with the age-old, but still valid, reasons for a hierarchical command structure. If your commander cannot do his mission without contacting somebody higher up the chain of command, then there is a structural problem, not just a technical communications problem. You either have the wrong commander (wrong skills, wrong training, both) or you have the wrong procedures. Communications are important. But it is not any substitute for proper preparation.

FINANCIAL AND ADMINISTRATIVE PREPAREDNESS

Responding to accidents, incidents, and disasters will incur an administrative and financial burden. The Incident Command System and similar schemes recognize this by including administration and finance as important staff functions in the staff structure. A common shortcoming in CBRN/HAZMAT planning is to focus on the first golden hour of response. Getting the right capabilities to the right place in time to save people really is terribly important. But as the hours drag into days, the paperwork and the bills start to pile up. The following are my suggested "best practices" for financial and administrative preparedness.

Be Honest About What Things Might Cost and Where You Might Get Them

There is no point in trying to "low ball" cost forecasts. Requests for extra funding at the last minute are likely to cause trouble. It is better to be honest about what you might need. If your plan will need you to buy things, have a source in mind. There is not much point to a plan requiring 400 gallons of bleach, if you cannot find a commercial source to sell it to you.

Have Finance and Administrative People Identified for Incident Command and Operations Center Roles

If finance and administration have a slot in the organization chart for incident command (and they do in ICS), then it is important that people be identified for these roles well before an incident. Actually, getting some support personnel into an operations center in a useful capacity may take some doing. Very often, the people who work in financial and administrative roles in emergency services organizations do not wear a uniform or respond to emergencies. Their jobs are more similar to their counterparts in civilian agencies and the private sector than they are to the responders that they support. Efforts to select financial and administrative staff ahead of time and to integrate them into the incident management scheme will pay off well.

LEGAL CONSIDERATIONS

The last thing any planner or responder wants is having lawyers tell them "you aren't allowed to that." It is even worse when it happens at the last minute. We live in civilized society and we have a framework of laws. The emergency response disciplines are part of the state apparatus and need to operate within the law. Problems will happen, however, when the law has not caught up with the circumstances of the modern day. I could (and maybe will, someday) write an entire work on CBRN/HAZMAT and the law. That level of detail is not important in this book. But it is useful to highlight some important considerations about the law and emergency response that may be relevant.

Problem Areas

The most important thing that I can do is to identify key areas that either have posed legal issues in the past or have the potential to cause issues in the future. If planners can identify the problem areas that have potential to affect their ability to protect a major event, and can do so sufficiently early, then the political leadership can be made aware of the problems. Legal advice can be obtained from competent legal advisors. It is possible with enough support from the political process that laws and regulations could be modified to accommodate the realities of CBRN/HAZMAT preparedness. Because of proactive (if not always effective) legislation and litigation, many of these issues have spawned a body of case law in the United States. The same may not be true in other countries. The following sections highlight the key areas of legal difficulty that I have encountered over 20 years, and I hope that by playing "devil's advocate" for a few pages here that you can avoid trouble later. These are the kinds of issues you need to refer to your legal advisors early in the planning process.

Litigation and "good faith" lifesaving actions: Are you going to get sued for your actions, even though you acted in good faith to save lives or property? You might guide a person from an incident site and take them to the decontamination station. Decontamination may result in them being cut from their clothing and sprayed from head to toe with water and then re-dressed in a blanket or a hospital gown. Because it is a high-profile terrorist incident, the media may film the entire process. The process can be seen as a degrading assault on human dignity. Is your agency going to be sued for this, even though it is a lifesaving action? In the USA, gaining certifications and declarations under the DHS SAFETY act can serious limit liability in many potential circumstances.[2]

What is a CBRN weapon? What is HAZMAT?: We can and already have discussed what CBRN/HAZMAT is for the purposes of planning and response. But what are the legal definitions that apply for the purposes of arresting people? The actual legal definition of what constitutes a chemical weapon is often hazy. A person can legally buy a gallon of a strong acid for use in a swimming pool. The same acid can be used to produce toxic gases. Can we detain him for carrying that same legally obtained bottle of acid while standing in line to buy tickets for a major event? What, if anything, has caused the miraculous transformation from pool chlorination product to weapon? How do we articulate this in front of a magistrate without looking like an idiot? Legal definitions need to be understood, and security personnel need to know where the law is clear and where it is unclear.

Health and safety regulations: Most countries have regulations and laws about the type and nature of equipment that is required to protect workers from hazardous substances. Very often, the protective equipment that emergency responders use falls under these same laws and regulations. Sometimes there are strong differences between what the military is allowed to use and what civil responders are required to use. Does the equipment you propose to use comply with the law? If not, then perhaps responders have legal grounds to seek compensation or refuse to work.

Collection and preservation of evidence: Can the legal system accommodate the collection and preservation of evidence that could be inherently dangerous? If a terrorist uses a gallon of homemade "Sarin" nerve agent, is there a laboratory that can process forensic evidence that is contaminated with the nerve agent? Is there a process by which a sample of the nerve agent can be collected in a way that can be defended by prosecutors in court? Is your process for handling evidence subject to attack by a competent defense attorney?

Search and seizure: Can a CBRN detector be used as basis for a legal search and seizure of a suspect's person or property? If police are driving a mobile radiation detector down the street and it alerts on a vehicle, do they have the legal authority to stop the vehicle, question the driver and if need be, search it? What about fumes escaping from a ventilation duct at a suspect's premises? Can these be sampled covertly? Is a warrant required?

Emergency medicine: Laws and regulations heavily govern emergency medical care, particularly at the pre-clinical level when provided by paramedics and EMTs. Much of the existing legal environment is based on conventional situations and commonly seen medical problems and traumas. Do the emergency medical protocols in use have provision for CBRN/ HAZMAT situations and mass casualty incidents?

Environmental regulations and laws: Is an environmental regulator going to investigate you or levy a large fine on your organization for taking actions to save lives? If you change the air-handling system on a building to vent fumes to the outside in order to reduce the level of hazard (and thus admitting a toxic gas to the atmosphere), are you going to be prosecuted for violating a clean air law? If you operate a decontamination operation that ends up flushing toxic liquids down a storm drain, will you be fined for water quality violations?

Role of the military: The use of military personnel in civil environments is an interesting legal environment in most countries. There are numerous questions that could be asked. We will discuss civil–military relations in greater depth in Chapter 12. Planners need to be aware that asking for and receiving military support can put a large number of issues into play. What sort of tasks can soldiers perform in emergency situations? What tasks are prohibited? Can civil responders legally use military protective equipment? Can a military medic use his full scope of training and equipment on a civilian patient, or does he have to abide by civil emergency medical protocols?

LANGUAGE SUPPORT

Major events do draw people from all over the world. The diversity of the public at a major event will, of course, vary greatly depending on the nature of the event. However, it should be safely assumed that there will be people attending or even working in support of a major event that will not have a fluent grasp in the prevailing language. Even people who might normally speak the prevailing language as a second or third language may not use it so well in a crisis situation when they are stressed or fearful. Several articles discuss this problem in some detail.[3,4]

Consideration of foreign language users is essential to emergency planning. Protective actions such as decontamination, evacuation, and sheltering in place will require communication with the affected people. If people do not understand what is happening, then it is possible that more people could be injured, that more emergency services personnel are needed to deal with the situation, or that the overall response is delayed. Simply by having some communication materials available in commonly encountered foreign languages is useful, as are the availability of interpreters. Incorporation of graphics could be useful, such as visual aids describing an emergency decontamination process using pictures and pictograms. Obviously, this needs to be done ahead of time. In some situations, translation programs and apps could be useful.

LABORATORY AND SCIENTIFIC PREPAREDNESS

Many emergency response scenarios in the CBRN/HAZMAT domain will require some degree of expertise from the scientific community. Response efforts may result in samples that need to be analyzed by laboratories. There may be incident scenes where the situation is ambiguous or complex and will require interpretation and analysis by external experts. The two most important preparedness measures in this sector are laboratory readiness and the development of "reach-back" schemes.

Laboratory Readiness

The support of qualified laboratories will be a component of effective CBRN/HAZMAT response as illustrated in Figure 13.2. Laboratories and their staffs are important to the major event planner for several reasons:

- Confirmation of results obtained by field detection and identification techniques
- Definitive analysis of materials to determine the presence of CBRN materials, usually to a level of confidence and detail suitable for forensic purposes
- Analysis of medical samples from patients who may be ill from exposure to CBRN materials
- Forensic analysis of conventional evidence that is contaminated by hazardous substances
- Testifying in court about their findings and defending their procedures against cross-examination

In the United States and several other countries, much work has been done to institutionalize a laboratory network. There is excellent guidance for laboratories on the Centers for Disease Control (CDC) website.[5] The US government has devoted much funding and labor to establish and maintain the Laboratory Response Network (LRN), which was established in 1999.[6] The LRN is a large network of well over one hundred laboratories across the United States that are trained and equipped to provide laboratory support in chemical and biological terrorism scenarios. My own experience interacting with LRN laboratories has been generally positive. The LRN is the largest laboratory readiness effort to date and is well suited to support major events. Therefore, this section of the book scarcely applies to planners and responders in the United States. Seek out your local LRN laboratories and most of the work has been done for you.

Identifying the Laboratories

Outside the United States or in remote areas, it may be necessary to do some work to identify which laboratories are capable of providing the necessary technical assistance. For chemical analysis, the OPCW's network of laboratories is a good start. It is quite likely that you will need to find separate laboratories for chemical, biological, and radiological sample analysis. This is generally the case in the United States, and the LRN actually has different networks for chemical and biological analysis. Public health and environmental health laboratories are both good starting points. However, do not ignore private or university laboratories; in many places, they have resources superior to their government alternatives.

Proximity

Distance to the laboratory is important, but it is not the most critical selection factor. A superior laboratory at a marginally greater distance may be a better option. Additionally, it is not very helpful to have a laboratory that is so close to the major event that it may become affected by an incident. I have been evolved in emergency planning situations where our laboratory support was conveniently right down the street. In hindsight, I think that it might not be wise to rely on a laboratory close enough to be part of the problem. If you cannot get the sample to the laboratory, the laboratory is evacuated, or key staff cannot get to the laboratory, much of the readiness planning will be negated.

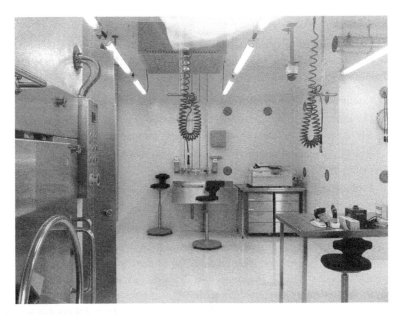

FIGURE 13.2 Laboratories are important to CBRN response operations.

Understand the Capabilities and Timelines

It is very important to take the time to meet with the laboratories in question and to define exactly what it is that they can do and cannot do. There are important questions to be asked and it simply will not do to wait until a crisis to ask them. What types of samples can be processed? What number of samples can be processed? What analytical procedures are used? It is also important to obtain an understanding of the time required for the various types of procedures in question. All laboratory procedures will take time, but the higher levels of management at a major event cannot always be expected to have a higher understanding of this. If it is going to take 24 hours for a definitive test for plague or anthrax, this needs to be understood well ahead of time so that the expectations of commanders and managers are not unrealistic. It is also important to understand which capabilities are available during normal duty hours and which ones are available at night or on weekends. What is the procedure for recalling key staff? How long does it take to get the laboratory up to full capacity at night or on a holiday?

Legally Adequate Procedures

As we have discussed elsewhere in this book, there is a difference between scientifically accurate procedures and legally and forensically correct procedures. The latter must always be the former, but the former is not necessarily adequate as evidence. The best-equipped laboratory in the world may be useless for obtaining a conviction in court if forensic procedures are not followed. Developing and fielding proper procedures can take months to implement, so it should be an early priority.

Communication Plan

Although I have done it, simply turning up at the door of the laboratory with a bag of powder is generally considered a poor policy. Communication protocols should be established to govern how the laboratory is to be alerted, both during normal business hours and after hours. Likewise, there needs to be a plan to govern the communication of the results. When the laboratory has results of its tests, whom are they going to call? I have witnessed situations where the laboratory sat on results simply because they did not know who to send them to. Backups and redundancy need to be built into the communication plan.

Packaging and Arrival of Samples

How is a sample going to be collected, packaged, and screened prior to movement to the laboratory? There may be a divergence of expectations between field personnel and the laboratory. Generally speaking, a laboratory will expect a properly packaged sample. While a biological laboratory may be well prepared to examine a sample for anthrax, they probably want some assurance that it is not an explosive, radiological or a chemical hazard as their laboratory safety procedures is not likely to account for such risks. Likewise, a well-equipped radiological laboratory may not be equipped for biological hazards and will want some assurance that the sample they are receiving is safe.

Screening Out Hazards

Laboratories will want to know what screening has been done in the field to rule out various categories of hazards. For example, while I took many samples of suspected biological origin to laboratories, I did not handle a sample that could have been an explosive device without it being cleared by an EOD technician to rule out the presence of explosives. Then I used a radiation detector to rule out the presence of radioactivity and a photoionization detector to see if there were any chemical vapors present. While none of this was conclusive, it did provide the laboratory with a degree of assurance. Consider adopting similar procedures.

Consider a Sampling Triage Plan

A laboratory will have a normal capacity to process samples. If you work with them to adopt emergency procedures, they will probably have the ability to temporarily increase their capacity. However, a large-scale incident can easily produce a volume of samples that can overwhelm even the largest laboratories. The 2001 anthrax terrorism in the United States provided such a large number of samples across the United States that public health laboratories could not keep up with the workload.[7] Even laboratories in parts of the United States well away from the incidents were overwhelmed by samples. It is possible that a laboratory may end up having to refuse samples, thus denying critical information to investigators and incident commanders.[8]

Serious consideration must be given to ways in which responders can make the laboratory's job easier and enable a faster response. Planners should consider having a procedure to prioritize samples in the event that the laboratory system becomes deluged with evidence. Qualified investigators and hazardous materials technicians can serve as "triage officers" and use field analysis methods and intelligence indicators to prioritize samples. As a very simple example, simple field screening techniques such pH paper, protein test kits, and non-specific DNA screening kits have been used to rule out the presence of pathogens in suspicious powders during presumed hoax or nuisance responses. Generally speaking, I have a relatively dim view of the operational utility of many of the biological detection devices marketed to the emergency responder market. However, such techniques could be used to triage samples into a lower priority category during a major incident. Another technique might be to use military assets to supplement the existing laboratory infrastructure. For example, I have seen situations in the United States where the Mobile Analytical Lab from a National Guard Civil Support Team would have been used to supplement existing laboratory capacity.

Exercise the Laboratory Plan

Like other response plans, the laboratory response should be tested by training exercises well in advance of the major event under both normal and stressed conditions, such as late-night hours when laboratory staff are off duty. Several training samples can be transmitted to the laboratory using the agreed protocols and the response can be observed. As a responder, I was surprised to find out how many lab response plans had not been exercised under realistic conditions.

REACH-BACK

An interesting development over the course of my career has been the development of "Reach-back" capabilities. Broadly defined, reach-back schemes are formal programs and projects designed to provide CBRN and HAZMAT technical support to responders. Reach-back schemes are generally organized as a pool of knowledgeable individuals who agree to place their technical, scientific, and/or operational knowledge and expertise at the disposal of emergency responders. On several occasions, before many programs were fielded, I have participated in some informal ad hoc reach-back schemes. Reach-back schemes may be government operated/funded or private. For example, several equipment manufacturers operate reach-back schemes to assist with interpretation of data produced by their detection equipment.

Reach-back can be a powerful resource if administered properly. Some excellent examples of situations where reach-back can prove useful include:

- Interpretation of misleading, unclear, or confusing data from sensors
- Diagnosis of casualties with unusual symptoms
- Assistance with hazard prediction and modeling
- Interpretation of data from shipping papers or container markings in a HAZMAT incident

Planning Considerations for Reach-back

No field responder, incident commander, or operations center watch-stander can possibly be expected to be an expert in all aspects of CBRN and HAZMAT response. Therefore, some sort of reach-back scheme is necessary for support to a major event. The value of reach-back needs to be tempered with some realism. It is important to remember that reach-back is not a solution for every problem and even the smartest expert is not on the scene looking at the problem directly.

Not All "Reach-Back" Schemes are Equal

Similar to the proliferation of operations centers, the variety and number of reach-back programs has increased over the last decade, particularly in the United States. Which reach-back scheme is best to use under which circumstance? The answer is not always readily obvious to the field responder. Major event planners need to establish specific guidance for responders as to which circumstances should require reach-back and which reach-back schemes to use. There may be little operational utility in three different responders asking three different reach-back centers the same question. The likely result is three different answers, which will only muddle the response. The only real answer to this is to do "comparison shopping" and work with the reach-back scheme that gives you the most confidence.

Garbage-in/Garbage-out

The subject matter experts of a particular reach-back scheme operate in a relative situational awareness vacuum. They do not see what the responder sees. They are not holding the detector in their hands. While distance from the problem will give the experts some clinical detachment from the problem at hand, they are dependent on information inputs to have enough information to provide useful input. If the reach-back experts are given incorrect, misleading, faulty, incomplete, or ambiguous input, then do not be surprised if strange advice is the result. The major event planner should interact with the reach-back provider(s) to figure out exactly what the reach-back scheme is expecting to receive from the field. Then the planner can field specific guidelines on what to report and how to report it.

Same Question/Different Answer

Specialized education and experience often coincide with strong opinions and parochial outlooks, particularly in the CBRN field. While a reach-back scheme will issue facts, it will often also issue opinions. Opinions, even well-reasoned ones based on sound logic and a good grasp of the facts at hand, can be wrong. I have seen situations where the same question and same information was passed to different experts and different advice was given. This is fundamentally part of the nature of expert advice and responders and incident commanders need to be aware of the fact that opinions from experts may vary.

Availability

Not every reach-back scheme is as truly available as it says on its web page. There are certain questions that should be asked: Is a particular reach-back scheme available twenty-four hours a day? Or does it actually require staff to be recalled or contacted at home during off-duty hours? What are alternative ways of reaching the reach-back hotline? The heated moments during an incident are not the time to be asking these questions. I would estimate that many reach-back customers are under the belief that their reach-back provider(s) are sitting in an operations center waiting for the phone to ring. Generally, these experts all have a regular job and get reach-back calls very infrequently.

Commercial Reach-Back

There is always the possibility that reach-back schemes operated by equipment manufacturers may be the strongest when dealing with the company's own products. I do not know of a specific instance of a problem occurring, but many response teams will use a wide variety of equipment from different manufacturers. Will company X give the most useful and informative interpretation of data from company Y's detector?

Training and Evaluation

If a reach-back scheme is going to be relied upon for technical support during a major event, it needs to be tested and evaluated. The best way to do this is to include the reach-back scheme(s) into a training exercise. Give the scheme some questions and input and see what information gets provided by the reach-back scheme.

LESSONS LEARNED: THERE'S ALMOST ALWAYS AN INTERNATIONAL CONTEXT

A CBRN incident or a major HAZMAT incident at a major public event will almost certainly take on some international aspect. Spectators, participants, journalists, and/or visiting dignitaries from some other country will be present in some degree or another. Even a nation with close ties to your own will want some reassurance that their nationals are well looked after and that an investigation into an incident is being undertaken correctly.

With events that have visiting political leaders, you are likely to have protection details with them. This will add complexity and require advanced planning. Indeed, it is even possible that such delegations might be a target of terrorism.

It is also important to understand that some events occur near national boundaries and a major CBRN/HAZMAT incident could cross a boundary. Cities like Detroit, San Diego, Basel, and El Paso have all hosted major events at some point. Major events in places like this have the need for international cooperation "baked in."

REFERENCES

1. BBC. London 2012: Mobile phone network 'under huge strain.' BBC Online, 23 September 2011. http://www.bbc.co.uk/news/uk-england-london-15039431.
2. United States Department of Homeland Security. *Safety Act.* https://www.safetyact.gov/, accessed 22 July 2022.
3. Ogie, R. et al. (2018). Disaster risk communication in culturally and linguistically diverse communities: the role of technology. *Multidisciplinary Digital Publishing Institute Proceedings* 2 (19): 1256.

4. Majid, A.M. and Spiro, E.S. (2016). Crisis in a Foreign Language: Emergency Services and Limited English Populations. *Proceedings of the International ISCRAM Conference* https://idl.iscram.org/files/amirahmmajid/2016/1363_AmirahM.Majid+EmmaS.Spiro2016.pdf.

5. Centers for Disease Control and Prevention. *Emergency Preparedness for Laboratory Personnel.* https://emergency.cdc.gov/labissues/index.asp, accessed 24 March 2017.

6. Centers for Disease Control and Prevention. *The Laboratory Response Network Partners in Preparedness.* https://emergency.cdc.gov/lrn/, accessed 10 April 2019.

7. (a) Center for Counterproliferation Research, National Defense University (2002). *Anthrax in America: A Chronology and Analysis of the Fall 2001 Attacks.* Washington (DC): US Government.(b) See alsoUnited States General Accounting Office (2003). *Bioterrorism: Public Health Response to Anthrax Incidents of 2001, Report GA-04-152.* Washington (DC): US Government.

8. Shelton D. Testing for anthrax has overwhelmed state labs: officials in Missouri and Illinois say they might have to have to start denying requests. St. Louis Post-Dispatch, 2001

PART III

Response

CHAPTER 14

The First Hour

The first stages of a serious incident will be extremely important in saving lives, preventing damage to property, and preventing damage to the environment. How you respond in the first few hours of a serious CBRN/HAZMAT incident may also set the tone for how a longer-term follow-on response evolves. The nature of disasters and emergencies include a high degree of unpredictability, confusion, and chaos. In such situations, it is far easier to descend into chaos and disorder from a position of authority and calm than it is to do the reverse. If you start out organized, you can always end up in a mess just by letting events take their natural course of action. If you start out disorganized, it is an uphill struggle for you to get a handle on the situation. The actions and responses taken in the first phases of an incident will dictate a successful outcome versus a poor one.

WHAT TO DO IN THE FIRST HOUR

It's not possible to say how you should most profitably spend your first hour in a major incident. They will all be different. Different types of incidents will unfold at a different rate. The nature of some threats is that they will affect people slowly over time. Other incidents will be manic; you will have to spend early minutes to save lives, often at great risk to yourself. However, here are some general principles for what to do in the first hour of a quickly emerging CBRN/HAZMAT incident. (A quick reference sheet summarizing these immediate actions is in the appendices to this book.)

Safety

Keep a safe distance from the scene. Approach from upwind, upstream, and uphill if possible. Perform initial reconnaissance from a distance. Don PPE and stage all response assets away

CBRN and HAZMAT Incidents at Major Public Events: Planning and Response, Second Edition. Daniel J. Kaszeta.
© 2023 John Wiley & Sons, Inc. Published 2023 by John Wiley & Sons, Inc.

from the problem, upwind if possible. Appoint a safety officer and give them the job of keeping an eye on responder safety. Consider the presence of secondary devices.

Establish an Incident Command Framework

Take charge of the initial response by starting your incident command framework. Start passing information up to your chain of command. Assess the incident. The GEDAPER process described on page 177 is helpful.

Establish Scene Control

Try to establish an initial perimeter. Start recording the names of responders on scene. Keep unequipped responders from getting dirty. Prevent the public from entering. If people leave, try to get their name. (If it is a slow-acting agent, you might need to get into contact with them later.) Get police on scene to keep public order. Establish hot, warm, and cold zone. Try to corral ambulatory victims so that they can be decontaminated. Remember, the perpetrator(s) may be among the victims.

Consider Emergency Rescues

Determine if there are live victims present. Consider emergency rescue efforts. Only do this if you can do it safely within established guidance. Stick to a "scoop and go" tactic until the situation can be assessed. Heroics may only get more people hurt.

Consider Emergency Temporary Mitigations

Is there a high-gain, low-risk tactic that could reduce the spread of the hazard and/or prevent further harm? For example, if the hazard is inside a building, can you cut the HVAC system to prevent further circulation of a respiratory threat? Can you put a few large buckets over a puddle of liquid agent?

Establish Emergency Decontamination

Start decontaminating any victims who are grossly contaminated. Use large volumes of low-pressure water, and plain soap if available. If it is cold, mind the hypothermia risk. Set up separate decon for responders.

Medical Care

Initiate emergency care for victims who are out of the immediate danger area. Start your mass casualty plan. Initiate triage and appoint a triage officer. Start basic life support measures—ABCDD. Remember, it is respiratory distress that will kill most quickly.

Keep dirty victims out of clean ambulances. Let the hospital system know what is happening.

Call for More Assistance

Based on your first assessments of the situation, call for additional help. Just as important, call off assets and responders that you don't need. You or someone else might need them later.

Start a More Detailed Follow-on Response

Based on your assessment, start appropriate public protective actions based on your best guess as to where the hazard will travel. These can include blocking traffic, evacuation, sheltering in place, and other tactical options.

Start Thinking Forensically

Don't destroy or damage evidence early on in the process if you don't have to. Put trained investigators at the head of the decontamination line to collect evidence and gather critical information. If you've managed to do even some of this in the first hour and not mess it up too badly, you're doing well. From the standpoint of being the incident commander, though, the most important thing is to assess the overall situation.

ASSESSMENT OF INCIDENTS

When an incident occurs, it is critical that information be collected and assessed about the nature of the incident. In some cases, many of which will be illustrated in the practical scenarios later in the book, it may not even be apparent at first that you are dealing with an "incident." Regardless, some kind of scheme or strategy or technique to assess incidents is necessary. In Chapter 5, we talked about the need to have a multidisciplinary assessment team in your organizational structure. In this section, we talk about how to put that team to use.

Initial Reporting

Many times, the first opportunity to assess an incident will be at the communications center or operations center level. For example, people may call to a 911/112/999 call center to report an accident. Likewise, security or event staff on the ground at the scene may radio or call into the operations center. The person or persons receiving such initial reports of a potential problem have an opportunity to capture important information. A CBRN/HAZMAT checklist, containing relevant questions to ask, should be made available to call center and operations center staff.

Use Reconnaissance and Surveillance Assets to Help You

Even if you do not have any trusted observers at the scene of an incident in the first few minutes, you may have a large amount of information at your disposal. Live media coverage may show useful information on an incident. CCTV systems may have visibility over the scene. It may be possible, even likely, that a witness has uploaded a YouTube video, a Tweet, a TikTok video, or some other type of post onto the internet with some sort of useful information. If you have 60,000 people at an event, someone is posting a selfie. A security command center or operations center should have someone monitoring social media posts from the event. During a CBRN incident, such a person or cell could be a vita source of information.

ESSENTIAL ELEMENTS OF INFORMATION (EEI)

Your communications center, 911 dispatch center, and/or your emergency operations center need to have checklists and, preferably, a poster on the wall. This checklist and poster should contain the "shopping list" of all of the information you would like to receive during the initial stages of incident. Of course, it is highly unlikely that a single source of information will have all, or even a high percentage, of this information. But if you are watching an incident unfold on TV, or better yet, have someone on phone or radio who can answer questions, it is good practice to have a framework such as an "essential elements of information" (EEI) to give staff guidelines during what may be a stressful situation.

Local requirements will obviously take precedence, but an EEI may look something like this:

- Where is the problem?
- When did it occur?
- Who saw it?
- What is the weather like in the exact location of the problem?
- How many people are in the vicinity?
- What does it look like?
- Is it solid, liquid, gas?
- Is it indoors or outdoors?
- Is there a package or container involved in the incident?
- Does anyone in the vicinity appear to be ill or have an obvious medical problem or injury?
- How is the problem affecting people?
- Do we have any surveillance video that shows either the problem itself or the vicinity?
- Do we have any sensors at the location? What do they say?
- Who do we already have in the area? Police, surveillance assets, intelligence, local medic teams, etc.?
- What is the weather like? i.e., Do we know which way the wind is blowing at the incident?
- What else is going on at the current time or the near future in the vicinity of the problem? (VIP movements, public ceremonies, etc.) This may impact upon how you manage your response.

Don't Look to Turn a Small Problem into a Big Problem

Not everything that happens is a major incident. Some incidents are minor. The purpose of an assessment team and process is precisely that—assessment. Some situations need to be escalated; others need to be de-escalated. A common fault with CBRN responders is that CBRN incidents are rare, which can lead to a bit of "what I am doing is mostly pointless" thinking, which in turn leads to a subconscious urge to validate one's existence. This can, through no particular deliberate fault of any one person, result in escalation of incidents.

The mere presence of an emergency response may serve to escalate a situation. Sometimes this is helpful, but many times it is not. The public and the media pay attention to emergency responders. Indeed, emergency responders use things like lights, sirens, special vehicles, special parking areas, avoidance of traffic rules, and distinctive uniforms to make sure that they get notices. The bigger and more visible the response, the greater the public and media attention, and with major events there are more public and more media in the area. There is every possibility that a highly visible response to an intermediate scenario will serve to inadvertently escalate the situation.

SIZING UP THE INCIDENT

When the assessment team, or any other responder for that matter who is first on the scene, arrives, they will do an initial assessment. Even if it isn't called that, it is basic human nature to use one's senses to see what is going on. It is sometimes called an "initial size-up." The first concern is safety of the assessment team. This is why you always want to appoint a "safety officer" as a first step in incident management. Someone will need to keep an eye out for the safety of the rest of the team. In US incident command system-based models, this is usually mandatory. Regardless, it is a good practice.

Incident command, discussed in previous sections, is meant to be a flexible and adaptable system capable of addressing a wide variety of situations. They are the structure that enacts an incident management scheme. There are a wide variety of approaches to dealing with CBRN/HAZMAT incidents and I have received training in many. There are many different schemes contained in various books and training courses, and as far as I can tell they are all very similar. Three examples are the "8 Step Process" by Greg Noll and Jim Yvorra,[1] "GEDAPER"[2] by Dave Lesak, and "DECIDE" by Ludwig Benner, which are widely available online in many sources.[3]

Having attended several high-quality training courses under FEMA's aegis, I feel that Lesak's GEDAPER process is a very highly useful tool and Mr. Lesak was kind enough to provide permission to use it in the first edition of this book. By all means, adopt another tool if it suits you better.

The GEDAPER Process

GEDAPER is an acronym that stands for 7 steps of managing an incident:

G—Gather information
E—Estimate potential course and harm
D—Determine strategic goals
A—Assess tactical options and goals
P—Plan and implement actions
E—Evaluate
R—Review

G—Gather Information

Ideally, the incident command structure has started gathering information (our EEI) even before any responders have arrived at the scene. By their nature, both terrorist incidents and HAZMAT accidents are unexpected events (see Figure 14.1). It is not likely that we will know very much about the nature of the event. The responders will need to determine as much as they can about the scope and nature of the threats present at the scene. Information may be gained about the incident and local conditions. Various means are available:

Direct observation: The initial impression of the skilled observer is important because it tends to color the nature of the response until more information is derived. What can the responders see and hear when they show up on the scene? How big is the site? How many people are involved? What's the response of people on the scene? For example, there's a difference between people walking from a building complaining about a bad smell, people running from a building in fear, a situation with nobody leaving the building because everyone inside is overcome by the incident, or a situation where everyone is still at their desk, but is annoyed by a liquid dripping from a package in the corner.

Other's senses: The assessment team is not the only observer. What have people in and around the area observed? What has been seen on CCTV? What was the first report to the operations center or 999/911 call center? Witnesses or surveillance footage will have seen things that the response teams will not have been able to observe. This is where the EEI in the operations center is worth investigating. What have they been told. It is particularly important to gather witness information as soon as possible. Dispersal or evacuation of witnesses could mean that useful intelligence is lost or delayed.

Instrumentation: Many types of CBRN/HAZMAT hazards are completely undetectable or unquantifiable by human senses. Detection and identification equipment can provide useful information on the type and concentration of hazard present. The use of detection equipment in CBRN/HAZMAT operations is so significant that an entire chapter is devoted to it. Sensors may be your only means of identifying and measuring many types of hazards.

The situation and condition of people: Are people dead, injured, or ill? What signs are they showing? Can you speak to any victims? What symptoms are they experiencing? Can on-scene medical personnel provide any useful information, either directly or through the use of simple diagnostic procedures? Do not disregard effects on animals in the vicinity.

Containers and markings: Commercial and industrial chemicals usually are transported and stored in containers. These are usually accompanied by distinctive markings and paperwork. This may not apply to terrorist incidents. However, in a terrorist incident, we can still assess the type and size of container as an important indicator. Is it a tank or a barrel or a bag? The type of container will give an indicator as to the physical state of the material. What size is it? This will allow an estimate as to the maximum amount of material. Are the markings or labels? While markings may not be accurate portrayals of the contents, they could be useful clues later in the investigation.

Estimate the quantity: In many circumstances, it may be able to make a rough educated guess as to the quantity of threat material involved. The size and shape of a container or device, if known, can be used as a rough upper limit to an estimate.

Intelligence: Intelligence reports might provide useful information. For example, a reported theft a quantity of a commercial chemical in the same city during the week prior to the incident may be of note. Does the incident at hand have any resonance with any intelligence reports you may have received?

Existing local knowledge: Earlier in the book, I stressed the importance of site surveys and reconnaissance. In a major event environment, the response structure should have key information about the venue at their disposal. For example, the responders should not arrive on the scene wondering how many people are likely to be in a major venue or whether the air intakes for the HVAC are at ground level or on the roof. Somewhere, someone should have this information at hand. In the case of a commercial or industrial accident, many countries have legal requirements for reporting of inventories of hazardous substances. This information can be critical in some scenarios.

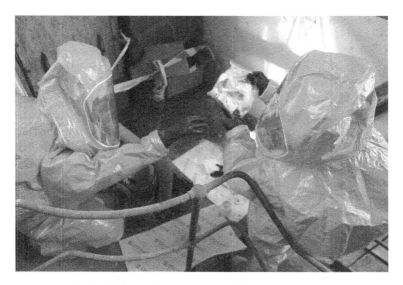

FIGURE 14.1   Gathering information may mean doing a reconnaissance entry to look at a problem.
Source: Photo: US Dept of Defense, public domain image.

Reference material: There are many excellent printed and online technical references available, both in print and in electronic format. This will be particularly valuable in HAZMAT scenarios, where you may well know the identity of the material.

Environmental Information: Information about weather and geography are very important. Which way is the wind blowing? One thing that I learned is that no matter how many weather stations are in service in the area, the wind at the site of the HAZMAT leak is going in a different direction than any of the sensors. What is the lay of the land? Look up and down as well—it is easy for someone back in the command center to take a two-dimensional view of the problem by looking at the map.

E—Estimate Potential Course and Harm

In the "G" step, you've tried to understand what the problem is and how much material may be present. Now, based on the information you gather, the "E" step is to try to make an educated assessment of where the problem is heading. Which way is the problem moving? Who might be in further danger as the situation evolves over time? Is the problem static (staying the same, in intensity and size) or dynamic (growing, shrinking, changing, moving) in nature?

The following concepts may be at your disposal to help estimate the potential course and harm:

Physical characteristics of the substance: Is the threat solid, liquid, gas, vapor, aerosol, or some mix? Is it heavier than air, lighter than air? Does a liquid mix freely with water or float on top? Is it flammable or explosive? All of these factors are significant in estimating course and potential harm.

Informed judgment and "Reachback": Based on technical expertise and operation experience, what do the responders think is going to happen? Some skilled HAZMAT practitioners have responded to hundreds or even thousands of incidents and have developed enough expertise to offer informed opinions of consequences in traditional HAZMAT situations. If expertise is not available on the scene, it may be found through "reach-back" schemes. In practice, "reach-back" schemes vary widely in the quality of information provided. But one should always remember that the reach-back expert is, by definition, not on the scene and cannot feel what is really going on at the scene.

Reference materials: Some references provide good default guidance on isolation distances, based on computer modeling. For example, the North American Emergency Response Guide contains many useful tables of isolation and protective action distances. While many of these resources are generic, they still have a place. If other information sources are lacking, these references can serve as a useful source of information to estimate how far a hazard is likely to travel.

Hazard modeling: We discussed the various advantages and disadvantages of hazard modeling at length in elsewhere. If you have a good modeling capability, a good operator, and (most critically) good information to put into the model, then hazard prediction technology can be useful in estimating the dispersion of CBRN/HAZMAT threats. One important factor is that the more information fields that an operator has to enter, the higher the potential is for mistakes. Therefore, check your data entry closely. Remember, these are tools that are better understood as "this is what might happen" as opposed to "this will definitely happen." Hazard modeling is also discussed in the next chapter.

Vulnerability assessment: Estimating the likely path of a hazard is only part of the problem. It will be necessary to analyze the vulnerability of everything in the estimated path of the hazard. In most situations, people in the open will be more vulnerable than people inside buildings. The nature of the buildings will also have a bearing on vulnerability assessment. You may need to make the difficult decision that keeping people in buildings in the likely path of the hazard is, overall, a lower risk endeavor than bringing them out into the open in the course of an evacuation.

The incident commander needs to use all the information and resources at your disposal to assess the various scenarios that will evolve. Responders should base their operations on what they believe the most likely outcome will be.

D—Determine Strategic Goals

The "D" stage is to determine what your desired outcome is. What are your overall objectives at the end of the incident? This can be viewed by the "mission statement" of the response, and the strategic goals should be prioritized. Strategic goals will vary greatly from scenario to scenario. Policy, established plans, and procedures, as well as guidance from upper management, will influence your goals.

Generally, in a major event environment, the strategic goals would be something like the following, in approximate priority order:

Save lives
Prevent further illness or injury
Protect critical infrastructure and property of cultural significance

Protect other property
Protect the environment
Restore public order
Allow the major event to resume or continue
Allow businesses

A—Assess Tactical Options and Resources

This step of GEDAPER gets you to consider the actions you need to take and the resources you require to start progressing toward the strategic goals identified in the previous stage. The strategic goals are the "what" and our tactical options are the "how." Realistically, this is a complex subject and the following chapters are largely devoted to tactical options and resources. These could be actions like rescue, medical care, decontamination, forensic investigation, and numerous other courses of action. The "A" step in GEDAPER can involve a fair bit of brainstorming. You may have a lot of choices, or very few, depending on the situation.

P—Plan and Implement Actions

This is the P in GEDAPER. This step is the implementation phase of incident management. The incident command team has established strategic goals, decided on which tactical options and resources are best, and have now undertaken to execute a tactical plan to manage the incident.

Plan of Action: This is, in HAZMAT terms, the "entry plan." It is the detailed description of who is going to do what, how they are going to do it, and what resources they are going to use to do it. US ICS procedures generally call for this to be a written document, and even if only on a form or a whiteboard. Complex situations will require many actions by many people; many subsequent or subsidiary plans of action will be needed. A large incident will require someone to prepare and monitor these plans as well as plan ahead. A good incident command scheme will accommodate the need for a plans officer in the incident command structure.

Existing procedures and guidelines: In dynamic incidents, there may not be time to concoct a new plan from scratch. Many aspects of a plan of action should not need to be spelled out if existing standard operating procedures (SOP) or standard operating guidelines (SOG) are in use. For example, the plan of action can cite a sampling SOP and decontamination SOP, rather than belabor the point.

Safety: Remember that responder safety as an important function. A safety plan is usually considered an integral part of this step of GEDAPER.

E—Evaluate

The second E stands for evaluate. This part of the GEDAPER process refers to the evaluation of tactical actions to see if they are working. The entire incident management framework benefits if you can be flexible enough to adapt to changes in circumstances and new information. Further, the incident command structure needs to make sure that the work that is occurring actually furthers the strategic objectives. Is the work that you are doing actually getting you closer to the objectives?

R—Review

R stands for review. GEDAPER only really works if you view it as a continuous cycle, rather than a single process. It is important to remember that incident assessment is not a single act at the beginning of an incident. Situations can be dynamic. Incidents and accidents will change. New information will emerge. Weather conditions usually change. Whatever incident management scheme is adopted for use at your major event, you must include recurring periodic reassessment of the situation in it. GEDAPER teaches us to make periodic review a standard part of the incident response.

Immediate Action Drills and SOGs/SOPs: Because time is often of the essence in a CBRN/HAZMAT incident, there is tremendous value in pre-existing "immediate action drills" and standardized operations that are, in effect, "cut and paste" from a playbook. During the initial stages of an incident, there will be a lot of things you can do that provide general value while you are still trying to work out the bigger picture using the "GEDAPER" process.

During your planning phases, spend some time working out some standardized responses to some scenarios and situations that are part of your design threat. You should not have to invent a procedure for "investigate a suspicious odor" or "set up emergency decon on the street" because you should have standing drills or guidelines in place for how to do this. You might have to adapt a SOP/SOG/drill to local circumstances, but at least it is a start.

This chapter is not the place to discuss such immediate action drills in great detail. However, the various practical training scenarios in the later part of the book will be a useful basis for you to develop your own immediate action drills.

WHAT NOT TO DO IN THE FIRST HOUR

Both experience and technical knowledge give us some examples of things that you could do, but which are not actually very helpful in most, if not all circumstances. As has been made clear in several other parts of this book, it isn't always possible to advise you what to do, but it is very easy to advise you on what NOT to do. Further suggestions in the "what not to do" category are contained in the practical scenarios section at the end of the book.

Panic

Don't panic. It might be the worst day of your life. But losing sight of reason and giving in to fear and chaos is not going to make it any better. Panic makes it worse. Worse, the people you are trying to help will see your panic. Rely on your training, experience, and procedures.

Assume Panic

Don't expect other people to panic. They might. In many circumstances, people do not panic. As discussed in Chapter 3, take firm, decisive steps to communicate clearly with other people, including the public.

Assume the Incident Is Smaller Than It Is

In some senses, this is the opposite of a panic. It is easier to take a large perimeter or cordon and make it smaller than it is to take a small one and make it bigger. My advice is to take an expansive view of where the boundaries of the problem are at the earliest stage. Being too conservative too early on in a CBRN or HAZMAT incident may make it more difficult in later hours.

WITHDRAWAL AS A TACTIC

Although emergency responders do not like to admit it, we must remember that withdrawal is a tactical option in some circumstances. Both terrorism and accidents can confront responders with situations that are beyond the scope of their capabilities. A HAZMAT team can do an entry into the incident site only to discover a potential secondary device. A radiological "dirty bomb" could have damaged a building and withdrawal is the best course of action due to the risk of structural collapse. It is better in some cases to withdraw and wait for more help than to charge into a situation and end up needing to be rescued.

LESSONS LEARNED FROM HISTORY

Shortages

You will be short of everything. In a major incident, at the beginning stages you will probably not have the resources you need to do everything that you need to do. Additional personnel and equipment will gradually arrive. You must, therefore, prioritize your actions and think about the things you can do right away with what you have.

FIGURE 14.2 A situation with multiple casualties may require a triage scheme to maximize the available resources.
Source: Photo: US Dept of Defense, public domain image.

Triage Exists for a Reason

If you have a lot of injured or sick people, you may not be able to help everyone at once as shown in Figure 14.2. Triage is discussed in other chapters, but it is during this first hour that triage becomes very important in many scenarios, particularly chemical ones and situations with conventional injuries. (Biological and radiological signs and symptoms largely have delayed onset.)

Getting People to Work for You, Not Against You

As discussed in Chapter 3, panic is not a foregone conclusion in the public. Nor is resistance to emergency response efforts. An important tactic may be to get the public to actively help out rather than merely be victims, a target for response activities, or bystanders in the way. If members of a crowd can be given work to do, or if people are given agency, even in a minor way, fear and anxiety are reduced and the likelihood of a crowd following instructions will increase.

There are many ways in which this could be implemented. In a mass casualty situation, there are a lot of things that untrained people could be instructed to do such as holding pressure on dressings, holding up an IV bag, or watching a monitor ("hey, yell if this pulse ox reading drops below 90"). In a day and age of ubiquitous smart phone ownership, getting someone to stand there and take some photos and send them to the operations center is, if not always critical, at least not useless and keeps someone busy. You can get some people to believe that they are part of the response and not part of the problem, and this will help manage the public.

REFERENCES

1. Noll, G. and Yvorra, J. (1995). *Hazardous Materials: Managing the Incident*. Stillwater (OK): Oklahoma State University Press.
2. Lesak, D. (1998). *Hazardous Material Strategies and Tactics*. New Jersey: Prentice Hall.
3. Benner, L. (1975). DECIDE in hazardous materials emergencies. *Fire Journal* 69 (4): 21–26.

CHAPTER 15

Characterizing the Threat

This chapter talks about tactics, techniques, processes, and technologies for characterizing the threat in the early stages of an incident. By using the word "characterizing," I mean detecting, categorizing, measuring, identifying, or otherwise collecting information on a substance, condition, or environment.

DETECTION, IDENTIFICATION, MEASUREMENT, AND IDENTIFICATION EQUIPMENT AND ITS USE

The previous edition of this book went into some depth in this field, but the state of the art in CBRN/HAZMAT instrumentation changes every year. The first edition of this book was in print for over a decade. Some things in the detection space have not changed at all. Other things changed considerably. If I get into specifics, whatever I write may not be useful in five years. Instead of cataloging the available equipment or trying to teach how to use a particular instrument, it is a more productive use of my time to explain some useful principles in how to use whichever equipment is available to you in the most productive way possible. For purposes of this book, I am going to use the word "sensor" to refer to the entire spectrum of detection instrumentation.

Far deeper discussions of the subject are available in standalone books on the subject. For the most useful discussions, I recommend Chris Hawley's *Hazardous Materials Monitoring and Detection Devices*[1] and Houghton and Bennett's *Emergency Characterization of Unknown Materials*.[2] Both are relatively recent, the authors keep their work updated with new editions, and they are approachable to the emergency services audience rather than academic specialists. Both should be in your library alongside this book.

CBRN and HAZMAT Incidents at Major Public Events: Planning and Response, Second Edition. Daniel J. Kaszeta.
© 2023 John Wiley & Sons, Inc. Published 2023 by John Wiley & Sons, Inc.

What Is the Purpose of Sensors?

Sensors are information tools used to make decisions. The purpose of a sensor, whether it is a one-cent piece of pH paper or a rolling suite of laboratory instruments in a large trailer, is to provide information. The value of the sensor is based on, and is only based on, the value of the information it provides to you. In turn, the value of the information provided by a sensor is only as good as the decisions that you make using that information. In other words, sensors are decision-making aides. If the information a sensor produces is not useful in making a decision, then it raises the valuable question of why you have it at all.

Let's look at a non-CBRN/HAZMAT situation. A thermometer can be a useful sensor. Assume that you live in a house. You can put a thermometer outside of the window. You can look at that thermometer to see how hot or cold it is outside. If you use that information to make a decision as to whether to put on a jacket, hand, and/or gloves, then that thermometer is being a useful sensor. If you don't look at the thermometer and just stick your head out the window to see how hot or cold it is and then use your own feelings to make a decision about apparel, then that thermometer is not being useful in that scenario.

Sensors that cause people to make bad decisions have negative value. So do sensors that waste time or resources. If a sensor says a situation is safe and a responder dashed into the room only to be met with injury or death, then the sensor has negative value because it lead to a poor decision. Likewise, a sensor that gives indeterminate readings after a long period of time may have wasted valuable time that could have been used for some other useful purpose.

Using Sensors in a CBRN/HAZMAT Context

Within a CBRN/HAZMAT context, you can use sensors for several purposes. It is important to state, categorically and firmly, that there is almost never an ideal sensor for every purpose. There are things that you may want sensors to do, but for which the technology and products

FIGURE 15.1 GC/MS systems, such as the Inficon Hapsite, provide high quality identification of many hazards.
Source: Used with permission from R. Mead.

simply do not exist yet. Since sensors are information tools to help us make decisions, this section provides a general description of the broad categories of ways in which you can use sensors to protect against CBRN/HAZMAT threats. The text boxes contain general information on the major categories of chemical, biological, and radiological sensors. Always use them in ways that give you actual actionable information. Otherwise, there is little point.

Sensible Missions for Sensors

Detect to warn: You can use sensors to attempt to warn you of an upcoming threatening situation. You can use sensor readings to help you make decisions. However, in major event environments, the kind of decisions you can make are often high penalty ones like event cancellation or emergency evacuations. Often, sensors cannot provide the level of information to make these decisions on their own. More often, you need to develop a tiered set of procedures whereby detect-to-warn sensors trigger a more detailed special investigation. For example, you could have radiation detectors in key areas. A detection even would not trip a major evacuation automatically, but would trigger a team going to investigate with handheld equipment to localize and identify a problem.

Field Chemical Sensor Technologies: A Basic Overview

Color-changing papers: These are inexpensive technologies to characterize liquids. Military chemical warfare agent detection paper, pH paper, and glucose testing strips are all part of this group.

Combustible gas indicators: These are commonly used in industrial and firefighting settings. They detect and measure combustible gases like methane or propane. They are often bundled together as part of a multi-gas meter. Useful for ruling things out in a CBRN environment and insuring that you are not operating in an environment with an explosive gas problem.

Colorimetric tubes: These are often known as "Drager" or "MSA" tubes after their manufacture. Glass tubes filled with specialty reagents, through which a sample of air is drawn with a specialty pump. These are often good for tests for specific types of gas or vapor. Time-consuming if you need to work your way through a battery of them. Some are prone to cross-sensitivities.

Wet chemistry techniques: There are various products which are basically portable chemistry kits that use a variety of reagent chemicals to test for specific substances or categories. Often inexpensive, but time-consuming to use properly.

Industrial gas detectors: There are a number of specific technologies for detecting specific substances like carbon monoxide and oxygen, as well as specific threats like ammonia, cyanides, or chlorine.

Photoionization Detectors: Good, fast, economical but generic volatile chemical gas or vapor detector. Often bundled with industrial gas detection and combustible gas detectors.

Ion Mobility Spectrometry (IMS)**:** Very good for rapid detection of chemical warfare agents. Inherent false positive risk for compounds of similar molecular weight.

Flame ionization detectors: Alternative technology for detection of chemical warfare agents. Sometimes used as alternative to IMS or as a supplemental to IMS to confirm a detection.

Fourier transform infrared (FTIR): Sophisticated and relatively expensive technology. Uses infrared energy to collect and identify a spectrum of a substance and identify it from a library. Good for a wide range of substances. Mixes/blends sometimes troublesome.

Raman spectroscopy: Interrogation of solids and liquids with a laser to identify them from a library. Useful in many environments for identification, although not every substance has a Raman response. As with FTIR, blends and mixtures can be difficult to ferret out.

Gas Chromatography/Mass Spectroscopy (GC/MS): A bit of a "gold standard." The technologies are often fielded in tandem for definitive analysis and identification of unknown substances (see Figure 15.1).

Stand-off detection: Several infrared technologies for detection of gases and vapors at a distance. At time of writing, still very troublesome and expensive in urban settings.

Detect to treat: Some types of sensors don't work fast enough to give you information ahead of time. But they might be able to work fast enough to give some type of information to be able to provide medical treatment to save lives. An example of this would be aerosol collection filters that sample the air at a venue. The concentrated air samples are then periodically removed and examined in a laboratory for evidence of biological warfare agents like anthrax. If anthrax was detected, there's a time window of several days to allow for everyone who was at the event to be tracked down and given appropriate lifesaving antibiotics.

Types of Biological Sensors and Technologies

Protein, sugar, starch, and pH testing: There are a lot of simple tools based on medical or chemical laboratory processes that can be used to rapidly characterize suspicious powders or liquids, usually to *rule out* biological hazards. For example, a powder with a high or low pH is unlikely to be a pathogen

DNA detection: There are field analytical techniques, such as test kits, to look for the presence of DNA. This could be used to rule-in or rule-out a particular sample as biological (and thus needing more analysis) or non-biological.

Hand-held assays: Believe it or not, the "lateral flow" test kits for COVID-19 have their early scientific roots in hand-held "immunochromatographic" assay tests for specific threats like anthrax and brucellosis. Such devices are small, liquid-based kits for specifically detecting a single threat material or, on occasion, a closely related group of threats.

Aerosol samplers: These aren't really sensors in the direct sense. They collect material from the air and concentrate it into a sample that could be analyzed by other means, either in the field or in the lab.

Particle sizing: Particle sizers examine the air and count the number of particles of various sizes. A spike in the size range of 1–10 μm *could* be an indicator of a biological attack.

Microscopy: A properly trained technician can use powerful microscopes in a field laboratory to look at a sample and make a qualitative judgement as to whether it is biological or not.

Flow Cytometry: These devices look for specific biomarkers of interest in a sample, such as the presence of the compound ATP, which is useful in making a "is it living or not living" determination.

UV Fluorescence: Some biological material gives off useful information when exposed to the right type of UV light.

PCR testing: PCR devices examine a sample for a DNA-match.

Screen people and materials: Detection technology can be used to screen people and materials upon entry to a venue. Often, this is generic technology, such as metal detectors and X-ray equipment. However, both could detect the means of dissemination of CBRN materials so they should not be discounted as CBRN sensors, even if the technology is not CBRN-specific.

Investigate a situation: Handheld or manpack sensors can be used to investigate suspicious circumstances, confirm or resolve alerts by other means, and provide more localized information.

Define a perimeter: After an incident involving CBRN/HAZMAT circumstances, sensors can be used to define a reasonable perimeter to help isolate the problem and protect the public.

Identify unknowns: Numerous sensors are useful to classify or even fully identify unknown materials. Many possible scenarios start with the presence of suspicious materials that might be dangerous. FTIR and Raman devices have a good track record of identifying unknown solids and liquids, while GC/MS devices are also good at gases and vapors as well.

Protect responders: Some types of sensors can be used to determine if a responder is operating in a safe or unsafe environment. One example might be a combustible gas indicator, which would alert if things like propane or methane are present.

Find evidence: Sensors can be used at a CBRN crime scene or at a HAZMAT incident to locate the places where environmental samples can be taken. This can greatly facilitate a forensic investigation.

Verify decontamination: Sensors can be used to verify whether or not emergency decontamination processes have reduced the level of hazard. This is illustrated in Figure 15.2.

Types of Radiological Sensors and Technologies

Dosimetry: These devices measure cumulative exposure to radiation. Some are "direct reading"—you can read them on the spot. Others need to be read after the fact by specialty equipment.

Count-rate and Dose-rate monitoring: Various types of detectors (gas-filled tubes such as "Geiger Counters," scintillators with solid material like sodium iodide, ion chambers) detect and measure the presence of radiation in real time. They are useful for making safety decisions, detecting the presence of possible threats, and measuring contamination on surfaces. Many such devices are described by the actual detection media used, such as "sodium iodide crystals" or "solid state" or "Geiger Mueller Tube." All will have relative advantages and disadvantages.

Gamma spectroscopy: Some devices collect the spectrum of gamma radioactivity and use this to tell you what isotope is emitting the radiation. These can be useful for telling the difference between natural background, medical isotopes, and dangerous threats. (Mind you, medical isotopes can be dangerous in the wrong context.)

Neutron detectors: These are specialty devices that detect neutrons. Neutron detection is more highly specialized and more expensive. It has some niche application.

Alpha Detection: Some devices are optimized for detection of alpha particles. These devices might be of use in some fringe scenarios where the threat material is principally an alpha emitter.

It should be noted that many instruments combine several of these technologies.

What NOT to Do with Sensors

Just as important as what you might plan to do with sensors is the concept of learning from other people's mistakes. Here's what NOT to do with CBRN/HAZMAT sensors:

Use them like a household smoke alarm: A number of times I've dealt with officials who acted as if CBRN sensors could work like a smoke alarm in a house. The alarm goes off, everyone flees, problem avoided. Unfortunately, they don't work that way. Even smoke detectors rarely work that way in large buildings, where they are simply one input in a complex system. If you are placing a lot of sensors in and around a venue, you want to look at the totality of the circumstances if one of them goes off. Look at the area on CCTV. Send a team to assess a problem. Call someone on the radio in the area. One sensor reading is a data point in a decision process, not a decision unto itself.

Let sensors overrule patient assessment: I witnessed a training exercise where the patient was simulating clearly nerve agent signs and symptoms, but the trainees would not treat the patient for nerve agent exposure because their detector didn't give a reading. (Many nerve agents, such as VX, give off almost no vapors, are primarily skin contact hazards, and may be hard to detect with vapor detection systems.) Sensors can be a useful indicator of some things, but in medical terms, follow your medical protocols.

Build procedures around a sensor: You don't want to buy something and then figure out how to shoehorn it into your operational processes. But this happens. "The boss bought ten of these and now we have to use them somehow" is not a way to develop response procedures.

Prove a negative: You should not try to provide that no hazards are present. No sensor detects everything. Furthermore, even operating within their scope, detection instrumentation cannot definitely prove that you have removed all of a particular hazard. Even the most sensitive sensors have a minimum detection threshold.

Use them outside their performance envelope: Sensors have specifications, and they cannot be expected to operate effectively outside of their specifications. If you use them the wrong way, you will not get good information from them. You might even get bad information. Specifications are discussed below.

SPECIFICATIONS OF SENSORS

Regardless of the instrumentation that you use, every one of them has a set of specifications. They were designed to do certain things. You need to understand their limitations.

Sensitivity and detection threshold: There is always a lower limit below which a sensor will not detect something or measure something. This is the "detection threshold." What is the minimum quantity of material that will give a reading on the detector? In most types of detectors, the sensitivity correlates with false positives. In other words, you get more false alarms if you set the instrument to be more sensitive.

Range of detection: "Detection range" refers, in the context of sensitivity and selectivity, to both the bottom and top limits of detection. As a simple example, think of a thermometer, with the bottom limit as −50 and the top limit as +140. This is the "range of detection" of the thermometer. It is important to note that the selectivity and responsiveness of a sensor may differ across its range. Knowing where the maximum limit of a sensor may be is critical to making safety decisions.

Selectivity: Selectivity is basically the ability of a sensor to sense what it is looking for without getting confused by other similar substances. It can be seen as the ability to reject interference. In practical terms, it means that when you are looking for chlorine, it is not giving you a chlorine reading when you encounter, say, ammonia.

Cross-sensitivity: A cross sensitivity is when a sensor reliably, consistent, and routinely reports an alarm for a substance that is not a threat. In lots of ways, cross-sensitivities are the opposite of selectivity. For example, many CWA detectors have a cross-sensitivity issue with a chemical called methyl salicylate (MS). MS is so reliable at causing alarms for H-series blister agents (the mustard agents) that it is used as a simulant for equipment testing. Cross-sensitivities are common causes of false positives. (See below.)

False Alarm/False Alert: The term "false alarm" is often used, but really there are two kinds of false alarm of importance to this discussion, the false positive and the false negative. No detection technology is completely free from false alarms. We should be precise when we are using language to describe "false alarms" because many instances of detector functions are not really false alarms at all. It's important to unpack some of the terminology here:

False positives: A false positive is when a detector says something is present, but it is not. An example would be a detector that got a sniff of diesel exhaust and gave an alarm for nerve agents. False positives are dangerous because they can cause erroneous responses to otherwise innocuous situations.

Some things are not false positives: Sometimes you really are detecting something, but it is not actually a threat. Some types of plastic emit small amounts of volatile chemicals in high temperatures or direct sunlight. Small amounts of chlorine or ammonia may be present due to chemicals normally used for sanitation. A radiation detector that senses radiation from a person who has received a benign radiopharmaceutical is NOT a false positive. The detector is functioning as intended in this circumstance. It is up to the operator to determine whether the circumstances warrant a response.

False negatives: A false negative occurs when a detector fails to detect something that it is meant to find. For example, a chlorine gas detector that receives a sample of chlorine gas but fails to properly report an alarm is a false negative. A detector that fails to provide an alert based on an amount of substance that is below its designed sensitivity is NOT providing a false negative, as the detector is operating within its published specifications.

Background readings: Many instruments are detecting, measuring, or identifying things that are in the normal background. For example, radiation detectors will detect the normal cosmic and terrestrial radiation in which we are bathed every day. Volatile organic chemical sensors, such as photoionization detectors (PIDs) will detect all kinds of low-level background chemicals in practically every household or commercial setting. It is important to take background into account. Use your sensors around the venue ahead of time to understand what the normal background is.

Environmental specifications: Any piece of equipment will have environmental specifications, such as temperature, altitude, humidity, whether or not it can withstand water, and related criteria. Using them outside of their performance envelope will make the information they provide less reliable to you or even give you false alarms.

Intrinsic safety: An instrument that might ignite combustible gas is not always a safe one to use, particularly if we are talking about HAZMAT environments where many substances are flammable. An instrument that is designed to be able to operate in a combustible gas environment is referred to as an "intrinsically safe" device. The intrinsic safety or lack thereof for a particular instrument is an important planning consideration.

PUTTING IT TOGETHER: WRITING SENSORS INTO OPERATIONAL PLANS

Remember that sensors are tools to accomplish a mission. Sensors should provide information to support decisions. Remember to apply the mission planning guidance contained in part II of this book. Apply the following thought process:

- Work out a good concept of operations (CONOPS) to achieve your mission. This should have no reference to specific instruments or sensors. It should be a statement of what you have to do in order to protect safety and health.
- Work out what tasks people and organizations will need to do to execute the CONOPS. In other words, develop a task list.

FIGURE 15.2 Sensors can be used to check the effectiveness of decontamination.
Source: Photo: US Army, public domain image.

- The mission, CONOPS, and task lists will drive what equipment you will need, including sensors.
- Work through some tabletop scenarios to see where, when, and how sensors will help you make decisions.
- Evaluate the available sensors to see if they contribute information to help make decisions that are needed in your CONOPS.
- Procure the sensors that make sense for your mission.
- Understand what they can and cannot do for you.
- Train with them until their use is familiar. Practice your training in exercises.

It sounds simple, but speaking both as a former responder and as a sensor salesman, many organizations do not even come close following this logical thought process.

No amount of advice that I can give will apply to every major event, situation, or scenario. However, the following guidelines may help to effectively implement:

- Select sensors based on the quality and quantity of their information output.
- Use the existing CCTV coverage to its fullest extent to look for signs of CBRN problems.
- Use different technologies ("orthogonal detection") to increase confidence in sensor readings. For example, combine FTIR and Raman for analysis of suspicious powders and liquids.
- Use simple bio/non-bio field techniques to rule out the presence of biological substances when dealing with suspect powders.
- Since you will definitely encounter them, use isotope identification to identify medical isotopes.

- Since false alarms are practically impossible to eliminate, develop procedures to verify alarms, particularly in situations where protective actions will have a high operational impact.

- Use instruments with low false-negative rates to rule out possible categories of threats.

- Buy sensors well ahead of the major event so that personnel can train and exercise with them.

- Consider independent testing of new sensors that you have acquired to determine their true operational envelope.

- When making decisions about public health and safety, by all means use the sensors that you have, but err on the side of caution if there is any doubt.

THINKING OUT OF THE BOX—DETECTION USING NON-SPECIALTY SENSOR TECHNOLOGY

Not every type of sensor that is useful in a CBRN/HAZMAT scenario is a specialty detection instrument. We can make use of other types of sensors as well. While sophisticated electronic sensors may have greater sensitivity and selectivity than some cruder or less specialized methods, this does not mean that the crude methods are without value.

The Human Eye

Human vision is a sense. Human senses are still useful in many ways. If looking at something with the human eye, either directly or through a camera lens, gives us "detect to warn" or "detect to treat" information, then the human eye is a CBRN/HAZMAT sensor as much as anything else. A good pair of eyes, looking at the problem through binoculars, is better than many sophisticated electronic instruments. Security personnel, medics, and event staff dispersed throughout venues may end up observing suspicious circumstances long before a sophisticated sensor analyzes a sample.

One person in a crowd having difficulty breathing is a medical problem. Ten in the same area is a CBRN/HAZMAT problem. Generally speaking, if multiple people in a crowd start having similar symptoms, something bad is going on. Security staff, first aid staff, and other venue employees can be an early warning network. This will, of course, require effective communication infrastructure. A standardized procedure for reporting suspicious incidents is also highly recommended.

Field Medical Observation

If event planners follow good special events medical services processes, there will be a network of medical personnel in and around the major concentrations of people at major events. Effectively, such medical personnel comprise a network of sensors that are looking for signs and receiving reports of symptoms. Reports from field medics may be the first signs of trouble in some scenarios. Sometimes, the most definitive information may come from signs and symptoms, not any particular specialty equipment. For example, a report of several people with dimness of vision and pinpointed eye pupils will be a far more definitive detection of a nerve agent incident than many other means.

CCTV/Video Surveillance

CCTV is an electronic extension of the human eye. Many CBRN/HAZMAT problems can present themselves in ways that are visible to the human eye. Because safety and security personnel cannot be everywhere looking at everything, video cameras can be used as extensions of the human eye.

Over the last decade, digital video has supplanted analog video systems. The advantages of digital video have far outweighed any disadvantages. Security video systems now require less power, less installation, less storage for old footage as the product is now digital, and are generally considered to be superior. One of the key advances in video surveillance has been in the field of image processing. As the output signal from a modern CCTV system is now a digital product, it can be easily subjected to analysis by modern software and shows great potential for use in antiterrorism efforts, including the CBRN/HAZMAT field.

Video signals once relied upon a human being to monitor a blurry black and white display, usually in conjunction with many others. Automated algorithms can now analyze such signals nearly instantaneously. Various software routines can be applied to the analysis of behavior in the video signal and alert surveillance operators if certain actions or behaviors are automatically detected. Some specific examples that would have utility in the CBRN/HAZMAT arena could include the automated detection of:

- Persons entering a restricted area
- Persons leaving bags, packages, or objects in a secured area
- Unattended vehicle
- Members of a crowd or audience suddenly moving from vertical to horizontal, or rapidly dispersing, as in a panic reaction
- Counterflow detection—people or vehicles moving the wrong way
- Appearance of a liquid puddle on a surface
- Sudden decrease in visibility, possibly indicating the presence of a vapor

CCTV surveillance is also a useful "force multiplier" for more traditional CBRN detection methods and devices. Emplaced CBRN sensors can be paired with dedicated camera systems. This allows a video image to be called up any time when an alarm or a fault is detected by the sensor. The use of video tied to chemical sensors has proven to be useful in many settings for the reduction of false alarms and the rapid analysis of potentially hazardous situations.

X-ray Systems and Metal Detectors

Do not discount the boring, old-fashioned conventional X-ray device. While security X-ray systems cannot detect or identify CBRN materials, they are very good at detecting liquids, suspicious areas within packages, and materials that may be part of the dispersal mechanism, such as explosives. A chemical bomb is just that—it is a bomb that contains chemicals. Things that might detect a conventional device may very well detect a CBRN device. Likewise, metal detectors can screen people for possible devices as many CBRN devices may have a metallic component. Many X-ray systems are capable of using backscatter technology to detect the presence of organic matter. While this technique was invented to detect the major categories of explosives, many CBRN materials are organic and might be detected by such means.

Conventional Explosive Detection

Conventional explosives are a valid and recognized means for dissemination of CBRN material. Under some circumstances, it may be easier to locate terrorist CBRN devices by detecting explosive material than by trying to sense the presence of the CBRN materials. Both electronic and canine means (detection dogs) have been of great use in this field. However, there is a concern that use of dogs around hazardous chemicals could damage their health or ability to smell.

HAZARD PREDICTION MODELS

I wrote at length about the pros and cons of hazard prediction models in the previous edition. While these are generally tools for responding to incidents, they need to be procured well ahead of an incident and integrated into procedures if they are going to be effective, so I am discussing them in the preparedness section.

There are many different types and methods of hazard modeling, and it is useful to look at an overview of the major categories. In general terms, there are two forms: templates and plume models. Templates are general outlines designed to be applied to a map. They are quick, simple, and generic and rely on very simple inputs. Plume models are computer programs that take all the available information about the incident and attempt to calculate where the hazardous substance will travel.

Approach the use of hazard prediction models carefully. Avoid the old cold war era military templates (the so-called ATP-45 models), as they are not suitable for use in the civil environment. Generally, my own experience is that you can get some useful outputs from hazard prediction models, but only if you have excellent inputs (good data on type and quantity of substance, weather data, and actual modeling of your terrain) AND good experience with using the model. If you can spend the time and effort to train with a good prediction model, by all means, give it a go. But I have found that they can cause as many problems as they create. The final word of wisdom—always react to the problem in front of you, not what the computer screen says. For further thoughts on this, seek out the first edition of this book.

LESSON LEARNED: LEVERAGING EXISTING DETECTION CAPABILITIES

There may be analytical capabilities at major events that were not set up for CBRN purposes. But some of the equipment and expertise available for other purposes might be adaptable to CBRN and HAZMAT situations. If you think hard through what might already be going on in and around a major event environment, you might have more capability than you realize. Three relevant examples will illustrate this point.

Many organizations, such as fire departments, public utilities, and private companies that use industrial chemical, will have commercial gas and vapor detection equipment, such as combustible gas indicators and PIDs. Depending on the scenario, much of this equipment will be of some use. A large percentage of toxic chemicals will be detectable on a PID. Some hazards might be detectable on other instruments, particularly when you take into account the many cross-sensitivities of the instruments. For example, a Dräger hydrogen chloride sensor has a significant cross-sensitivity to hydrogen cyanide.

Several types of equipment useful for drug and explosive detection, such as ion mobility spectrometry devices used in airport and border control missions. Rather a lot of the technology useful for explosive or narcotic trace detection is highly useful for detection of chemical warfare agents. Many of these devices actually have one or more operating modes that will do that. My advice would be to investigate the actual capabilities of hardware already in use.

One development of particular note involves illicit drugs and music festivals. It is a fact of life that large music festivals have a drug scene. Poor quality, tainted, or otherwise adversely affected lifestyle drugs have caused injury and death. At some music festivals in recent years, there have been efforts to set up testing sites so that drugs can be discreetly tested for impurities and hazards. Some technologies such as Raman spectroscopy and FTIR spectroscopy have been used for this purpose at music festivals.[3] As noted elsewhere in this book, both of these technologies (and often the exact same device) are highly useful for identification of unknown substances. With enough coordination, or in an emergency situation, a drug-testing operation in a festival tent could provide useful capabilities to identify chemical threats.

REFERENCES

1. Hawley, C. (2018). *Hazardous Materials Monitoring and Detection Devices.* Burlington (VT): Jones & Bartlett Learning.
2. Houghton, R. and Bennett, W. (2020). *Emergency Characterization of Unknown Materials.* Boca Raton (FL): Taylor & Francis.
3. (a) Gerace, E. et al. (2019). On-site identification of psychoactive drugs by portable Raman spectroscopy during drug-checking service in electronic music events. *Drug and Alcohol Review* 38 (1): 50–56.
   (b) Laing, M.K., Tupper, K.W., and Fairbairn, N. (2018). Drug checking as a potential strategic overdose response in the fentanyl era. *International Journal of Drug Policy* 62: 59–66.

CHAPTER 16

Medical Response

This chapter discusses the medical response effort, with an emphasis on interventions in the field environment. The author's medical training is at the basic field care level, so I am not in a position to advocate changes to existing doctrine on treatment. Both this chapter and Chapter 8, as well as the bibliography, point to excellent resources for both field and hospital medical treatment protocols.

DIVIDING THE PROBLEM INTO SYNDROMES

In the first phase of emergency response, the exact cause of illness or injury may be unknown, or at least not obvious. Many people get hung up on this concept. However, the medic's job is to treat the patients that they are presented with, not some archetype from a training scenario. Always treat the patient. The array of potential CBRN/HAZMAT threat materials that can cause casualties is bewildering. The easy way through this is to actually analyze how CBRN/HAZMAT causes acute injuries. By doing this, we can divide up the problem into manageable bit. We can look at the mechanisms of injury that are likely to make people acutely unwell at the scene of an incident and divide the problem into simple categories. First, it is likely to be chemicals that cause immediate problems. Biological and radiological injuries are likely to present themselves far later than any on-scene medical personnel are likely to confront.

Looking at the possible spectrum of threats, there's only a few pathways to go down. Despite the wide array of potential threats, the signs and symptoms observable to the field medic and the range of medical interventions available at the field level are fairly narrow. A "syndrome" is a bundle of signs and symptoms. At the clinical/definitive care level, it may be

CBRN and HAZMAT Incidents at Major Public Events: Planning and Response, Second Edition. Daniel J. Kaszeta.
© 2023 John Wiley & Sons, Inc. Published 2023 by John Wiley & Sons, Inc.

another situation altogether, but at the field level it is easy to narrow the problem down into some basic categories. The medical responder at the CBRN/HAZMAT incident is likely to be confronted with the following syndromes.

Nerve Agent Syndrome

The nerve agent syndrome is the so-called cholinergic crisis in the human nervous system caused by an exposure to nerve agent through the respiratory tract. The signs and symptoms are fairly distinctive, with little scope for differential diagnosis in severe cases. A patient with miosis (pinpointed eye pupils in both eyes) and lots of bodily secretions is likely to be a nerve agent patient.

Cyanide Syndrome

Cyanides, which include the warfare agents hydrogen cyanide and cyanogen chloride, as well as some other chemicals that could contaminate food or drink, produce a distinctive syndrome unlike any other class of chemicals. It is important to note that when these agents produce mild symptoms, the exposure is sub-lethal and a full recovery is expected. Victims either die very quickly or will survive with minimal problems.

Skin and Eye Irritation

Many chemicals, both CWAs and HAZMAT, cause various kinds of irritation to exposed tissues, such as skin and eyes. This can range from mildly annoying to debilitating. However, the eyes are very sensitive and the degree of pain and irritation is not an indication of severity of the overall problem. Some riot control agents, for example, are extremely painful but mostly harmless, while mild irritation of the skin and eyes from some vapors is actually indicative of a fatal concentration. There's also some materials that cause irritation right away and others (like the "mustards"), which cause irritation after a delayed interval.

Respiratory Distress

Many substances will produce respiratory distress. Chlorine and ammonia are examples, often cited for its availability and long history of causing injury. A wide variety of chemical substances will cause respiratory distress. Unlike nerve agents and cyanides, which have a discrete and recognizable set of signs and symptoms, it may be neither possible to identify the causative agent based on observation of the victims nor absolutely necessary, as the field medical interventions are basically the same.

Unknown Substance Inhaled or on Skin

Many possible CBRN/HAZMAT scenarios will not have immediate signs or symptoms. In these scenarios, victims will present with powders or liquids on their skin or clothing, or they may report having been exposed to an unknown gas, vapor, or aerosol without immediate effect. An explosion that provides a lot of dust or liquid contamination, particularly out of proportion to the size and nature of the explosion should arouse suspicion among responders.

Conventional Injuries

In many situations, such as an explosive dissemination of CBRN material, there may be a large number of casualties with conventional injuries. In some circumstances, such as an RDD, the only likely acute casualties will be conventional. Blast and fragmentation will cause a wide variety of injuries. Crowd behavior may cause a wide variety of conventional traumas if large numbers of people panic and stampede.

Psychological/Behavioral Issues

The stress and anxiety of a terrorist attack may cause episodes of acute anxiety. Many signs and symptoms of anxiety can be mistaken for various mild or moderate chemical agent exposures. Shortness of breath, which can lead to hypoxia, as well as rapid heart rate, can be caused by a wide variety of causative agents, including anxiety. Nerve agents can cause strange behavior, cognitive issues, and memory loss.

MANAGING THE INCIDENT—BEING REALISTIC IN CHEMICAL SCENARIOS

There is a practical limit to what responders can really expect to do in a field setting, particularly when confronted by a mass-casualty incident. The major event environment is not like a modern battlefield, where every soldier has a mask and three nerve agent antidote autoinjector kits for self-aid or buddy-aid, backed up by a combat lifesaver at squad level and a platoon medic, all trained in treating nerve agent casualties. The reason why modern armies go to such lengths is that such a level of care is the only way to realistically save people in the event of a nerve agent attack. Field care may take ten minutes, half an hour, or longer to arrive. What can we reasonably expect in major event situations?

Nerve agents and blood agents will kill people very quickly from systemic poisoning that will lead to respiratory and circulatory failure. Many respiratory irritants will do so if the victim's ability to breathe is compromised. Patients without obvious nerve syndrome, cyanide syndrome, and respiratory distress may very well need serious medical intervention, but there may not be much that can be done, other than decontamination, in the field setting. It is worth examining some of the major categories of chemical exposure.

Nerve Agents

Nerve agents can kill very quickly, but the severity of symptoms is dependent on both dose and route of entry. Systemic poisoning from dermal exposure will take some time, sometimes many hours, as witnessed by the Skripal poisoning. On the other hand, inhalation of nerve agent will cause medical problems rapidly. Medics may be confronted with a range of victims, with symptoms ranging from mild to severe. Decon, clean air, and perhaps a single dose of antidote may be sufficient to stabilize mild exposures, but patients presenting with severe exposure will be very demanding.

Nerve agent exposure can require a number of medical interventions. For severe exposure, these interventions include the following, in accordance with commonly found guidelines, generally in the following order:

- Establish and maintain airway, through intubation if necessary
- Control secretions through suction

- Ventilate with oxygen, using bag-valve mask
- Monitor pulse, commence compressions if pulse stops
- Administer atropine, pralidoxime (or other oxime, in accordance with local protocols), and diazepam
- Decontaminate any possible skin exposure
- Establish IV access to allow further atropine
- Administer additional antidotes as required
- Move to definitive care
- Constantly reassess airway, breathing, and circulation

I have done this in an exercise setting, both as an exercise participant and as an exercise controller. Even with well-trained medics, these tasks take at least six trained hands, in other words at least three people. Additional difficulty is added by the fact that these tasks often need to be done by medics wearing PPE. One or two victims in such a state of distress may be manageable at an incident scene, but is it realistic to expect to treat thirty? I can definitely see a scenario where we have six medics to throw at two people, but I do not foresee many scenarios where we have ninety medics to throw at thirty patients. If field care arrives after 15 minutes, the unfortunate situation is that many people with severe exposure will have deteriorated and "self-triaged" into the expectant category simply because there are not enough trained hands to help them.

Blood Agents/Cyanides

The scope for field interventions in cyanide cases is limited to the first few minutes after exposure. Effective antidotes exist, but skilled providers must administer them quickly if there is to be any value. Rapid intervention with nitrites and sodium thiosulfate are lifesaving. Amyl nitrite can be administered through a bag-valve mask as a temporary measure. Sodium nitrite and sodium thiosulfate require IV access, and IV medications are fairly useless if the patient has no pulse or circulation. This means that cyanide victims are labor-intensive.

As a practical matter, I wonder how much any of this will matter in a major event environment. The medical interventions required to make a difference are fairly advanced and must be done quickly. Cyanides kill within 5–10 minutes. Victims who are seen by medical providers after this point need no help because they are either dead or will pull through on their own. My own advice is to not spend a huge amount of time on cyanide scenario.

Blister Agents

The most likely blister agents (i.e. the mustard family) will not provide many immediate symptoms, if any. Exposure to blister agents will require decontamination to prevent the spread of contamination and may reduce the scope of injury, but medical interventions will likely be done at the clinical level. Lewisite and phosgene oxime will likely require supportive care for respiratory distress.

Phosgene

Phosgene and diphosgene are an interesting case. Most phosgene victims will be asymptomatic at time of exposure, as the pulmonary edema that is the fatal mechanism of injury takes some time to develop. Indeed, patients with symptoms within a few hours after exposure have a very poor prognosis. People exposed to phosgene should be evacuated as litter patients, even if they are apparently well and asymptomatic. Experience from actual cases shows that this is a discriminator between good and bad patient outcomes.

General Respiratory Distress

The thousands of substances that provide varying degrees of inhalation hazard and respiratory distress are generally managed similarly at field level. While the authorities on the subject discuss the various mechanism of injury at length, intervention at the field level is really all about the ABCs. Maintaining airways, provision of oxygen, and assisted ventilation may be needed to keep victims with serious distress alive.

FIELD CARE—REMEMBER THE ABCDD

The most important thing to remember about medical treatment of CBRN casualties is that the basic principles of emergency medicine still apply. General patient care and solid assessment of patients are still the most important aspect of basic life support. In a CBRN/HAZMAT incident, the medical interventions that are most important at the field level are best understood using the mnemonic ABCDD—Airway, Breathing, Circulation, Drugs, and Decon, in that exact order of priority.

Airway

Regardless of the mechanism of injury or exposure to threat materials, a patient with no airway will surely die, as surely as if you are choking him to death. All the antidotes and decon in the world will not help. Establishing and maintaining an airway is as important in CBRN/HAZMAT scenarios as it is in conventional trauma or illness. Therefore, one of the most important medical skills is to establish and maintain an airway while wearing appropriate PPE. Suction to maintain an airway may be necessary in many cases, such as nerve agent exposure. Consideration as to how that will be accomplished under near-battlefield conditions is important. Even placing victims in recovery position is better than nothing. Using an endotracheal (ET) tube can be difficult and is usually considered an advanced life support skill. Inserting an ET tube while wearing PPE will be much more difficult. Some value may be had from using a "Combitube ™"[1] or similar simplified device, which is often considered a less advanced life support skill.

Breathing

If we are not breathing, we die in minutes. This fact remains true in CBRN environments. Despite all of the other things going on in a nerve agent poisoning, the actual mechanism of death is almost always respiratory failure.[2] Ventilation will be required in many circumstances. In the case of nerve agent exposure[3] or exposure to respiratory irritants,[4] ventilation has a high priority. For obvious reasons, old-fashioned mouth-to-mouth resuscitation is not advocated in CBRN/HAZMAT environments. Devices such as bag valve masks are preferred and are effective tools in a dirty environment.[5] Although filtered resuscitation devices certainly exist, they may not be available at the incident scene.

Many people ask what may be the point of ventilating in contaminated environment. It is important to note that the concentration of vapor present at the site of treatment may not be the same as when the injury occurred. Concentrations of vapors could be much lower by the time the medic is on scene to provide ventilation. It is more likely that the concentration has gone down rather than up since the incident. While not ideal, ventilation with potentially contaminated air is far superior to the alternative of no air at all. Additional exposure *may* harm the patient; lack of oxygen will certainly kill the patient.

Circulation

If blood does not reach our tissues, we eventually enter a state of shock, our essential organs will not receive oxygen, and we will die. Again, this also remains true in CBRN environments. Antidotes can be administered, but if there is no circulation, the medicine will stay at the point of administration. Nerve agent victims who are pulse-less and not breathing are unlikely to survive, but aggressive resuscitation did save one such victim in the Tokyo subway attack.[6] CPR and defibrillation are worth attempting if equipment and training are available and local protocols allow.

Drugs

In the case of the nerve agents and cyanides, effective antidotes are available and can be lifesaving if administered quickly and correctly. Atropine, various oximes (2PAM in the US, but others are in use), and diazepam are considered vital for nerve agent treatment and are often fielded in autoinjectors for easy administration. Amyl nitrite, sodium nitrite, and sodium thiosulfate are potentially lifesaving in cyanide poisonings, but only if administered very quickly after exposure. The cyanide antidotes require more expertise to administer than nerve agent autoinjectors. As mentioned above, amyl nitrate is administered through a respiratory route by crushing an ampoule into the bag of a bag-valve-mask, while the other two drugs require an IV line.

Decontamination

Decontamination (see Figure 16.1) serves several purposes, as discussed in the following chapter. But it should also be considered as part of medical care. Decon will not take priority over the ABC basic life support measures. Unfortunately, I have seen training exercises where casualties were carefully decontaminated before any emergency life support care was rendered. Decon considered a lifesaving measure in nerve agent exposure, but not in priority before airway, breathing, and circulation. But in other situations, decon is not likely to be needed as an immediate lifesaving measure. With chemicals that are causing direct injuries to skin, like corrosives, decon reduces the scope and extent of injuries. Decon is desirable as it removes the source of exposure. For the purposes of basic life support, decon may be as simple as cutting off clothing or a quick flush with water. Nerve agent patients will simply continue to get worse if they continue to absorb agent.

PRACTICAL INCIDENT MANAGEMENT MEASURES

I have described what is, to many readers, a very grim scenario. The challenges are very stark indeed. However, planners and responders need to try to manage this problem in a way that

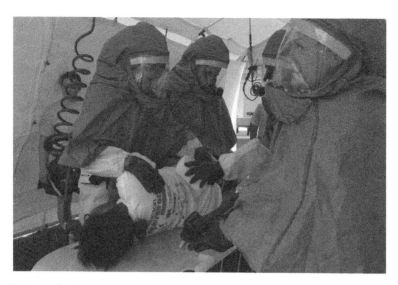

FIGURE 16.1 Patient decontamination is different than normal personnel decontamination.
Source: Photo: US Dept of Defense, public domain image.

saves lives and reduces suffering. Certain key considerations must be analyzed and incorporated into the incident management process.

General Patient Care—Prioritize Airway and Breathing

The mechanism for prompt lethality in chemical scenarios is almost invariably respiratory failure. Drawing up doses of drugs and washing skin when somebody isn't breathing is practically the same as holding a pillow over their mouth. The principal effort should be to get airways established as quickly as possible and keep as much ventilation going as possible. If nothing else, this will buy precious minutes to get other interventions started. The fact that some antidotes exist for a few of the more clearly defined chemical warfare agents has often clouded the minds of many and obscured the fact that airway and breathing are the key to survival. Antidotes are fine, but they may not exist in the right place and right quantity. Ventilation, airways, and suction, while less glamorous, will save lives.

Treatment in the Hot Zone?

The issue of treatment in the hot zone is even more contentious than rescue from the hot zone. Some response agencies, such as the UK HART program, seek to provide some degree of medical care in a hot zone environment. Equipment, such as special ventilators, has been developed for precisely this mission. But providing medical care in the contaminated zone is controversial. My own opinion is that provision of emergency treatment in the hot zone is noble in principle and difficult in practice. From an incident management standpoint, treatment in the hot zone needs to balance two different perspectives.

Why we should—the humanitarian perspective: There are situations where people will die if they do not receive help. There is a limited period of time in which people can be saved, particularly in nerve and cyanide scenarios. Clearly, some lives can be saved by aggressive medical interventions applied quickly by skilled medics in many scenarios, such as nerve and cyanide incidents. There are no insurmountable technical barriers to providing care in the hot zone. Paramedics and physicians can be trained and equipped for this task. The issue is not possibility, but practicality.

Why we should not—the risk management perspective: The capacity and capability to perform medical interventions in the hot zone requires a significant commitment in training and equipment. Medical interventions carried out in a hot zone are more difficult and more time-consuming. Advanced interventions, such as inserting IV lines or intubation, are more difficult when done while wearing PPE. Many invasive procedures risk the possibility of further introducing the contaminant into the victim's body. Furthermore, aggressive interventions tie up several providers. From the perspective of what is best for the overall management of the whole incident, the trained and equipped personnel who are in a position to take the time to enter a hot zone and render medical care may be better utilized outside of the hot zone, where they can work on more patients in the same period of time. Furthermore, time will be lost upon exit from the hot zone due to contamination. Simply put, the same doctor or paramedic can do more, for more people, more efficiently and at less risk to their own health and safety by staying out of the hot zone.

The compromise position: The best approach, in my opinion, is to split the difference. There's little operational penalty for having the capability to do some medical intervention, and then deciding not to use it. My advice would be to have a team that could do hot-zone care and rescue in small scenarios where they might make a difference, possibly using the UK HART as a model. However, the incident commander needs to use the same assets more effectively in a large incident, where the same skills can help more people outside the hot zone. In medium and large incidents, "scoop and go" by responders in PPE is probably the best course of action.

Triage

The key to managing large numbers of casualties is triage. Once an incident has breached the "mass casualty" point (which will depend on local policy), then implement triage and adhere to it. Develop and use triage categories, even if you have to adapt them from reference books. For chemical warfare agents, triage is heavily discussed in Chapter 15 of *Medical Aspects of Chemical Warfare*.[7]

DEFINITIVE CARE

Packing the patients off into ambulances and getting them away from the incident scene may seem like the end of the problem for the incident command on-site, but in reality, it only extends the incident site to the hospitals. While I am not really qualified to comment on the technical content of medical care provided at the clinical level, much of the success or failure depends on logistical and operational matters as much as it does on clinical interventions.

References

Many references exist for the hospital-based provider. The American College of Emergency Physicians,[8] the American College of Surgeons,[9] the US Centers for Disease Control,[10] and many others have resources available online for hospital managers and practitioners. One of the most thorough documents available is the "best practices" document[11] published by the US Occupational Safety and Health Administration (OSHA) and the hospital manager could do worse than to digest this exhaustive document.

Managing the Flow of Patients

There is little point to overwhelming one hospital while others remain un-burdened. The hospitals near the incident site are the ones most likely to get overwhelmed by "walk-in" or self-referring patients. Patients who were stabilized in the field can probably get safely transported to hospitals further away from the incident site. The incident management system needs to monitor the flow of casualties and disperse them among available facilities. Such systems already exist as part of MCI plans in many areas, but not others.

Apply Incident Management Systems at the Hospital Level

Incident management procedures and policies are as useful inside the hospital as they are in the field. The same principles, such as unity of command, interoperability, and scene control, have applicability in the clinical environment. Incident management systems, such as ICS, have had less history of use in hospital settings than in the field. This is beginning to change, as there is a recognition that there is value in such systems and that, in a large event, the various hospitals actually need to be part of a large incident command system. As one example, the state of California has developed its own Hospital Incident Command System,[12] which is a useful template.

Assume That Patients Will Self-refer and Self-transport

No incident plan can assume that all victims will be decontaminated at the incident scene. As discussed in other chapters, it can be assumed that some number of people, both injured and "worried well," will turn up at medical facilities seeking assistance.

Decontamination

Decontamination plans, procedures, and equipment are required for both ambulatory and nonambulatory patients, including those with special needs. Despite the best efforts, decontamination is likely to be needed at the hospital-level, and many hospitals have derisory decontamination processes.

Do Not Disregard the Other Patients

Just because there is a terrorist incident does not mean that people will stop having heart attacks, strokes, and car accidents. Hospitals will need to be able to cope with their "normal" patients as well.

Hospital Security and Scene Control

A primary imperative for hospitals in all CBRN/HAZMAT situations is to defend the hospital from being over-run. There are many potential scenarios whereby members of the public-seeking assistance can overwhelm a medical facility. If such a situation occurs, there are many ways in which contamination can be spread into a hospital. This could have the effect of taking the hospital out of service. Defending a hospital's perimeter is key in managing such a scenario. Hospitals have many entrances and exits, and scene control in a CBRN/HAZMAT incident poses some challenges and may require a significant effort by security personnel. Effectively, it requires scene control in reverse. The "hot zone" is outside, a "warm zone" is the entrance through a decontamination process, and the "cold zone" is the interior of the hospital. Maintaining scene control in a CBRN incident at a hospital will probably require a fairly assertive security "lockdown" response.

Investigative Issues

Hospitals are likely to be the recipients of evidence. People will arrive with potentially useful evidence on their clothes or belongings. Mobile phones and cameras that are isolated as contaminated goods at the beginning of a decontamination process may hold useful video footage. Many of the people who arrive seeking care may be useful witnesses, who saw or heard something of significance to an investigation.

Triage Considerations at the Hospital Level

Triage applies as much in the hospital as it does in the field. Triage is routinely practiced in hospital emergency departments. The same general triage categories apply, and most hospitals have a disaster plan or MCI plan that provides for triage. Patients arriving by ambulance should be re-assessed, as their triage category may have either improved or deteriorated since they were seen at the incident scene.

PPE and Detection

CBRN/HAZMAT situations will require some hospital staff, such as triage and decontamination teams, to operate in PPE. Generally speaking, unless the incident occurs at the doorstep of the hospital in question, the amount of contamination arriving at a hospital is likely to be significantly less than that at the incident scene itself. This likelihood, however, should not be taken for granted. Sensors should be used to establish the presence of any hazards. In all likelihood, Level C PPE is all that is needed for most hospital decontamination operations, but safety laws and regulations need to be followed.

LESSON LEARNED: TREAT THE PATIENT NOT THE SCENARIO

It sounds almost too simple, but the most important thing in providing emergency medical aid is to treat the patient, not the incident. Assess every patient and treat accordingly. I have witnessed far too many exercises where highly trained medics made treatment decisions based on assumptions and detector readings rather than the signs and symptoms of the patient. I also have seen training exercises where simulated atropine was administered for a patient who clearly was in respiratory distress, but had no unique symptoms of nerve agent, clearly out of some belief that a nerve agent attack had occurred. There was another situation where there was one nerve agent victim (simulated) and a lot of trauma casualties. Guess what? The participants decided to give everyone atropine, whether they needed it or not. Treat the patient in front of you, not a belief in what happened. You would not do that with a cardiac patient or a trauma patient, so why would you do it with a CBRN patient?

MINI-CHAPTER: LESSONS FROM A PANDEMIC

There is no point denying the obvious fact that this second edition was revised during the worldwide COVID-19 pandemic. Indeed, pandemic-related issues have caused delays in the production of this edition.

Does the COVID-19 pandemic qualify as the sort of incident relevant to this book? Is it a CBRN incident? Is it a hazardous materials release? The answer is not actually terribly clear. First, let me dismiss the obvious. My own view is that the COVID-19 pandemic is not deliberate nor is SARS-Cov-2 a biological weapon. I stand by my views that I expressed in the *Washington Post*.[13] This pandemic started in nature. Did it escape from a laboratory environment through some sort of mishap? This is a now long-standing allegation and many people making this claim are doing so in ways that are disingenuous and/or unhelpful. By this point, it is just speculation and, barring some stunning revelations out of China, a laboratory escape is probably an unprovable but plausible hypothesis.

The more I think about it, there is a sensible way to look at a regional or global pandemic from the viewpoint of this book. This book does what it says on the cover and focuses on "major public events" and CBRN/HAZMAT threat. The "major events" I cover are basically pockets of human activity above and beyond the normal baseline in a particular point in time and space. A pandemic will sweep across these major events like a wave from the sea washing across pebbles on a beach. All of society is affected. Therefore, major events will be affected. Because everything is affected.

The point of this mini-chapter is not so much to tell you that you can prevent a pandemic. That sort of thing is beyond the scope of the book. But what I can do is to point out some things that we have learned during this pandemic that are relevant to protecting major public events in different scenarios. We can learn from this dreadful pandemic and take pandemic lessons and apply some of them in non-pandemic scenarios. There is a reason why this bit of the book is placed here. Many of the mitigations that have been used to combat the COVID-19 pandemic are useful in other contexts.

Understanding the Threat Vector

In a CBRN terrorism attack or a traditional HAZMAT accident, there's usually a definable "source term"—a point of origin for the dangerous materials, such as a leaking tanker truck or a terrorist's device. In a pandemic, the source term is people. With some diseases, it can be obvious who among the public is a possible source of infection. Chicken pox, for example, is generally quite obvious. But COVID-19, influenza, and similar diseases can easily be unobservable to human senses or so generic as to not be distinguishable. Furthermore,

COVID-19 and many other illnesses have a period of time where asymptomatic contagion can occur. From an event standpoint, it is like knowing that the threat is people with knives, but the knives are invisible. In a pandemic situation, people are the threat and you likely will not know which ones.

"Super-spreader" Events

We are now all familiar with the concept of super-spreader events. Major events, which concentrate a lot of people in one place at one time, can become a multiplier for the spread of aerosol pathogens. There is also the possibility of reputational damage. Pushing forward with an event, without mitigation, and becoming a source of infection might be permanently damaging to an institution.

Event Postponement and Cancellation

If you can postpone the Olympics and cancel Glastonbury festival, then the precedent is pretty well set for other events. Cancellation and postponement are, in fact, valid mitigation strategies. If your event is going to make the overall problem worse rather than better, then a serious consideration of postponement or cancellation is a viable course of action. There are penalties associated with this. But yet another thing that we have learned from the current pandemic is that contracts, terms, and conditions associated with event management are quickly evolving in this area. One lesson that we learned during the COVID-19 Pandemic is that cancelling events is not actually the end of the world.

Ventilation

Ventilation at interior events is an important aspect of defense against respiratory threats, whether they are chemical, biological, or radiological in origin. The exchange of fresh air with air that might contain hazards has already been mentioned as a good mitigating tactic. With a CBRN incident outside of a building, one would want to limit the exchange of air from outside, thus using the physical barriers of the building to buy time and provide a degree of protection. With respirable aerosols emitting from people, the opposite is the case. You would want more fresh air from outside into the event, thus diluting hazards. Interestingly, this is a phenomenon that can be measured. Humans emit carbon dioxide when they exhale. In an occupied space, measuring how much of this carbon dioxide has built up can give some indication as to the relative level of risk from aerosol pathogens.[14] You can now use some relatively inexpensive measurement techniques to see how good your ventilation strategy might be.

Outdoors Events

Moving an event from indoors to outdoors can provide a degree of mitigation against pandemic threats. But in terms of vulnerability to CBRN/HAZMAT threats, an outdoor event venue can be more vulnerable. You will need to balance the probabilities and balance the "pandemic threat" against the CBRN terrorism. Moving some threats outdoors may make them more vulnerable to conventional attack. We need merely to look at the 2017 shooting in Las Vegas, where a gunman killed 60 people and wounded hundreds while firing at the crowd at a music festival.[15]

"Social Distancing" and Queues

Queue management becomes an important consideration. I have always worried about people congregating outside major events, long before I ever heard of COVID-19. COVID-19 has forced people to study the effects of queueing in the context of a respiratory pandemic.[16] Having a bunch of people standing in long queues, often snaking past each other in serpentine queue arrangements, is not conducive to mitigating the spread of respiratory illness. While queuing might be necessary, review your arrangements and think of ways to cut down the queueing. More staff to process entry reduces the length of queues and the time spent in them. Having a straight queue instead of a zig-zag means that people in a queue are not facing each other. Staggering entry times with timed tickets may mean that fewer people are in a queue at any particular moment.

Passes, Passports, and Apps

The COVID-19 pandemic has yielded a wide variety of administrative control measures such as vaccine passes, virus passports, track-and-trace apps, and similar tools, both in paper and software form. The usefulness of these tools has been hotly debated. While I await the inevitable study of which ones were or were not useful, it is necessary to admit that such tools might be useful in the future in major event environments. No doubt some clever people are working on improvements to such tools for use in future pandemics.

Rapid Diagnostics

The speed with which rapid diagnostic technology was fielded, such as PCR tests and "lateral flow" antigen tests, was astounding. In all likelihood, in future pandemics, we will see similarly fast, or even faster, deployment of technology to test for a pathogen in a human population. If I think about it, the first edition of this book has been in print for ten years. It is not unreasonable to think that, within the print lifetime of this edition, there may be rapid diagnostic capabilities for many pandemic illnesses that could be used very quickly at entry points into events as a screening tool.

The Antisocial Resistance

The rise of anti-vaccination, anti-masking, and anti-lockdown agitators and conspiracy theorists during the COVID-19 pandemic is troublesome, to say the least. Indeed, some have even concocted fake CBRN attacks as an excuse for their own illnesses to avoid admitting that COVID-19 was a real phenomenon, as witnessed by the "activists" in Texas who, erroneously, claimed that their illnesses after a conference were the result of an anthrax attack.[17]

Conspiracy theories and belief in unorthodox alternative narratives may hinder emergency response at a major public event. An anthrax attack, for example, will require a mass campaign to ensure antibiotics are taken, possibly by many thousands of people. If this were to happen now, I would almost guarantee you that some refuseniks will try to sabotage the effort.

Some Ideas for Next Time

Everyone planning a major event from about March 2020 onward was caught out by surprise by the speed at which the world came to a halt. But have we learned? I think that we have. If you are working on events with long planning horizons, such as months or years in advance, you may able to build some contingency plans into your event planning. If you build some pandemic contingency plans into an overall safety and security plan, and are ready to activate them, you may be able to avoid cancellation or postponement. Or, at a minimum, you may be able to offer the overall management team some options to consider.

Some of these options to consider might be any of the following:

- Consider, in theory, what your infectious disease and vaccination policy are going to be if a pandemic threat crops up during your planning timeline. Have a draft screening plan that you can activate if needed that allows time for any of the following:
  - Scanning an app
  - Scanning some sort of "passport" or "certificate"
  - Scanning body temperature
  - Inspection or provision of masks
- Review the movement, congregation, and queuing of the public with a view toward increasing distance and dispersion.
- Understand your ventilation. Consider carbon dioxide monitoring as a proxy indicator of how good your ventilation is.
- Develop a cancellation or postponement policy that accommodates infectious disease concerns. Have some idea what milestones or decision-points ahead of a major event might be significant in a decision to postpone or cancel an event.

REFERENCES

1. Agro, F. et al. (2002). Current status of the Combitube™: a review of the literature. *Journal of Clinical Anesthesia* 14 (4): 307–314.
2. Tuorinsky, S.D. (2008). *Medical Aspects of Chemical Warfare*, 174. USA: Government Printing Office.
3. Ibid, 181.
4. Ibid, 363–364.
5. Schumacher, J., Weidelt, L., Gray, S., and Brinker, A. (2009). Evaluation of bag-valve-mask ventilation by paramedics in simulated chemical, biological, radiological, or nuclear environments. *Prehospital and Disaster Medicine* 24 (5): 398–401.
6. Medical aspects of chemical warfare, 342.
7. Ibid, 511–548.
8. American College of Emergency Physicians (2001). *Disaster Response and Biological/Chemical Terrorism Information Packet*. American College of Emergency Physicians.
9. American College of Surgeons. Disaster Response. https://www.facs.org/for-medical-professionals/membership-community/operation-giving-back/resources/disaster-response/, accessed 22 July 2022.
10. Centers for Disease Control. Chemical Emergencies. https://www.cdc.gov/chemicalemergencies/index.html, accessed 4 April 2014.
11. US Occupational Safety and Health Administration (2005). *OSHA Best Practices for Hospital-Based First Receivers of Victims from Mass Casualty Incidents Involving the Release of Hazardous Substances*. The Administration.
12. California Emergency Medical Services Authority. Hospital Incident Command System – Additional Resources and References. https://emsa.ca.gov/hics-additional-resources-and-references/, accessed 22 July 2022.
13. Kaszeta D. No, Coronavirus is not a biological weapon. *Washington Post* 2020; 27.

14. (a) Vouriot, C.V.M. et al. (2021). Seasonal variation in airborne infection risk in schools due to changes in ventilation inferred from monitored carbon dioxide. *Indoor Air* 31 (4): 1154–1163. (b) Bidilă T., et al. Monitor Indoor Air Quality to Assess the Risk of COVID-19 Transmission. *2021 23rd International Conference on Control Systems and Computer Science (CSCS)*. IEEE, 2021.

15. Pearce, M. (2018). The most comprehensive look yet at how the Las Vegas concert massacre unfolded. *Los Angeles Times* 19.

16. Perlman, Y. and Yechiali, U. (2020). Reducing risk of infection–The COVID-19 queueing game. *Safety Science* 132: 104987.

17. Press-Reynolds, Kieran. (2021). Far-right influencers suggest widespread illness following an event may be an 'anthrax attack,' despite similarity to COVID-19 symptoms. *Insider*, 23 December 2021. https://www.insider.com/qanon-influencers-far-right-anthrax-attack-reawaken-america-tour-2021-12, accessed 13 September 2022.

Decontamination

Decontamination, decon for short, will be an important part of response plans for a major event. Many of the potential threats that we have discussed will provide contamination of personnel, buildings, and equipment. Threat materials that become deposited on a surface are "contaminants" and anything with contaminants present on it or in it becomes "contaminated." I define decontamination as any process for removing or destroying contamination.

WHY DO DECONTAMINATION?

Although the reasons for decontamination are usually taken as an article of faith in the CBRN/ HAZMAT community, such a level of understanding does not always permeate through the other sectors of the emergency response community or the public. The specialist may need to articulate the reasons why decontamination may be necessary.

Decontamination Is Often a Medical Necessity

People may need to get CBRN materials off their skin or else they eventually will suffer ill effects. In some cases, decon is a life-saving intervention, whereas in other situations decon will reduce the severity of injuries and reduce discomfort. With some industrial chemicals, decon will provide immediate relief from pain and irritation.

Reduces the Spread of Dangerous Materials

Decon can be seen as a component of scene control, because it controls what comes out of an incident scene. Even very simple decon measures serve to keep material from spreading too far

CBRN and HAZMAT Incidents at Major Public Events: Planning and Response, Second Edition. Daniel J. Kaszeta.
© 2023 John Wiley & Sons, Inc. Published 2023 by John Wiley & Sons, Inc.

outside the incident scene. Decontamination near the event will help to prevent contamination spreading to ambulances, hospitals, and peoples' homes.

Returns Personnel and Equipment to Service

Responders who are dirty cannot simply drop what they are doing and do something else. Decon can allow specialized personnel and specialized equipment, both of which are likely to be in short supply, to redeploy to other roles at an incident.

Provides Psychological Assurance

In a CBRN/HAZMAT incident, many people physically unaffected by the incident will be convinced that they need help. Decon can provide some assurances to such "worried well" as it does no additional harm but is a firm measure that is clearly meant to provide help.

Assists Investigation Efforts and Forensics

A decontamination corridor can and should serve a useful role as a controlled exit to an incident scene. After a CBRN incident, valuable evidence may be on the skin and clothing of victims. People waiting for decontamination may be important witnesses to the incident that just took place. Placing trained investigators at the beginning of the decon process to collect evidence and interview witnesses will aid in any resulting criminal investigation.

FIGURE 17.1 Emergency decontamination needs to be done with whatever assets are available early on in the incident response.
Source: Photo: US Army, public domain image.

Prompt Decon Saves Resources

Decon helps to keep the size of the problem from growing any bigger than it already is. If people are covered in powder or liquid, the incident scene, by definition, is where they happen to be. As contaminated people move about, the size of the "hot zone" gets bigger and bigger. One does not have to be a specialist to understand that a larger hot zone will take more resources to deal with than a smaller one.

DECON AT MAJOR EVENTS

The purpose of this chapter is not to teach responders how to perform decontamination or to specify which products to use. I have strong opinions on these subjects, but it is likely that most of the readership lives and works in areas where the response authorities have procedures for decontamination. There are many different types and methods of decontamination, and vitriolic arguments have arisen over whose method is superior. These arguments will not be rehashed here. Initial decon, emergency decon, gross decon, technical decon, and other terms can mean different things even within the firefighting disciplines, let alone between rival operational disciplines. There are good primers available to provide an overview of CBRN/ HAZMAT decontamination,[1] as well as some excellent books and training courses that describe decontamination in great detail,[2] and the reader would be well served to make a serious study of the subject.

Size

Decontamination at a major event poses problems of scale and scope that make it a fundamentally different problem than that faced on the battlefield or at a "normal" civilian HAZMAT accident. Most people with practical experience of decontamination are used to dealing with small groups or at most an army battalion. HAZMAT accidents rarely involve the requirement to decontaminate large numbers of people. Even large military decontamination operations are generally planned for the battalion level at the largest, with many hundreds of people and a few hundred vehicles, but not thousands of victims.

Scope and Variety of Victims

Events with large numbers of the general public will mean that there will be a wide variety of people with special needs, such as the elderly and disabled. A military decontamination operation is usually dealing with healthy soldiers who are wearing chemical protective gear and are accustomed to taking direction. A HAZMAT incident in an industrial setting may only have "normal" adult victims drawn from a typical industrial workforce, but if you have 20,000 members of the general public, statistics do not work in favor of the "normal" victim population as a planning basis. It is likely that the incident commander will be confronted with children, the elderly, people in wheelchairs, and people who do not understand the local language, not to mention epileptics, people with guide dogs, the deaf, and people with objections to removing clothing for religious reasons.

Lack of Precedent

The major event planner does not have much historical experience to work in the decontamination realm. While excellent theories exist and many products have been developed, the truth is that nobody has had to put decontamination plans into operation at a major event. Nobody has had to decontaminate a stadium full of 75,000 people, so we do not know how it would have worked. The lack of useful precedent has left ample room for speculation as to what will work and what will not.

CATEGORIES OF DECONTAMINATION

Authorities differ on how to classify decontamination into categories. I am not sure that the exact categories that exist in military manuals (such as the US Army's FM 3–5[3]) or the fire service (such as the US's NFPA 472[4]) are immediately useful for major events. The most useful way I can think of to categorize decontamination is through use of examples as follows:

Emergency Decon

Responders show up at an incident scene and as an emergency measure, they perform a quick wash-down of victims and responders that they encounter. This can be at the very early stages of an incident or in the event of an emergency evacuation of an entry team. Generally, an emergency decon in a civil setting is done with plain water, usually by firefighting crews. This is illustrated in Figure 17.1.

Mass Decon

Responders decontaminate victims who are affected by the accident or incident. This is done both to limit the spread of contamination and as a vital first aid procedure to minimize health threats posed by contaminants. Mass decontamination has three subsets of particular interest.

Ambulatory: Ambulatory decon is when the victims are capable of walking under their own power and are capable of following simple direction. Such people may or may not be injured.

Non-Ambulatory: Decon for victims who are incapacitated, such as those on stretchers or unconscious. Also called "litter decon" or "patient decon" in some references.

Special Needs: Victims such as the elderly, physically and/or mentally disabled, people who do not understand the local language, or children who will require additional support. Bear in mind that some substances may affect people in ways that hinder their faculties, so in many situations many victims will have "special needs." For example, people exposed to small amounts of nerve agents may be behaving in a strange or confused manner.

Responder Decon

Responders establish a formal decontamination lane to process their colleagues out of the hot zone in a way that systematically removes all contamination from the exterior of their PPE. This is illustrated in Figure 17.2. Responder decon may be performed with more detailed care than mass decon, as the PPE may need to be reutilized for further entries.

FIGURE 17.2   Responder decon is usually more meticulous than mass decon. *Source: Photo: US Air Force, public domain image.*

Equipment/Technical Decon

Responders undertake the detailed cleansing of mission critical equipment to ensure that it can be reutilized. While this type of decontamination is very resource-intensive in terms of time and labor, it is necessary in many cases. Many types of response equipment may be needed for other incidents and will have to be decontaminated in order to return to service. In addition, detailed decontamination may need to occur repeatedly during investigations at an incident site for the purposes of evidence integrity and prevention of cross-contamination.

Sensitive Equipment Decontamination

Some equipment, such as electronics, is sensitive to the brutal nature of many standard decontaminants. Computers and radio equipment generally cannot withstand hot soapy water or bleach solutions, for example. Therefore, special care will be needed for the decontamination of mission-critical-sensitive equipment. Special products are made for this sort of decontamination.

VIP Decontamination

Dignitaries may require separate and segregated decontamination due to their security requirements. Protective teams will be in an agitated state after a terrorist event and will require special handling well away from the public.

Hospital Decon

People will self-evacuate and may seek treatment on their own. Hospital decon is any decontamination process that occurs away from the incident site at a medical treatment facility. In a sense, hospital decon is really just mass decon done at the hospital entrance. In a perfect world, all decon will be done at the scene, but we all know that that is unlikely. Worried well will turn up at the hospital, but there is also every possibility that grossly contaminated individuals will show up as well, particularly with agents that have long latency periods.

Building Decontamination/Remediation: Buildings and their contents can become contaminated. Some materials might remain dangerous for a very long period of time. Building decontamination is the process of cleaning buildings and (often) their contents. This is a growing line of inquiry in many countries, but it is beyond the scope of this book, as I am generally not discussing long-term recovery issues.

METHODS OF DECONTAMINATION

There are many different approaches to decontamination. The relative merits of the different approaches depend on many variables, such as objective, resources, and time available. There are others, but the following are the basic categories that I feel are relevant for major event purposes. Please note that in some cases there is overlap between categories. For example, some chemicals degrade in water (a process known as hydrolysis), so a water wash-down may also count technically as neutralization. Selection of a decontamination method will depend on many factors, such as the threat material, logistical considerations, time, and the number of victims.

Weathering/Aging

Natural processes and the passage of time will affect some contaminants. Radioactive isotopes have half-lives, and some are short enough to be of tactical significance. Some chemicals degrade, evaporate, or diffuse to a point at which they are no longer a hazard. Many biological agents are sensitive to the ultraviolet light in natural sunlight and are inactivated relatively quickly under the right conditions. While weathering and aging are not the primary methods of decontamination in most circumstances, it is important to understand that in a major event the scope of potential contamination may be so widespread that not everything can be decontaminated immediately by other means, and some things may have to be left to age.

Removal

The most common technique is removal. Water, with or without soap, is very commonly used to remove contamination from skin, clothing, and equipment. Most decontamination procedures in the civil emergency services around the world are based principally on water wash-down techniques. It is very important to remember that removal may dilute or move hazards, but it may not necessarily inactivate or neutralize hazards to health and safety. Water used to remove a toxic chemical may become a solution of toxic chemical. Decontamination with soap and water may often save lives while transferring the hazard from an immediate health and safety problem to a longer-term environmental problem. However, decontamination with water is by far the most affordable and logistically sustainable approach to large contamination problems. In some cases, water may actually create nearly as much of a problem as the threat material, as some chemicals will react to water, and other chemicals may have decomposition products that are nearly as problematic as the chemical itself. Removal of the outer layer of clothing is often the most useful and effective step for many types of personnel decontamination. It should be noted that removal is usually considered the viable technique for radiological decontamination.

Dry Decon

Not every contaminant is a liquid. In radiological situations, the contamination may take the form of dry dust and debris. In such circumstances, contamination may be removed by dry techniques such as the use of vacuum cleaners and brushes. Use of a high level of filtration (i.e., HEPA) in vacuum cleaners is necessary for this technique to have any effectiveness.

Absorption and Adsorption

Some decontamination techniques seek to assist the removal of contaminants by adding an absorbent (a material that will soak up a contaminant) or an adsorbent (a substance that will stick to a contaminant, easing its removal). A classic example is Fuller's earth, a clay-like mineral that easily absorbs liquids. Some absorbents/adsorbents may also have some neutralization effect due to the chemical properties of the material.

Neutralization/Disinfection

It is also possibly to attack the contaminant by physical or chemical means. For example, acids and bases have chemical properties that react with many chemical or biological substances to render the threat substance less harmful. For example, chlorine ions in a solution of household bleach act to kill microorganisms. Acids or bases may act on chemicals or biological materials. Decontamination by neutralization is often accomplished by applying some type of decontamination solution to the contaminated surface. Very few neutralization techniques are instantaneous; most will require some contact time. While chlorine solutions were considered the primary decontamination solution for many chemical situations, they are considered quite obsolete for many purposes. Chlorine solutions can be dangerous to skin and therefore may not be suited for personnel decontamination unless diluted. However, weak bleach solutions have a valuable role to play in many biological situations. Many commercial products have been designed for decontamination by neutralization. Relevant examples include Reactive Skin Decontamination Lotion produced by RS Decon and the various solutions produced by the Cristanini company in Italy.

TACTICAL CONSIDERATIONS

Executing decontamination can be challenging. The following paragraphs represent the planning considerations that I feel are important for the major event safety and security planner.

Not Every Situation Requires Decontamination

Remember the difference between exposure and contamination. Decontamination is a remedy for contamination, but it may be useless to remediate exposure. It is very important to remember that not every CBRN/HAZMAT situation requires decontamination as part of its resolution. Some hazards do not provide contamination of personnel or equipment. For example, hydrogen cyanide is largely a respiratory problem. It is extremely volatile in liquid form, so even if encountered in liquid state (unlikely), it would evaporate before it could be decontaminated. While decontamination might be the most logistically burdensome part of a response operation, it should be remembered that it would not be needed in every single scenario.

Decontaminate in the Field, Not in the Hospital

One of the oldest operational imperatives in military decontamination is to decontaminate as far forward in the battle zone as possible. This maxim applies in the major event environment as well. When translated to the civil response context, this means providing for decontamination as close to the incident scene as possible. Part of the reason for decontamination is to reduce the spread of contamination. It is far better to transport clean patients in clean ambulances. Placing dirty patients into clean ambulances will likely result in mission critical patient transport assets becoming unserviceable, and most emergency plans will not sustain loss of too many ambulances. Most local authorities understand this and do not rely on decontamination at the site of hospitals as their primary plan, but the point bears repeating.

Logistics

Decontamination is likely to be one of the most logistically demanding portions of a major event CBRN/HAZMAT plan. Mass decontamination will require a significant operational footprint in terms of equipment, manpower, water, decontamination solutions (if used), and related resources. Decontamination can be labor-intensive, will require crews to wear protective equipment, and is likely to require personnel to work in shifts. Does your major event security plan adequately address the resource requirements for extended large-scale decontamination operations?

Decontamination Can Be Dangerous

While decontamination is necessary in many circumstances, it can also be harmful to people in some instances. Even relatively mild weather can provide the hazard of hypothermia when you strip clothing off of people and put water on them, particularly with some populations at higher risk from exposure, such as the elderly. Temperature is a key planning consideration. People who are ambulatory and not in visible distress are probably not in an immediately life-threatening position. However, if you soak them in cold water on a cold day, a minor problem can be turned into a major problem. Traditionally, this has been addressed with various techniques, such as warm water for decon, heated tents, and ample provision of towels and dry clothing for re-dressing, all of which increase the logistical burden of a decontamination operation. Mass decon, if done properly, often creates a mass care situation. In a cold weather situation, vapor pressure and volatility work in our favor. If only cold water is available on a cold day, perhaps the best approach with lightly contaminated people is to get them to quickly shed their clothing into plastic bags and re-clothe them in clean garments. In many circumstances, merely removing outer layers of clothing achieves 90% of the value of the decontamination process.

Decontamination Solutions Versus Water?

In the early 1990s, personnel decon was largely an argument of bleach versus plain water, or in some European militaries, Fuller's earth. The subject of decontamination has since become more complex. There are many possible ways to conduct decontamination. Numerous products have been developed for use in decontamination. Most are useful. While I personally believe that many of the products available on the market are useful and effective, and have various scientific studies supporting their claims, I think that cost will make their use in mass decontamination situations problematic. The various decontamination products cost money, whereas warm soapy water is relatively cheap. Whether one believes in the effectiveness of decontamination solutions or not, there is no question that trying to stockpile enough product to decontaminate thousands of people is a costly proposition. In addition, it is important to remember that few of the products available are truly polyvalent—able to counteract every hazard. Depending on the threat, warm soapy water may still be the best solution, even if you have three cargo containers of a specialty product. I believe that soap and water are going to be the cheap and practical solution for mass decon. The various specialty decon substances available are useful and effective if used in accordance with the manufacturer's guidelines. They have a role to play, but having large stockpiles of them just for a major event will be very expensive and it may be difficult to quickly deploy them in the right place. Consider reserving the specialty items for responder decon and equipment decon.

How Big Is the Problem? How Many People Will We Have to Decontaminate?

The most difficult part of planning process will be to calculate the scope of the problem. The gut reaction of many would be to plan based on the number of people at a venue. This can make the size and scope of the problem assume astronomical proportions. I have sat in meetings where discussions of this nature have had a chilling effect on planning. But how do we realistically assess this scenario? If a chemical device detonates in the middle of a stadium holding 50,000 people, does this mean that we are going to have a queue of 50,000 people who require decontamination? This is the equivalent of three or four army divisions. There is no easy answer to these questions at present, but it seems clear to me that the answer is something less than 50,000. For what its worth, the US Army tells us to expect a five to one ratio of uncontaminated to contaminated victims at an incident,[5] although it is unclear how they arrived at this figure. Until we get a better answer, the answer for now is that you work out a planning threshold for each venue and if more people turn up than you plan for, you don't lose your head, you muddle through.

Triage

Decontamination takes time, and some people will need it more than others. Many authorities suggest using a triage scheme. As one example, the US Army provided guidelines in January 2000, suggesting that the victims requiring ambulatory decon should be decontaminated in the following order:

> "The highest priority for ambulatory decontamination are those casualties who were closest to the point of release and report they were exposed to an aerosol or mist, have

some evidence of liquid deposition on clothing or skin or have serious medical symptoms (e.g., shortness of breath, chest tightness, etc.). The next priority are those ambulatory casualties who were not as close to the point of release, and may not have evidence of liquid deposition on clothing or skin, but who are clinically symptomatic. Victims suffering conventional injuries, especially open wounds, should be considered next. The lowest decontamination priority goes to ambulatory casualties who were far away from the point of release and who are asymptomatic"[6]

This certainly seems like someone worked hard on this, but it seems impractical to me. Short of a color-coded ticket system, a squad of bouncers, and a miraculous amount of docility among the victims, I do not think I could ever implement this scheme, nor does it make a huge amount of sense to me. In practice, I wonder how well any triage scheme will actually work without the presence of large numbers of security staff to maintain order. I think that on-scene personnel are going to have to do the best they can to triage people for decon, but in large-scale situations any attempt to implement anything more than a "if you are having trouble breathing you go to the front of the line" triage policy will be difficult.

Worried Well

Many people who do not need decontamination may turn up anyway. The more agitated and anxious people are, the more likely it is that there will be problems maintaining order at the decontamination site. Indeed, some people may require decon only for reassurance purposes.

Site Selection

Site selection is important, but it is very difficult to make recommendations in the abstract. The site survey process should identify primary and backup areas for decontamination at your major venues. The following selection factors are most important:

Wind direction: Ideally, the decontamination area should be upwind from the problem. Because winds will change, this factor demands a degree of flexibility among responders.

Proximity to likely mass evacuation routes: Is it going to be easy or difficult to funnel evacuees into the decontamination area? If, as I have seen, a decontamination area is on one side of a building and the emergency exits are on the other, this can make for an interesting operation. If poorly executed, such an operation runs the risk of losing scene control, as victims may wander off in every direction, thus increasing the size of the contaminated area.

Surface: Is it gravel? Dirt? Asphalt? Grass? Concrete? Are contaminants going to seep into the ground and then desorb later on?

Water supply and drainage: Most mass decontamination plans will involve large quantities of water. We need to have a source of water sufficient to the requirements and we need to have an idea where the water is going to go after we are done with it. Dirty water will go somewhere. Having that water available and having some idea of where the dirty runoff will go is critical. Drainage also poses a potential contamination problem in itself. Dirty water leftover from decontamination efforts can cause environmental damage.

I think that while having a good idea where to set up a decontamination area ahead of time is very useful, it is important to have flexibility. Have a backup plan. The situation may be very different on the day of the incident than on the day the site survey was done. I have been through a few planning meetings where some manager decreed: "Decontamination will happen in X area" without regard for changes in wind direction or consideration of the drainage.

Queuing and Public Order

Mass decontamination is likely to involve a lot of people standing in line to endure an experience that they are likely to find unpleasant and possibly degrading. People will be suffering from fear and anxiety. Crowd dynamics will become an important factor in the emergency response. Therefore, it is important that security and public order be a part of the planning for mass decontamination. It seems likely to me that it will be necessary to use police and security personnel to enforce order at a decontamination site. Training exercises (which generally use relatively docile role players, such as Red Cross volunteers) have not been particularly useful in simulating this aspect of decontamination. While I do not think that it is likely rioting victims will over-run decontamination lines in a terrible stampede, neither do I believe that everyone will form an orderly queue and wait for their turn. The reality is likely to be somewhere in the middle. I believe that there is scope for more research on the crowd dynamics of mass decontamination.

Public Information

It is not that easy to provide good, clear, and well-understood direction to members of the public through the facepiece of a self-contained breathing apparatus. Planning efforts need to consider ways in which the decontamination responders can communicate effectively with the crowd. Bullhorns and PA systems will be needed. At events with international attendance, some of the victims will not understand commands given to them.

Not Everyone Will Be "Normal"

Not everyone or everything that needs to be decontaminated will be in normal categories. Major events will have a broad category of people who may require decontamination. Studies have shown that such populations can be accommodated if appropriate measures are incorporated into planning and exercises.[7] In some places, response agencies now include people with special needs into decontamination exercises. This is a practice that I highly recommend. Any planning effort should account for the following special categories:

Elderly
Disabled
Children
Speakers of other languages
Obese victims
Dead bodies
Prisoners
People with religious garments or head coverings

Decontamination Should Not Willfully Destroy Evidence

Decontamination procedures must not be set up in a way that all useful evidence is destroyed. Some law enforcement authorities have been known to make statements that were, effectively, anti-decontamination.[8] We must acknowledge the grain of truth in this viewpoint, because decontamination can destroy evidence and the fire service crews who are the most likely practitioners of decon are not approaching the problem from an investigative viewpoint. Efforts must be made to ensure that items of interest in forensic investigations are preserved. As discussed in the forensics section, positioning trained evidence collectors at the beginning of a decontamination line may be useful in collecting evidence and obtaining statements from witnesses.

Custody of Personal Effects

Decontamination procedures must account for the fact that everyone will have a wallet, keys, passport, digital camera, and/or other valuables. Custody of the personal effects of twenty people is easy. For 10,000, it is a large task that will require significant resources. Assurance that personal effects will be looked after and eventually returned will help with crowd control. Furthermore, some smartphones and cameras may, in fact, contain useful evidence.

Perpetrators

A terrorist may be in the decontamination line. We must be aware of the fact that persons who perpetrate a terrorist act may end up as victims of their own actions. A terrorist may not have been able or have wanted to escape after employment of CBRN materials. A dissemination device may have not functioned as intended. The materials involved may not have immediate effect. There are many situations in which a perpetrator may turn up seeking decontamination.

Video

Should decontamination procedures be recorded? There are good reasons both for and against this practice. Video evidence taken in the immediate aftermath of an incident may prove useful in the investigation of the evident, particularly if it is possible that the perpetrator(s) may be present. Bear in mind that in this day and age, many people will probably be filming the decontamination process on their smart phones anyway. There will probably be public agitation and resistance if video surveillance is performed too overtly, as people are likely to be disrobing as part of the decon process. It is relatively easy, given current digital video technology, to place several small digital video cameras in unobtrusive locations. Privacy and data protection laws will probably apply to any video recording and need to be taken into consideration.

Drain on Fire Suppression Assets

Many countries rely heavily upon fire departments and firefighting vehicles to supply adequate water for mass decontamination operations. While fire engines and firefighters may indeed be the best suited for the provision of decontamination, are extended operations going to erode the ability of the local fire services to conduct traditional fire suppression? One argument for military support in decontamination is that it frees up fire service personnel and equipment.

LESSON LEARNED: LARGE VOLUME DECONTAMINATION IS POSSIBLE

What do you do if you have entire rooms full of contaminated surfaces and objects? Or the back of an ambulance? Or the cockpit of a helicopter? Or your operations center? What is going in with a bunch of water or bleach simply isn't feasible because of the prospects of water damage?

Fumigation is a possibility. Long used in industrial settings for biological decontamination and sterilization, it is possible to use substances like hydrogen peroxide and hydroxide radicals in a fog inside enclosed spaces to decontaminate both chemical and biological hazards. Several companies (including the firm Steris) used vaporized hydrogen peroxide for such work. The Italian firm Cristanini uses hydroxide radicals and has successfully demonstrated this approach in aircraft cockpits.[9]

Major event plans should consider such tactics and techniques for ambulance decontamination and possible decontamination of sites of special artistic or historical interest.

REFERENCES

1. Maniscalco, P.M. and Christen, H.T. (2011). *Homeland Security: Principles and Practice of Terrorism Response*, 172. Burlington (MA): Jones and Bartlett See also: C. Hawley, op cit.
2. Medema J., op cit.
3. United States Army Field manual 3–5:(2000). *NBC Decontamination*. United States Government.
4. National Fire Protection Association (2002). *NFPA 472 Standard for Professional Competence of Responders to Hazardous Materials Incidents*. Quincy (MA): National Fire Protection Association.
5. Lake, W.A., Fedele, P.D., and Marshall, S.M. (2000). *Guidelines for Mass Casualty Decontamination During a Terrorist Chemical Agent Incident*, 3. United States Army.
6. Lake, W.A., Fedele, P.D., and Marshall, S.M. (2000). *Guidelines for Mass Casualty Decontamination During a Terrorist Chemical Agent Incident*, 15–16. United States Army.
7. Bulson, J., Bulson, T.C., and Vande Guchte, K.S. (2010). Hospital-based special needs patient decontamination: lessons from the shower. *American Journal of Disaster Medicine* 5: 353–360.
8. For example, in the context of radiological contamination: where evidence cannot be decontaminated without compromising evidentiary value, the examination must take place without decontamination. *FBI Laboratory Annual Report*, 2007. http://www.fbi.gov/about-us/lab/lab-annual-report-2007.
9. Cristanini Decontamination Systems (22 March 2016). *Decontamination Systems for Aircraft*. Army Technology https://www.army-technology.com/contractors/nbc/cristanini-spa/pressreleases/pressdecontamination-systems-aircraft2/.

CHAPTER 18

Public Affairs and Crisis Communication

The relevance of this book is based on the fact that disasters and catastrophes can adversely affect members of "the public" or "the community." One of the reasons why people are reading this book is that they want to protect the public. Officials have a duty to protect the public and we all have ethical and moral duties as human beings to help our fellow humans when they are in trouble. We also have a duty to communicate to public officials. Part of protecting and helping will be communicating.

As we discussed earlier in Chapter 3, we cannot get very far in planning for responding to emergencies by treating other people like inanimate objects or just passive recipients of a service. At some point, we need to talk to people as part of the process of responding to an incident or, at a minimum, reassuring the public who are not involved in the problem that the right people are doing the right things. But we will often want people to take certain actions, even if those actions are simple like answering a "does anyone recognize this person?" question about a possible perpetrator or "avoid the XYZ stop on the subway." In order to effectively communicate in a crisis situation, there are certain practices that will help.

Several useful books are good for additional guidance in this area. Kathleen Fearn-Banks *Crisis Communications: A casebook approach*[1] is a well-used text in this subject. The American Hospital Association's *Crisis Communications in Healthcare*[2] is useful in healthcare settings. Although neither address CBRN/WMD scenarios specifically, they are good general guidelines with practical examples. With specific regard to the US framework, there is an official document, the *National Incident Management System Basic Guidance for Public Information Officers*[3].

CBRN and HAZMAT Incidents at Major Public Events: Planning and Response, Second Edition. Daniel J. Kaszeta.
© 2023 John Wiley & Sons, Inc. Published 2023 by John Wiley & Sons, Inc.

THE AUDIENCE: WHO ARE WE TALKING TO?

When communicating outside of one's own group in a crisis, it is particularly important to figure out WHO you are talking to. An excellent article by John Drury, David Novelli, and Clifford Stott[4] draws a distinction between "the public," "the community," and "crowds." Communication to the different categories may have to take a different approach. To this categorization, I would add "experts" and "the media" as communications audiences. A one-size fits-all communication strategy is likely to not convey everything that needs to be conveyed in the best possible manner. Let us look at this briefly across the categories.

The Public

Communications to "the public" are, in this age of social media and a 24-hour global news cycle, communications to the world. If you are talking to "the public," then you are talking to the world and have to be broad in your outlook and expectations. "The public" comprises billions of people, only a small fraction of whom are directly affected by the crisis at a particular time and place. Communications intended for the global or national public will largely be informational.

The Community

Talking to a "community" is very much a subset of talking to the global public. This would be the city or region where an incident is occurring. The community in question could be affected by the incident or event. There would be need for more specificity about how the incident affects the local area and local residents. Such communications will have scope for more than just provision of information. Communications to "the community" will need to be more tailored. In addition, such communications can be directive as well as informational. You are more likely to ask people to do things or at least not to do things. Community communications may be needed for protective actions such as evacuation, sheltering in place, or reporting of relevant information.

The Crowd

We could also call this "the spectators" or "the audience." "The crowd" is a convenient term for the group or groups of members of the public who are directly involved at a specific location at a specific time. Communication to this audience needs to be very local and very specific. It is likely to be far more directive ("we need you to exit the stadium toward the parking area") than merely informational.

Experts

We talk to experts in a particular field in ways that are different from we communicate to people without specialty knowledge. A set of briefing slides or emails destined for subject matter experts will be formulated differently from communications to the general public. An important aspect to mention, however, is that such materials often leak into the public domain. Bear in mind that what you think is internal correspondence may end up being published.

The Media

When we talk to the media, we may be talking to any of the above, all at the same time. A press conference to talk about a major incident is going to have an audience that likely combines all of the above elements. But it also important to understand that journalists working for various media outlets are, in themselves, also a bit of an audience. Because they may be in the room in front of you, they will communicate more directly with you than a more distant audience, who will be reading the newspaper or looking at a post online.

PRINCIPLES OF COMMUNICATING IN A CRISIS

Use Professionals for this Role

There are public affairs and crisis communications specialists who do this sort of thing as a job. Rely on them, even if they are not the one always answering the questions at the lectern. If officials and managers are answering questions, make sure that they get briefed by the communications professionals as to how to do this. It is easy to get out of one's depth. A fire department chief may be used to talking about a major warehouse fire, but not a CBRN attack. You should have competent professionals in the public affairs roles in your organizational scheme. Just as importantly, these positions need to be integrated into the incident management structure.

Tell the Truth

The ultimate defense against ignorance is truth. Always ensure that your statements are truthful. Avoid deception. There may be things you do not want to talk about. Leave the "no comment" comment for Hollywood. Be prepared to address things that you are not ready to talk about yet. You can say things like "that is under investigation right now" or "we have not heard that report." "No comment" is asking for trouble. Also make sure that you always keep your word. If you say "we'll find out and get back to you" then you should always do that.

Be Consistent

Lengthy response operations may mean that, over the course of time, different people will be giving media briefings. While different spokespersons may have different styles, try to ensure some continuity and consistency, lest different persons send different messages. Use the same terms and definitions consistently over time.

Be Fast

From a public communications standpoint, it is a bit of a disaster to have the press turning up and asking questions of responders at an incident site. Get some sort of information operation open and running as fast as possible. By having someone able to deal with the press ready to go and integrated in the incident management infrastructure, you can hope to achieve this.

Keep a Regular Briefing Schedule

Make sure you establish and maintain a regular briefing schedule. You can always add additional briefings to an established schedule to deal with emergent issues. But it will help your liaison with the media if you can maintain some degree of regularity.

Keep Your Word. Trust That Is Lost Is Hard to Regain

It is highly likely that CBRN/HAZMAT incidents at major public events will require information to be passed to the general public. Therefore, engagement with the media should be embraced as one of the most efficient routes for doing so. If you need to announce general protective measure that the public need to take, put out information about a suspect that the police are looking for, or to put out a call for witnesses to an incident, then the time-honored press conference is a very good way to do it. Having information like this to put out at a press conference is a good way of maintaining the initiative and keeping control of the agenda.

Explain Jargon

CBRN/HAZMAT is a highly technical subject, and many aspects have both complexities and subtleties that vex experts, let alone laypersons. Credibility can be lost in seconds if someone in authority makes technical errors or gross oversimplifications. Assume that every time you use a technical term, someone at home and back in the newsroom is looking it up online. If you refer to cyanide as a "deadly nerve gas" (it is not a "nerve gas") or to bubonic plague as a virus

FIGURE 18.1 Establishing a point of contact for media and public information is important, even in a small incident.
Source: Photo: US Department of Interior, public domain image.

(plague is a bacteria) will immediately sabotage your credibility. Do not use a word unless you know what it really means.

A good practice is to have the technical content of a public or media presentation thoroughly vetted by technical experts. There is widespread precedent of getting a subject matter expert to speak on any necessary details. It is good idea to have such people within the response or support organizations identified well ahead of time. Get your organization's media relations person(s) to work with them on how to make effective public statements. It is likely that if you do not put your experts forward, the media will find "experts" of their own, who will probably not be familiar with the complexities of the situation at hand. Some may be supremely unhelpful and make your problem worse.

Consider How People Get Their Information

Communications plans need to consider exactly how people get their information. It is no longer 1980. Printing a full-page advertisement in a local newspaper is an old approach and not a terribly useful one to reach everyone. Nor is a news bulletin on television news. Serious consideration of the channels people actually use in their daily life is necessary. This is not to say that you should not bother to put things in newspapers or on TV, but that you need to embrace a broader set of dissemination channels than you would have used in 1980.

JOINT INFORMATION CENTERS

The last thing that an incident response effort needs is multiple public and media relations efforts, each putting out slightly different information. There have been numerous incidents in the past when different response agencies provided contradictory information to members of the press. Nor does any response effort need lots of media wandering about independently collaring responders for comment. One tactic that is useful is to establish a "joint information center" as a consolidated point of contact for media relations efforts. A joint information center should be set up as early as possible during an incident and should serve as the public "face" of crisis and consequence management efforts such as shown in Figure 18.1.

Many major events will establish a joint information center as a matter of routine business, regardless of whether an incident occurs. Provision should be made for alternative locations, as an existing media center for the major event in question may be part of an incident and thus would be unsuitable. The joint information center should serve as a focal point for all media enquiries, and it gives individual responders and commanders somewhere to refer media representatives. A joint information center also serves as a consolidated single point where public communications specialists from the major entities in your response effort, e.g., the police department, hospital, and fire department public spokespersons, can work together.

WHAT NOT TO DO

Give Unescorted Access to a Disaster Site

Scene control is important for many reasons, including safety and integrity of a criminal investigation. Never grant or permit unescorted access to a disaster scene.

Avoid Excessive Speculation

Do not get drawn into open-ended speculation. Stick to hard facts. Although "I don't know" and "I'm going to have to talk to my people and get back to you" may be hard to say, they bear less of a potential penalty than speculation that turns out to be incorrect or outright falsehoods. These are not ideal statements, but in some circumstances they are the least bad option. Avoid speculation if possible. Speculation on possible courses of action or outcomes may mysteriously transform into "expert opinion" over the course of a few hours. Speculation about what might be the circumstance, who might be the perpetrator, how long it might take to know the answer, or other unknowns provide room for trouble to arise.

Permit Unauthorized Spokespersons

One reason you want to control access to a scene is to prevent the unauthorized, untrained, or simply exhausted responder to become your unofficial spokesperson. Your incident management structure needs to specify who can talk to the press.

Interfere with Legitimate Media Affairs

Trying to interfere with the normal journalistic duties of the media will likely be counterproductive. Don't do it.

Breach Confidentiality

Do not give out information on victims to the media. There are many reasons for this, but respect the confidentiality of people. Sensitive data, on victims or anything else, can be misused for malinformation attacks, as described below.

THE PHENOMENON OF BAD INFORMATION

Crisis communication and public affairs can be seriously hindered by the generation and spread of bad information. As I write this, we are seeing this phenomenon all around the world during the COVID-19 pandemic. What I refer to as "bad information" (see Figure 18.2)

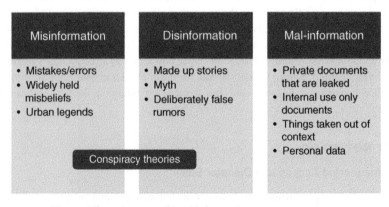

FIGURE 18.2   Three different types of bad information.

is considered by some as "information disorder"[5] and it comes in several varieties, all of which need consideration.

Misinformation

Misinformation is simply wrong information, but not wrong by intention. It may be due to ignorance, error, or even a simple typographic mistake. Misinformation lacks harmful intent, but it still can cause harm. It could range from negligible in impact (the wrong name of a photo caption of a fire department responder) to dangerous—mistakenly referring to a viral infection as a bacterial one, thus getting people to mistakenly take dangerous drugs as a possible treatment. The prevention of misinformation is through precision and due diligence at the point of origin. Primary countermeasures to misinformation are to provide the correct information as quickly and as authoritatively as possible and to get outside, respected voices to confirm and reinforce the correct information.

Disinformation

Disinformation is the deliberate spread of untruths, for malicious purposes. Because of its calculated intent, it is potentially more damaging than misinformation. Disinformation could be perpetrated by a number of actors for a wide variety of reasons. Individuals and groups with extremist or idiosyncratic agendas could take advantage of a crisis and use it as an opportunity spread disinformation out of a variety of motives. The perpetrators of a terrorist act could use disinformation to make the harm they caused worse and to hinder the response.

Malinformation

Truthful and accurate information can be misused. Malinformation is the use of legitimate, truthful information, but in ways that are unhelpful or harmful. For example, malinformation could be the deliberate leak of private or privileged information for some harmful purpose. The correct home addresses of emergency responders are a set of truthful information. But if this data gets leaked for purposes of harassment or to hinder a response, it is a bad thing. This would be a "malinformation" incident.

The categories can sometimes overlap. Sometimes, bad information of one or more gets woven into more elaborate constructs such as conspiracy theories or urban legends. A conspiracy theory may start out as some bit of truth, be warped by misinformation, and then get deliberately manipulated by malefactors who insert disinformation into the mix.

The advent of the internet and social media has not, by any means, started these phenomena. However, the internet and social media are fertile ground for starting and disseminating such constructs.

This book is not going to teach you how to combat bad information. The point I wish to make is that you will be confronted with one or more types of bad information. If there is a major CBRN incident, I can guarantee that bad information of multiple types will emerge. Dealing with bad information is complex and there is not yet a firmly agreed body of knowledge on how to deal with misinformation, disinformation, and malinformation in this modern era. However, I can suggest some basic best practices to follow:

Maintain Situational Awareness in the Social Media Space

Figure out which hashtags are relevant and watch them. You cannot react properly to information disorder by ignoring it or by not knowing about it in the first place. Also, early detection helps buy time to develop information strategies.

Share Information Internally

If you encounter disinformation or misinformation, let your team know about it. Someone might discover the bad information on their own more quickly than any monitoring effort.

Do not Feed the Trolls

Do not engage disinformation bad actors directly. This just amplifies them unnecessarily. React to misinformation and disinformation on its own and do not dignify the perpetrator or disseminator.

Maintain Good Data Security

Malinformation attacks are highly preventable with good discipline and data protection policies and procedures.

Report Dangerous Conduct

Some bad information can result in actual harm. Report dangerous misinformation and disinformation through your police, security services, and counterintelligence channels.

Keep Expertise Close at Hand

Some kinds of misinformation and disinformation will need technical, medical, or scientific debunking. Make sure that you have access to useful information and, more importantly, someone who can explain it coherently and simply.

Know Who the Neutral Fact Checkers and Debunkers Are

There are neutral, well-accepted fact-checking sites online. Learn them and point inquiries at them when relevant. You will not sway the hardened enthusiasts, but these sites can help mitigate damage among average people online.

Flood the Zone with Good Information

Make sure that every channel at your disposal is putting out good, accurate, and timely information. Where possible see if you can get it widely disseminated. Getting good information high up on search results on major search engines is highly helpful. While this will not sway hardened conspiracists, it will be helpful with the general public that may just be searching for information. Posts, sharing, and "retweets" by trusted figures can be helpful.

LESSON LEARNED: CONSPIRACY THEORIES AND DISINFORMATION WILL MULTIPLY IN A CBRN ENVIRONMENT

Recent instances involved CBRN materials have been a ripe operational environment for disinformation and conspiracy theories. Chemical warfare incidents in Syria and the attempted assassinations of Russians (Sergei and Yulia Skripal, Alexei Navalny) involving so-called Novichok agents have caused a proliferation of misinformation, disinformation, and conspiracy theories ranging from the sublime to the ridiculous. The causes for this are many and are beyond the scope of this book. You should expect disinformation efforts and conspiracy theories to turn up during and after an incident. Given the nature of social media, they will proliferate like mushrooms after a rainstorm, and there is little you can do about it.

It is important to understand that CBRN threats to human life and health are wrapped in mystique and are often invisible. There is often a large differential between the perception of risk and actual probabilities of harm with CBRN materials. Some things, like radiation, are less lethal than often perceived. Other things, like COVID-19, have been underplayed (the "it's just a bad cold" theme often peddled as misinformation) in terms of their actual risk. This general knowledge deficit of the actual risks posed by CBRN materials among the broader public provide the fertile environment for bad information to spread.

REFERENCES

1. Fearn-Banks, K. (2016). *Crisis Communications: A Casebook Approach*. Routledge.
2. American Hospital Association, Society for Healthcare Strategy and Market Development (2002). *Crisis Communications in Healthcare: Managing Difficult Times Effectively*. Chicago: Society for Healthcare Strategy and Market Development of the American Hospital Association.
3. Federal Emergency Management Agency (2020). *National Incident Management System Basic Guidance for Public Information Officers*. United States Government https://www.fema.gov/sites/default/files/documents/fema_nims-basic-guidance-public-information-officers_12-2020.pdf.
4. Drury, J., Novelli, D., and Stott, C. (2013). Representing crowd behaviour in emergency planning guidance: 'mass panic' or collective resilience? *Resilience* 1 (1): 18–37.
5. Wardle, C.C. and Derakhshan, H. (2017). *Information Disorder: Toward an Interdisciplinary Framework for Research and Policy Making*, 20–27. Council of Europe.

CHAPTER 19

Consequence Management and Other Response Measures

When I first entered Federal government service, it was all the fashion in emergency management in the Federal government to neatly divide response to terrorism or other disasters into the neat categories of "crisis management" and "consequence management." Both are valid concepts, but many catastrophic incidents do not neatly stop and then become just a set of consequences. This chapter is, by necessity, a mixed bag best described as "all those other bits you also need to worry about."

Dealing with CBRN/HAZMAT at a major event is likely to involve practically every sphere of human activity in the affected area. The majority of this book has been devoted to preparedness, initial response, and the crisis management phases of such incidents. But it is important to get a grasp on the long-term aspects of response. Long-term consequence management after CBRN/HAZMAT incidents is a very complicated subject, worthy of a book in its own right. Much of this subject is generic to disaster management as a whole and is not peculiar to major public events or CBRN/HAZMAT incidents. For general reference on disaster response and consequence management as broader subjects, there are numerous excellent references available.[1] This chapter steers the discussion toward the aftermath of CBRN/HAZMAT incidents.

GENERAL FRAMEWORKS AND REFERENCES FOR CONSEQUENCE MANAGEMENT

The United States has a broad general framework promulgated by FEMA. It is known, unimaginatively, as the National Response Framework (NRF)[2](see Figure 19.1), it is now in its fourth edition. While not particular to any scale or type of disaster, this framework does give a

CBRN and HAZMAT Incidents at Major Public Events: Planning and Response, Second Edition. Daniel J. Kaszeta.
© 2023 John Wiley & Sons, Inc. Published 2023 by John Wiley & Sons, Inc.

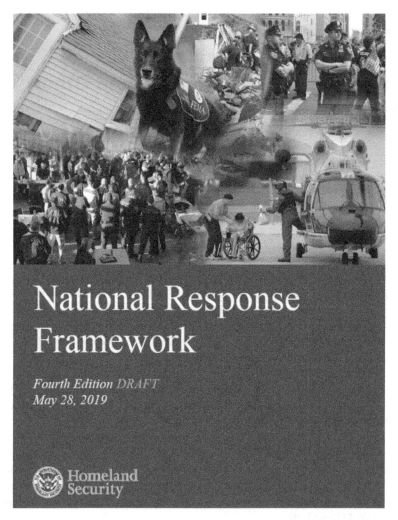

FIGURE 19.1 The National Response Framework gives broad guidance useful in managing CBRN/Hazmat incidents in the USA.
Source: Photo: US Department of Homeland Security, Public domain image.

useful way of thinking about consequence management. One of the more important features of the NRF and its predecessors is that of dividing up response into fifteen "Emergency Support Functions" or ESFs and several supporting functions (financial management, public affairs, international cooperation, etc.) Within the US Federal government, the NRF assigns responsibilities to different departments and agencies within the ESFs. Dividing up aspects of consequence management into the following ESFs is a very useful practice. This list of 15 ESFs is as logical a way to proceed in this subject area as I can imagine. It should be noted that there are a lot of overlaps between the ESFs, which are defined in broad categories.

The nature of CBRN/HAZMAT incidents means that I will address some of these in more detail than others. The FEMA website has explanatory annexes for all of these ESFs available online.[3] My suggestion is that you take a hint from the NRF and appoint a coordinator for each of the emergency support functions in your overall planning structure. If you are reading this book, you might already be the coordinator for ESF 8.

ESF 1 Transportation

Transport is crucial, both to dealing with an incident and in longer-term recovery. Sometimes, one part of incident response adversely affects another. Protective actions designed to help the populace may end up making other response efforts more difficult. Evacuation and relocation of affected persons from the area around a CBRN/HAZMAT incident may choke transportation routes, thus hindering the movement of personnel and equipment into the affected area. Transportation assets cannot be conjured out of thin air, so any reasonable emergency plan should provide for availability of transportation assets. This is an area where military support could be useful. Transportation planning also needs to consider the needs of mass evacuation. If hundreds or thousands of people need to move, transportation assets and routes will be needed to move them. Many areas rely on public transportation for such situations, but many issues and obstacles have been identified in using public transit systems in large disasters.[4]

ESF 2 Communications

This ESF covers both communications to deal with the emergency and restoration of communications infrastructure for the general use of society and the economy. Communications with regards to responding to the incident itself in the crisis phase are addressed in Chapter 13. Experience with major disasters and large terrorist events will strain communications capacity and capability. Communication failings were widely cited in the 9/11 response at the World Trade Center and Hurricane Katrina.[5] Major events and incident response at them require many different agencies and organizations to work together, making for a complicated communications environment. The issues and pitfalls of communications before, during, and after incidents are not unique to CBRN/HAZMAT. CBRN/HAZMAT environments mean that responders wearing PPE may have more difficulty communicating.

Consequence management efforts after a major incident are likely to result in lots of different departments and agencies responding to the affected area. Compatibility issues, interoperability issues, and radio frequency conflicts may adversely affect the ability to coordinate a broader effort. Some effort needs to be devoted to thinking about how to avoid such problems.

As a broader issue in consequence management, contamination of commercial telecommunications facilities or even physical destruction of equipment may cause communications problems in the affected area. However, the scope for such damage is less than with disasters like hurricanes and serious winter storms. The technical and procedural issues of communications at terrorist incidents are somewhat beyond the scope of this book, but should be considered.

ESF 3 Public Works and Engineering

The issue of what to do with contaminated public spaces and facilities will be an important one. CBRN/HAZMAT incidents pose both the potential of actual long-term contamination hazard (depending on the material involved) and perceived hazard. Actual contamination hazard is a measurable, quantifiable presence of harmful materials in excess of regulations or guidelines that would name a "safe" level of the material. Such guidelines often do not exist in a civil setting. Perceived hazard is more nebulous. It involves perceptions by responders, workers, visitors, and the general public. A hazard could be completely gone, but if people believe an area to be unsafe, they will not want to go there. COVID-19 may have changed some of that, as we have seen situations of people going to areas with clear biological hazard, as they perceive the threat to be low.

In any case, contamination of public works can have serious political and economic implications. Consequence management plans need to consider the problems of decontamination of essential public works, like roads, courthouses, shopping malls, transportation nodes, and other public spaces. Both the 2011 Fukushima incident and the 2001 "Amerithrax" raise the issues of decontamination. In Japan, the issue was landscape decontamination,[6] and in the Amerithrax situation, it was largely a matter of building interiors.[7]

Removal of contaminated debris is considered jointly between ESF 3 and ESF 10. If you have a situation where structural damage or even building collapse has occurred as part of a CBRN/HAZMAT incident, dealing with removal and safe disposal of building debris is a serious component of consequence management. It is worth considering the 9/11 experience in that an apparent non-CBRN/HAZMAT scenario nevertheless has resulted in a large number of illnesses and fatalities from hazardous materials over the decades following the attack. Non-CBRN incidents can end up having the features of a CBRN incident.

ESF 3 traditionally includes engineering and construction expertise, as well as provision of electrical supplies and water. Some types of CBRN/HAZMAT incidents may include situations with explosive yields, thus causing structural collapses. CBRN/HAZMAT incidents may occur in a context of a broader incident, such as conventional bombings or a large natural disaster like a hurricane that releases hazardous materials. Consequence management efforts may also require power and, especially, water in excess of what the existing post-incident infrastructure may be able to provide.

ESF 4 Firefighting

Firefighting as a support function to CBRN/HAZMAT incidents is addressed in significant detail in Chapter 10. It should be noted that traditional firefighting capabilities and capacities might be temporarily diminished due to the need to use firefighting personnel and equipment for decontamination and other specialized support. Consequence management plans may need to consider how to restore firefighting capacity that may be temporarily diverted to CBRN/HAZMAT response.

ESF 5 Information and Planning

This ESF was formerly called "emergency management," and it was a bit confusing as the whole framework is about emergency management. Chapters 4 and 5 cover this ESF in significant detail. However, one important feature of this ESF not mentioned elsewhere is a possible need for appropriate decisions and declarations to be made. In the United States and a number of other countries, many types of assistance and resources are contingent upon the right type of disaster declaration being made by an official of a certain rank or position, such as a state governor. Although this is basic emergency management at the core, you must understand what types of declarations and requests for assistance you are likely to need in the event of a large incident

ESF 6 Mass Care, Emergency Assistance, Temporary Housing, and Human Services

This ESF was formerly known as "mass care" and has been renamed to cover a broader range of activity. A CBRN/HAZMAT incident could cause serious disruption to people. Many incident response plans will call for isolation and evacuation of civilians. Some situations are more easily catered than other situations. Commercial and industrial accidents, occurring primarily in workplaces that are surrounded by other workplaces, are easy to manage. Workers can be sent home until the problem is sorted out. But what if the evacuation is kicking people out of their homes? Or what about hotels and public venues full of out-of-town visitors? The "I don't care where you go, just get going" tactic sometimes used by public authorities is only good for the first few hours of an incident. In a major event setting, there will be people who have nowhere logical to go. People who have been decontaminated might not have clothing or even identification documents. Eventually, people will need water, food, and a more permanent disposition, such as return to their homes. Emergency plans need to accommodate such situations. Many major events will have visitors from other countries, and there may be a need to coordinate with consular officials from other countries. Traditional emergency management and disaster recovery plans tend to have standing plans for activities in this ESF, and this is an area where the more generic emergency management plans may be adequate. The ESF annex for ESF 8 is actually quite thorough.[8]

Although not always expressly part of this ESF in US documents, it is also worth considering the mass care of responders. If your plans rely upon drafting in extensive external support such as personnel, not every department and agency will be able to logistically support them for long periods of time. Some of the same feeding and housing provisions for the affected population may need to be used to look after response personnel.

ESF 7 Logistics and Resources

This ESF, particularly when you read the relevant FEMA annex,[9] is very much a catch-all. Even in the detailed plans, it is very much a "make sure you have a way to have all the stuff that you need" provision. Relevant to CBRN and HAZMAT response, however, it is worth reviewing your ability to rapidly identify and procure goods and services that you might need in a contingency.

ESF 8 Public Health and Medical Care

Chapters 8 and 16 cover the more immediate aspects of public health and medical care in CBRN/HAZMAT situations at major public events. From a longer-term post-incident consequence management perspective, there are some significant considerations in this emergency support function that will need to be addressed in your planning processes. Three important aspects come to mind, although there will likely be others as well depending on the scenario.

Reconstituting Normal Medical Services

Mass casualty events might overwhelm both pre-hospital and hospital-level care in the area affected by a major CBRN/HAZMAT incident. Personnel, supplies, and capacity could be taken up by casualties. Infrastructure and equipment (e.g., ambulances) could be contaminated or damaged. The ability of the medical system to do its normal everyday job of looking after the health of the community could be compromised. Despite a major event, accidents happen. People need hospital care for many things. Good major event planning examines ways to make up for the possible shortfalls in capability and capacity and seeks to find ways to temporarily shift the burdens. Are there ways to get medical facilities in adjacent areas and regions to take up the slack? Are there mutual aid agreements where you can get additional EMS support? The time you need to be asking these questions is a year before your event, not 12 hours into an incident.

Considering Long-term Health Impacts

Many CBRN/HAZMAT situations involve dissemination of materials that may have long-term effects on human health. Dealing with just the immediate consequences of a CBRN/HAZMAT situation may, in fact, be just a small portion of the overall health effects of such an incident. There are numerous possible scenarios where people are apparently healthy, or at least not seriously ill enough to require immediate medical care, only to see these people get sick weeks, months, or even years after the incident. It is very much worth having a plan to investigate and document the presence of possible hazards as soon as possible after an incident. Attempt to develop a registry of people (including responders) who may have been affected by the incident, even if you do not know what hazards are actually present. Work with environmental health and toxicology professionals to identify possible sources of long-term harm.

Mental Health

The harm caused by a CBRN/HAZMAT incident may be more than just physical harm. Mental health care in the weeks and months after an incident is also worth considering. Major terrorism incidents can take a serious toll on mental well-being. Historically, this aspect of medical preparedness has been under-resourced or overlooked. Also, ESF 8 traditionally considers the overall safety of food and water, which overlaps with ESF 11.

Mortuary Affairs

CBRN/HAZMAT incidents may be lethal. This means that there may be human remains to deal with. Some human remains may be contaminated and existing procedures for their handling in conventional circumstances may be inadequate. In many situations, bodies of victims and perpetrators will be considered to be evidence and must be handled correctly. Plans, policies, and procedures for dealing with mass fatality incidents are well established in most of the developed world, but the extent to which these are adapted to situations that produce contaminated bodies varies widely.

Several references have been written, including an excellent and thorough article by Dr. David Baker and his colleagues in the United Kingdom,[10] and a US military policy document on mortuary affairs,[11] which examines the issue of CBRN contamination of casualties in its

Chapter 8. The latter document has, in theory, been rolled into a different broader document on joint logistics as an appendix.[12] I recommend both documents for any seeking to grapple with these complex issues.

Human remains will need to be recovered and identified, and doing so in a CBRN/ HAZMAT environment poses certain issues. Existing procedures will need to be modified, and additional equipment, such as special body bags and PPE for recovery team personnel, will be needed. The recovery team will likely need to operate in PPE. As with crime scene response, it is undoubtedly easier to spend the effort to train recovery staff in wearing appropriate PPE than it is to train CBRN/HAZMAT specialist in body recovery. As with any of the response functions involving entry into hazardous environments, operating in PPE will reduce the amount of time that recovery teams can effectively function, due to heat stress, endurance of PPE, and related issues. Therefore, a protracted recovery effort will require more personnel than a comparable non-CBRN/HAZMAT situation. Some of the traditional victim identification measures may not be possible due to degradation. For example, corrosive materials may hinder fingerprint identification.

Human remains may be contaminated. Neither contamination nor a lack of contamination should be presumed during recovery effort. Detection equipment should be used to establish the presence of hazards that will require protection greater than the normal bodily substance precautions typically used in human remains recovery operations, such as volatile chemicals or radiation. Not every CBRN/HAZMAT situation that kills people will produce human remains that are contaminated. Others might be heavily contaminated, and speculation will not result in firm information.

It should be noted that human remains from conventional disasters pose health and safety risks, and the presence or lack of CBRN contamination needs to be weighed in the context of this fact. Decontamination need not be exceedingly thorough and should aim to reduce contamination to a level consistent with normal precautions. Some guidance from Interpol is available on this subject.[13] Victim recovery and identification need to be closely coordinated with other aspects of crime scene response, as discussed in Chapter 18.

Existing mortuary capabilities may not be able or willing to accept contaminated human remains. It is common practice in many areas to establish a temporary mortuary for the management of fatalities from a disaster, such as a CBRN/HAZMAT incident. Temporary facilities for storage of contaminated casualties will likely become contaminated themselves, if not in actual fact, then in the public consciousness. Therefore, plans to establish temporary morgues in buildings that might require re-occupancy should be considered carefully.

ESF 9 Search and Rescue (SAR)

Traditional disaster plans discuss search and rescue largely in the context of either natural disasters (hurricanes, floods, earthquakes, tornadoes) or conventional terrorism (buildings collapsed after bombings). We have already discussed the oft-neglected issue of CBRN/ HAZMAT rescue in previous chapters. Traditional (i.e., non-CBRN/HAZMAT) SAR can take the form of maritime SAR, land SAR, or structural collapse urban SAR. Numerous technical references for these disciplines exist but are outside the scope of this book.

The important aspect to emphasize, with regard to SAR in the types of situations envisaged by this book is that you may easily have situations where traditional search and rescue is required in addition to specialty CBRN/HAZMAT rescue situations. For example, you could have a natural disaster such as a flood or tornado causing a serious hazardous materials release. This could lead to situation where both "normal" and specialty search and rescue is required.

ESF 10 Oil and Hazardous Materials Response

In many ways, this entire book is an ESF 10 plan for major public events. I have not addressed oil spills in this book, but they are a category of HAZMAT incident and should not be ignored. Particular to the consequence management after CBRN incidents, the US EPA has some resources online.[14]

ESF 11 Agricultural and Natural Resources

Much of this ESF is outside the direct purview of this book. Human life and health are the first priority in managing a CBRN/HAZMAT incident. However, we must be aware that any situation involving the release of dangerous materials could have adverse impact on the environment. Having a due regard for preventing or mitigating damage to the natural environment should be part of your priorities.

It should be noted that numerous CBRN threat materials could have an adverse effect on agriculture, either directly or through mechanisms of action like contamination of soil and water. If your major event is near anything remotely agricultural, you should consider the possibility of damage to plants, animals, or agricultural infrastructure.

Interestingly, at least in the American context, the phrase "natural resources" includes "cultural resources" and "historic properties." If your area of operations contains such assets, consider how you will address them in consequence management plans. Or at least think about who to ask for help. The ESF 11 annex has some resources on this for American-based readers.[15]

ESF 12 Energy

This ESF, defined as producing, storing, refining, transporting, generating, transmitting, conserving, building, distributing, maintaining, and controlling energy systems and system components, is largely addressed in conventional disaster plans. Beyond ensuring adequate power supply for your operations, this ESF is beyond the scope of this book.

ESF 13 Public Safety and Security

Much of Chapters 9 and 18 are devoted to this subject. From a consequence management standpoint, it is important to understand that a major CBRN/HAZMAT incident will be likely to tie up police and security resources normally devoted to keeping the public safe under normal circumstances. It should be noted that in a disaster situation, the homes and families of police staff could easily be involved in the overall disaster, thus abstracting police manpower from available duties.

There are numerous incidents in history of criminals and opportunists using the diversion of police during disasters to exploit the situation. Furthermore, properties that have been evacuated or temporarily abandoned could represent a golden opportunity for property crimes. Looting in post-disaster situations is not unheard of and is a frequently cited reason for deployment of National Guard troops even after relatively small disasters. Consequence management plans need to not only just accommodate the actual CBRN/HAZMAT incident but also need to encompass general policing and security needs in the post-incident environment. This will likely entail a need for more personnel, whether it be police, National Guard, or even private security staff. Once again, mutual-aid agreements with surrounding areas and support from higher levels of government may become an important aspect of your plans.

ESF 14 Cross-Sector Business and Infrastructure

This ESF was formerly called "long-term community recovery." It can be broadly termed as "getting the community back on its feet again" and encompasses many aspects of consequence management. It can be the longest leg of a multidisciplinary response. Once again, consequence management in this area is generally generic and not particular to CBRN/HAZMAT incidents. One aspect that deserves mention for CBRN/HAZMAT incidents is the effect on local businesses. Many aspects of your plans may rely on procuring goods or services from local businesses. However, if the effect of an incident is such that these businesses are closed or severely diminished in scope, you may have to consider procuring goods and services from further afield. For example, if your evacuation plans hinge upon renting buses from a local company, but the situation is such that the company's staff cannot even get in to drive the buses, you would have been better served by having an arrangement with a company a bit further away from the problem.

ESF 15 External Affairs

By external affairs, US documents use this ESF to mean providing "accurate, coordinated, timely, and accessible information to affected audiences, including governments, media, the private sector, and the local populace, including children; those with disabilities and others with access and functional needs,; and individuals with limited English proficiency." [sic][16] In other words, crisis communications. Chapter 17 discusses this subject in some detail and no plan is complete without consideration of these issues.

Support Issues

The US NRF also highlights several supporting areas that are not ESFs in their own right but which are worthy of consideration. These are financial management, public affairs, international coordination, worker safety and health, volunteer and donations management, and tribal affairs. Public affairs as a subject does overlap significantly with ESF 15, even by the admission of the official documentation. This area is covered in some detail by Chapter 17. Although tribal affairs are specific to US relationships with native American entities and a bit beyond the scope of this book, the other aspects are worthy of remark. Volunteer management is discussed in Chapter 11.

Financial Management

Eventually, crisis management and consequence management will cost money. Managing the detail of this is beyond the scope of this book. The Financial Management Support Annex to FEMA's NRF[17] is a good 34-page introduction to the subject. FEMA has numerous additional resources online.

International Coordination

CBRN/HAZMAT threats can pass across international boundaries and affect people of multiple nationalities. Even in the vast land area of the United States, there are sometimes complexities requiring international coordination. Aspects of consequence management in an international coordination context may include, but are not limited to:

- Management of contamination or environmental effects that cross borders
- Mutual aid between countries in terms of equipment, logistics, aid supplies, response teams, expertise, and other forms of assistance
- Sending samples abroad for independent analysis
- People seeking refuge across an international border
- Repatriation of casualties or deceased persons to their home country
- Looking after your own country's nationals affected by an incident elsewhere
- Consular assistance, such as verification of identity and replacement of documents lost due to contamination or decontamination

Worker Health and Safety

Just because the incipient phase of a crisis is over or a CBRN/HAZMAT incident is stabilized, it does not mean that a situation is somehow now safe. Some hazards may shift from acute hazards (i.e., able to hurt you quickly within minutes or hours) to chronic hazards (ones that build up risk over time). The long-term health issues and premature deaths associated with response to the 9/11 attacks are only one example of this. It is also important to note that long-term consequence management efforts necessitating lengthy times wearing CBRN/HAZMAT PPE may increase a lot of other health and safety risks. An injury from loss of dexterity or peripheral vision from CBRN PPE is highly possible, and we dare not treat the site of a CBRN/HAZMAT incident as just a regular workplace.

LESSON LEARNED: VICTIMS FROM OTHER COUNTRIES

The very nature of major public events is that we must always assume that there will be some type of international context if there is a CBRN/HAZMAT incident. The fluid nature of international travel and the general diversity of major metropolitan areas all over the world mean that it is exceedingly improbable that there will be no foreign nationals at some kind of major event.

Of particular concern in CBRN/HAZMAT events is decontamination. You may very well end up with nationals of 20 or more other countries, who have been deprived of their identity documents because all of their clothing was cut off during a decontamination process. Some will have nowhere to go and serious difficulty in communicating their plight. Enlisting the aid of the consular sections of embassies and other diplomatic missions will be crucial in helping such people.

REFERENCES

1. (a) Larsen, J. (2013). *Responding to Catastrophic Events: Consequence Management and Policies.* Springer. (b) McEntire, D.A. (2015). *Disaster Response and Recovery: Strategies and Tactics for Resilience.* John Wiley & Sons.

2. Federal Emergency Management Agency (2019). *National response framework*. United States Government https://www.fema.gov/sites/default/files/2020-04/NRF_FINALApproved_2011028.pdf.

3. Federal Emergency Management Agency *Emergency Support Function Annexes*. United States Government https://www.fema.gov/emergency-managers/national-preparedness/frameworks/response#esf.

4. Transportation Research Board of the National Academies (2008). *The Role of Transit in Emergency Evacuation*. Washington (DC): National Research Council.

5. Manoj, B.S. and Baker, A.H. (2007). Communication challenges in emergency response. *Communications of the ACM* 50 (3): 51–53.

6. Evrard, O., Patrick Laceby, J., and Nakao, A. (2019). Effectiveness of landscape decontamination following the Fukushima nuclear accident: a review. *Soil* 5 (2): 333–350.

7. Kaszeta, D. (2016). Decontamination of buildings after an anthrax attack. *Journal of Terrorism and Cyber Insurance* 1 (1): 47–60.

8. Federal Emergency Management Agency. *Emergency Support Function #6 – Mass Care, Emergency Assistance, Temporary Housing, and Human Services Annex*, June 2016. https://www.fema.gov/sites/default/files/2020-07/fema_ESF_6_Mass-Care.pdf

9. Federal Emergency Management Agency. *Emergency Support Function #7 – Logistics Annex*. June 2016. https://www.fema.gov/sites/default/files/2020-07/fema_ESF_7_Logistics.pdf

10. Baker, D.J., Jones, K.A., Mobbs, S.F. et al. (2009). Safe management of mass fatalities following chemical, biological, and radiological incidents. *Prehospital and Disaster Medicine* 24 (3): 180–188.

11. Joint Staff, US Department of Defense Joint Publication 4-06(2011). *Mortuary Affairs*. Washington (DC): United States Government.

12. Joint Staff, U Joint Staff, US Department of Defense Joint Publication 4-0(2019). *Joint Logistics*. Washington (DC): United States Government.

13. Interpol (2018). *Guidelines for Dead Body Management and Victim Identification in CBRN Disasters*. Lyon, France: https://www.interpol.int/content/download/5763/file/E%20DVI_Guide2018_Annexure16.pdf 1 February 2022.

14. United States Environmental Protection Agency. *Chemical, Biological, Radiological, and Nuclear Consequence Management*, 6 January 2022. https://www.epa.gov/emergency-response/chemical-biological-radiological-and-nuclear-consequence-management

15. Federal Emergency Management Agency. *Emergency Support Function #11 – Agriculture and Natural Resources Annex*, June 2016. https://www.fema.gov/sites/default/files/2020-07/fema_ESF_11_Ag-Natural-Resources.pdf

16. Federal Emergency Management Agency. Emergency Support Function #15 – External Affairs Annex, June 2016. https://www.fema.gov/sites/default/files/2020-07/fema_ESF_15_External-Affairs.pdf

17. Federal Emergency Management Agency. *Financial Management Support Annex*, June 2016. https://www.fema.gov/sites/default/files/2020-07/fema_nrf_support-annex_financial.pdf

CHAPTER 20

Forensics and Investigations

Most of the types of incidents that this book covers should be the subject of detailed investigations. Some incidents may require significant investigation after the fact to determine what actually happened. Incidents with a human cause will require investigation to determine responsibility. Many of these investigations will be criminal investigations to determine the nature of the crime committed and the identity of the perpetrators. Fatal incidents may be investigated as homicides. However, conducting forensic investigations involving dangerous substances such as are likely to be found at CBRN/HAZMAT incidents becomes more complex than more conventional investigations. Many countries around the world are simply unprepared to adequately conduct investigations in a CBRN/HAZMAT environment.

CBRN FORENSICS AS A SPECIALTY

Historically, the specialists in CBRN and HAZMAT response came out of the military and fire services, and had little or no background in forensics. Most law enforcement agencies have little operational history with CBRN forensics. Laboratories trained and certified to work with CBRN materials are not likely to have much experience in conventional criminology and the conventional forensic labs may not be allowed or equipped to operate with CBRN materials on their premises. These are clearly areas where personnel from different backgrounds are going to have to work together.

For the purposes of this book, forensics and investigation can be reasonably divided into three phases.

Collection

This phase involves collection of information and the physical and virtual collection and preservation of evidence, which can take a number of forms ranging widely in size and form.

CBRN and HAZMAT Incidents at Major Public Events: Planning and Response, Second Edition. Daniel J. Kaszeta.
© 2023 John Wiley & Sons, Inc. Published 2023 by John Wiley & Sons, Inc.

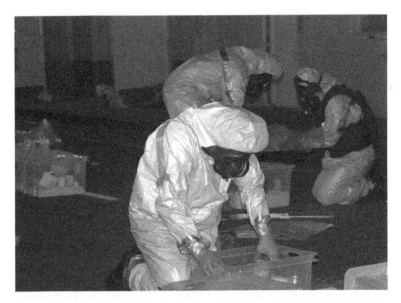

FIGURE 20.1 Exploitation of a CBRN crime scene must be done safely and meticulously.
Source: Used with permission from R. Mead.

Historically, this might have occurred entirely in the field such as illustrated in Figure 20.1 above. In an increasingly technological age, evidence can be virtual and exist only online or in electronic format, so it is impossible to say that collection only happens at the incident site. Collection also includes taking witness statements, which may happen at the scene or after the fact. The object of the collection phase is to assemble pieces of information that can be further analyzed and items of evidence, which can have information extracted from them.

Analysis

Analysis includes all of the various processes, both technical and non-technical Although there are some analysis tools which might be used at an incident site, most analytical work will take place elsewhere, such as in laboratories and offices. The purpose of the analysis phase is to extract all possible information, assemble it into a narrative that explains what actually happened, and package this information in a way that public authorities can use it.

Exploitation—Using the Collection and Analysis

The assembled information can be used in many ways. In conventional criminal investigations, the aim is usually to arrest perpetrators and try them for criminal offenses in a judicial process. However, there are other purposes as well. Forensic evidence can be used to identify exotic substances, thus leading to better medical treatment. National governments may make policy actions such as sanctions or military action based on evidence. Evidence may also be useful in civil proceedings or regulatory actions as well as criminal trials. Depending on the disposition of evidence, it may be challenged in court procedures and the evidence, as well as the personnel and processes used to collect and analyze it, may be called upon to defend this evidence.

For the purposes of this book, we will emphasize the first. Much of the components of post-event analysis, judicial processes, policy decisions, and other long-term reactions are outside the immediate scope of this work.

COLLECTION AND PRESERVATION OF EVIDENCE: GENERAL CONSIDERATIONS

The perpetrator(s) of terrorist acts involving CBRN/HAZMAT need to be caught. This will involve the collection, preservation, and analysis of evidence. This evidence will need to be examined and analyzed in a correct forensic manner. It may be needed for use in a criminal trial, in which case it is likely to be examined closely and cross-examined and questioned by defense attorneys. Evidence that is placed in doubt can imperil successful convictions.

Defining the "Crime Scene"

Much of modern forensic work involves defining and processing "crime scenes." In the major event environment, a CBRN/HAZMAT situation poses many possible crime scenes. Possible crime scene locations could include such diverse locations as the following examples:

- The scene of the detonation of an improvised device
- A room in a hospital
- A morgue
- The region around a wrecked truck or derailed rail car.
- An area of seating in a stadium or arena where people became ill
- A house or industrial building where a suspected terrorist may have been manufacturing toxic substances
- A delivery van containing a package that is leaking a toxic substance
- A hotel room that was used by suspected perpetrators of an attack
- The waiting room of a local hospital
- A car that was used to transport a terrorist device
- The cargo hold of an aircraft that contained baggage that leaked a strange substance

In many training courses and exercises "the crime scene" is obvious and simplistic. It was an area around where a terrorist device had functioned. Yellow "crime scene" tape was put around it, just like on television. Imagine, however, that you have the roll of tape and you have to figure out where to put it. In the real world, we may need to do some work to figure out where to put the yellow tape.

A crime scene may travel and expand. The latent nature of many CBRN/HAZMAT materials means that victims (or perpetrators) may turn up injured, ill, or dead at locations nowhere near the point of dissemination. Evidence may very well be found anywhere along the trail of the perpetrators and/or victims. In the 2006 poisoning of Alexander Litvinenko, traces of Polonium were found in many different places along the trail. At least a dozen were openly identified in the media[1] and there were others that were not publicly disclosed.

Assisting Victims

Evidence collection does not take priority over saving lives and helping the injured. Occasionally at exercises, I've heard casual talk of "victims just getting in the way" and such, but this should not be policy and is a heartless attitude. It is not defensible to let someone die while evidence is being collected. Indeed, it could be considered manslaughter under laws in many places. Rescue, emergency medical assistance, EOD render-safe procedures, firefighting, and related lifesaving activities take precedence. This does not mean that such procedures should take no regard for evidence, nor does it mean that some evidence collection cannot take place while some of these activities are still underway. There will be ways in which collection and preservation of evidence can happen alongside lifesaving and rescue activity. At a crime scene, with lots of victims needing rescue and decontamination, there will likely be some loss of evidence. But no crime scene forensic effort ever captures ALL evidence. We must understand this and accommodate it in our plans. However, also bear in mind that a victim who survives may provide useful evidence later on in the investigation.

Safety

Responder safety is paramount. The investigators at the crime scene will have to be trained and equipped to a level appropriate to allow entry into this environment. This will likely require PPE as well as appropriate detection and monitoring equipment. As a standard practice, any forensic team should appoint a safety officer whose primary task is to ensure that safety procedures are followed by all of the investigative personnel at the crime scene. Unknown hazards may be present. Some, like secondary devices, may be intended specifically to harm police personnel. Clandestine labs may include many precursor chemicals that may be hazardous in ways that are different from the end product.

The Transient Nature of CBRN/HAZMAT Evidence

Some of the important evidence in CBRN/HAZMAT situations can have a very short lifespan. CBRN evidence is subject to factors like volatility and degradation. By the time a trained and equipped investigator arrives at the scene, the evidence may have taken a different form or have degraded. Many of the possible threat agents will disappear or change their characteristics in a way that makes collection of evidence more difficult. For example:

- Liquid agents may evaporate into a vapor
- Aerosols may disperse or deposit onto surfaces
- Gases or Vapors may have dissipated
- Fine powders may have been carried away by wind or air conditioning
- Biological pathogens may have been degraded by exposure to the environment
- Chemicals may react to water or air and change their composition

This potential for attrition of evidence means that speed may be an operational imperative in processing crime scenes, particularly when chemical or biological substances are suspected. When balanced against the needs of rescue efforts, which take priority, the crime scene response is crushed between two operational imperatives. Good planning, policies, and procedures can be used to find a way of managing these conflicting imperatives.

INTEGRITY OF THE EVIDENCE

If evidence is going to be used for any greater purpose, such as prosecution, it must be protected and defended. Evidence that may end up in court is likely to be scrutinized by defense lawyers. Any defense lawyer, even a bad one, knows to attack the evidence. You do not want the embarrassment of guilty terrorists walking free because the evidence was tainted, lost, or mishandled. Evidence may also be used for other purposes, outside of judicial proceedings. Political officials are well within their rights to question the integrity of the evidence as well, if they are supposed to use it to make policy decisions. Integrity of the evidence, both physically and administratively, is of paramount importance.

Controlling the Crime Scene

Once you have established a perimeter, you need to control access. Your crime scene perimeter may or may not match other perimeters and cordons established for other emergency response purposes such as rescue, decontamination, or hazard mitigation. However, management will be much easier if it is. Maintain a single point of entry and exit, which is also good for managing decontamination. Keep a log of everyone entering and exiting the crime scene. Such a log may serve many uses. In the event of the discovery of a highly infectious disease, it will be useful in tracing contacts.

Chain of Custody

Any forensic evidence collection effort should have proper chain of custody of evidence. This means a trail of paper that shows where the evidence has been. Such a chain should have no breaks from the moment that the technician collects it, all the way through laboratory analysis and legal proceedings. Any break in the chain of custody means a lapse of accountability and represents a potential point in place and time where evidence could have been altered, stored improperly (and thus degraded), or even been substituted.

Cross-contamination is the Enemy

Cross-contamination occurs when a substance inadvertently gets introduced into a sample where it should not have been. Use of proper procedures will reduce the risk of cross-contamination. Cross-contamination will ruin evidence and can make it harder to process a crime scene. For example, a spoon or scoop used to collect a sample of powder should not be used to collect evidence from a second puddle. Any hint of cross-contamination will make the laboratory's job harder and a defense attorney's job much easier. Any tools that are needed for the collection of multiple samples will have to be painstakingly decontaminated before re-use.

Sterility

The integrity of tools and containers is very important, and is usually expressed most starkly in the concept of sterility. If the integrity of a sample container is compromised, the evidence is ruined for the purposes of any legal proceeding. You may be called upon to prove beyond a reasonable doubt that the bottles, tubes, vials, bags, etc. that were used, as well as any tools that were used to collect the samples, were sterile. For example, a defense attorney can argue that the soil sample that was collected behind a suspect's house was collected in a dirty bag. If the bag was dirty, how do we know if the sample was actually collected near the house, or was it residue that was in the sample bag already?

Document Everything

Document everybody and everything. Take names of everyone that was at the scene. Make a detailed log of every single item or sample collected as evidence. Take photographs of everything. For example, you should mark where each and every sample was taken and take a photograph of the spot. Use markers with numbers to help identify photographs and correlate them to a log. While this task is laborious, copious documentation of every step of the crime scene response will help you defend the legal integrity of your actions. Digital photography has made this much less onerous than it would have been twenty years ago.

Blanks, Blanks, and More Blanks

Because you must assume that a crafty defense lawyer is going to challenge your processes and procedures, there are some preemptive measures that you can take to help prove the integrity of your processes. Because cross-contamination is such a pernicious hazard, and one that easily takes advantage of lax processes, it is necessary to introduce extra control measures into your procedures as an additional safeguard.

Blanks are a procedural safeguard to prove sterility and to establish that cross-contamination has not occurred. The legal and procedural requirements for collection of evidence will require a comprehensive strategy for equipment and containers. If the sterility of a container is in doubt, so is the evidence that is in the container. A similar principle applies to tools and procedures. For example, a shovel that is used to collect multiple samples can be a source of cross-contamination. While I have strongly urged the proper documented sterilization of equipment, we should also provide control measures that can reinforce the integrity of our procedures. A commonly used strategy involves *blanks*. It is important that I dwell on this for a minute, because neither my military CBRN sampling training nor my fire service-based HAZMAT training covered the use of sample blanks in any great detail. Both Drielak and the US Environmental Protection Agency (EPA) have more detailed explanations.[2] I will discuss three kinds of blanks are useful in this setting:

> Control Blanks: Control blanks are samples of equipment of the exact type and lot used at the scene, sterilized by the same procedures (or bought sterile and kept that way), but kept secure and separate from the equipment used at the scene. For example, the crime scene team procures a case of sample jars. A few of the jars are selected at random and sterilized by the exact same documented procedure as the others, but are numbered and packaged and stored separately. They will not be taken to a crime scene, but will remain secured. They will then be submitted as evidence to the laboratory as well as samples from the crime scene. This will also help prove that no contamination of the sampling media took place prior to the incident.
> Trip Blanks (also known as Field Blanks): Trip Blanks are control blanks that are taken to the crime scene, but remain sealed and unused. They go with the technicians to the

sampling site. They are submitted for laboratory analysis as if they were used. They are a control measure to prove that transportation to and from the crime scene did not cause any contamination to be introduced to samples. Drielak suggests two trip blanks for every ten live samples.[3]

Equipment Blanks: Some sampling tools, particularly large and/or expensive ones, will need to be reutilized at the crime scene. For example, non-sparking shovels or soil sampling augurs will likely need to be cleaned and re-used for multiple samples. In this situation, an *equipment blank* is used to validate the cleanliness of the tool and rule out the possibility of cross-contamination introduced by the tool. For example, after being cleaned several times, distilled water is poured over a tool and collected as evidence. This clean, distilled water is collected, labeled, and documented as evidence as the equipment blank. The laboratory can analyze the water sample for contamination and, if shown to be free of contamination, proof that the tool is not a source of cross-contamination.

The consistent use of all types of blanks will be one of the most powerful tools to defend the integrity of the sampling process.

POSSIBLE TACTICS, TECHNIQUES, AND PROCEDURES

A crime scene contaminated with CBRN/HAZMAT will require some tactics, techniques, and procedures somewhat at variance from conventional crime scenes. It will take more time, more people, and more equipment. I highly suggest following procedures similar to those in the references cited above, or developing your own based on such references.

Types of Samples

A full treatment of sampling techniques and methods is best left to the likes of Drielak's textbook or various procedural guides. This chapter's purpose is to start the thought process in the correct way, not to serve as a textbook on sample collection in its own right. A crime scene response effort needs to be prepared, both procedurally and logistically, to collect the several types of samples.

Solid: Solids include powders, soil samples, core samples, and other types. Numerous techniques can be used. Remember to look for air filters from HVAC systems or protective masks, which could contain evidence.

Liquid: The majority of chemical hazards are liquids. However, they may not be in obvious locations. Remember to check drains (u-bends in pipes are brilliant for this) and things that can absorb liquids. Carpets, upholstery, paint, clothing, and soil can all absorb liquids. In small bodies of water, some liquids will float on top of water. Others will sink. There are specific techniques that can capture such samples.

Aerosol/Vapor/Gas: These can be tricky to collect. Evacuated cylinders, Tedlar bags, and desorption tubes are all described in various references as useful techniques. Remember vapor density. Most of the hazards are heavier than air and will collect in low lying areas. Some (e.g. hydrogen cyanide) will be lighter than air Wherever practical, use detection instrumentation to

Surface wipes: Not every CBRN/HAZMAT threat will be visible. Some may be an invisible residue on surfaces. These can often be obtained by swabbing wet and/or dry wipes on surfaces. Some of the nerve agent evidence in the Skripal case was found this way.

Dermal: In some situations, trace evidence may actually be on human (or animal) skin. It may not always be easy to obtain such samples, but some creative thinking can address this. Swabbing hands with sterile water before a person goes through decontamination is one option.

Biomedical: Largely, biomedical evidence will have to be collected in clinical settings or in post-mortem analysis. This can be from both humans and animals. Biomedical samples from both humans and animals have been useful in investigating chemical warfare use.

Rescuing Victims

Rescue and treatment of victims takes priority over investigative concerns, a point that bears repeating. Victim rescue can take place in ways that accommodate a forensic approach. Document the rescuers entry into the zone. Photograph the situation. If clothing is cut off a victim, collect it as evidence. If necessary, consider defining an exclusion zone that doesn't cover where active rescue and medical assistance is taking place, and continue forensic efforts there until your perimeter can be expanded.

Collection of Evidence at the Decontamination Line

Contamination is potential evidence. This evidence will not necessarily be collected in a proper manner, as most decontamination procedures are more worried about safety and health than evidence. Position a team of investigators up-stream in the decontamination line. If there are victims who are contaminated with toxic substances or residue, then this residue can be useful evidence. It is also possible that the perpetrator will be among the victims undergoing decontamination. By positioning some investigators at the head of the decontamination line, valuable evidence can be siphoned off before it is damaged or destroyed or lost during a decontamination line. This would include clothing and other personal effects that might have valuable residue of a hazardous substance, debris of a device, liquid samples from a decontamination pool, or similar valuable evidence. Likewise, there may be some useful witness statements that otherwise might take days or weeks to collect. Finally, a skilled forensic technician may be trained and equipped to collect *dermal samples* of CBRN materials.

Send Investigators to the Hospital(s)

Send some investigators to the hospitals supporting the medical response. It is very possible, indeed almost certain under some circumstances, that contaminated victims may have been transported to hospitals bearing evidence. Other individuals may have self-referred themselves to medical care, while bearing useful evidence about their persons and possessions. Persons who have relocated to hospitals may even have useful evidence in the form of video or photographs on smartphones and cameras. Useful statements from witnesses may be obtained during the first critical hours of an evolving investigation. Furthermore, the perpetrator(s) and or accomplice(s) may have been injured in the incident and evacuated to medical facilities. It may be possible to collect dermal samples here as well.

Initial Reconnaissance

One or more entries will need to be made into the scene for the purposes of initial survey and reconnaissance. Photos and/or video are useful at this stage and will establish a baseline state before the systematic removal of evidence. Other purposes of these reconnaissance entries will be to identify and verify the level of hazard in the crime scene and to develop a forensic strategy—the plan for sampling and collection. For example, photoionization detectors can be used to identify where higher concentrations of volatile organic chemicals may be found.

Reassessment of Scene Safety

Once a site has been stabilized and secured, there needs to be a definitive determination of the hazards posed at the site. The key imperative will be to define what level of PPE the crime scene investigators will require. It is possible that different zones of the site will require different levels of protection or special procedures, such as confined space entry or intrinsically safe instruments. A significant amount of information about the hazards at the scene will likely be available from earlier responders, but this information should always be verified as conditions may have evolved. Liquids may have evaporated, powders may have re-suspended in the air, and containers may have leaked. Radiation readings should always be taken, even in seemingly obvious chemical or biological scenarios, if only to rule out the possibility of a combined threat.

EOD Considerations

Many possible crime scenes might involve explosive dissemination of materials. Not every device may have detonated or detonated as intended. The crime scene may have energetic materials present in a form that poses a threat to responders. We must also consider the possibility of secondary devices and booby traps. Devices may have been deliberately emplaced to imperil responders or to hinder a response effort. If there is any doubt whatsoever, have an EOD team survey and clear the site.

Make an Evidence Collection and Sampling Plan

The next step in the process is to develop and execute a sampling plan. A sampling plan is a documented procedure that describes what samples will be collected. Drielak and other references, both CBRN and conventional, go into great detail on how to do this, so I will not repeat their advice. However, during the collection of evidence at CBRN/HAZMAT incidents there are some significant considerations.

Being in a Hurry to Catch Fragile or Short-lived Evidence

Some types of valuable evidence at CBRN/HAZMAT scenes will be fragile or transitory. Fine powders may blow away in the wind, volatile liquids may evaporate, or vapors may be dispersed by a building's ventilation system. While the technical response and rescue effort may have made some effort at product control, this cannot be assumed and it may be necessary to take emergency steps to safeguard evidence that is fragile in nature. A good established crime scene SOP should have established methods for ad hoc preservation of evidence until it can be firmly processed. For example, plastic sheeting can be placed over powders, buckets can be placed over puddles of liquid to reduce evaporation (or at least trap a sample of the vapor), and the air handling systems of a building can be shut off. Sterile materials should be used whenever possible, in order to maintain integrity.

Conventional Criminology

All too often, the CBRN responders will focus only on CBRN evidence. However, you will have conventional evidence as well. You will have things like footprints, fingerprints, body part residue from a suicide bomber, DNA evidence, fibers, and every other conceivable type of conventional evidence. You may have conventional evidence and CBRN evidence co-mingled. How do you deal with a smartphone with nerve agent on it? But that smartphone might be the single most important bit of evidence. This becomes problematic in the laboratory. The lab qualified to work with nerve agents may not be trained or equipped to do a forensic exploitation of the smartphone. This conundrum remains unresolved in much of the world.

LESSONS LEARNED

Don't Wash the Evidence Down the Drain

I attended a major exercise in the Czech Republic in 2008 which simulated a chemical terrorist device. The responders, who were emergency medical staff and firefighters, literally flushed the remnants of the explosive device down the drains. No forensic effort was made whatsoever. You may need to do decontamination to save lives and prevent the spread of hazards, but try not to botch the forensics this badly.

Think Like a Defense Lawyer

One must assume that a terrorist put on trial for a heinous CBRN attack will have very good lawyers defending him. But even poor defense lawyers know to attack the evidence. When developing and executing your forensic evidence processes, it is helpful to imagine the questions that a really sharp defense advocate would ask. Ask yourself the hardest questions about containers, sterility, documentation, and processes. Better yet, find a lawyer to probe your processes or observe an exercise.

Labs Need to Adapt

Almost always, the room comes to an awkward silence when I ask about the smartphone covered in nerve agent. As noted above, the labs who can deal with CBRN materials are often highly specialized and hold international certifications to do that work. They do it well. However, they are generally NOT the authorities on conventional criminological work. Both kinds of lab need to work together. More work is needed to bridge this gap. In your planning stages, ask yourself about the contaminated smartphone that may have been in the hands of the terrorist perpetrator. Where do you send it and who gets the information off of it?

REFERENCES

1. Oliver M. Radiation found at 12 Sites in the Litvinenko case. The Guardian, London, UK (30 November 2006).
2. (a) Drielak, S.C. (2004). *Hot Zone Forensics: Chemical, Biological, and Radiological Evidence Collection*, 40–41. Springfield (IL): Charles C. Thomas. (b) US EPA. *Region III Fact Sheet: Quality Control Tools: Blanks* (rev.1, 2009), http://www.epa.gov/region3/esc/qa/pdf/blanks.pdf
3. Drielak, S.C. (2004). *Hot Zone Forensics: Chemical, Biological, and Radiological Evidence Collection*, 41. Springfield (IL): Charles C. Thomas.

Practical Scenarios

Introduction to the Practical Scenarios

The best way to learn is by doing. However, CBRN and HAZMAT incidents do not happen very often. We try to learn by holding training exercises that simulate how bad things are happening, and we hope to learn from this experience. The First Edition of this book provided training scenarios that were (and still are) valid tools for learning. Feedback from readers was that this was the most helpful part of the book. These scenarios are either completely new, or significantly modified. Readers interested in the older scenarios should contact the publisher or the author, as they can still be made available. If in doubt, contact the author.

Rather than full scripts, these scenarios serve as outlines for training and as a teaching method to impart lessons. There's value in simply reading them. They will have more value if you use them as part of a table-top discussion. These scenarios are written primarily with the view that they will be used in tabletop training exercises. With some work and resources, they can provide a basic framework for an actual dynamic exercise. Feel free to use these in whichever way helps your organization best.

The scenarios are largely adapted to a North American audience. This decision is based on the fact that this book is being published by a North American publisher and the bulk of global sales of the previous edition were in North America. However, I have worked hard to ensure that the various "Americanisms" are replaceable with a little bit of creativity. Feel free to make whatever substitutions or adaptations you need to in order to tailor it to your own operational environment.

BEST PRACTICES AND WHAT NOT TO DO

Each scenario describes some "best practices" as guidance for those kinds of situations. But no book can always tell you what to do. There are too many uncertainties and variables in every scenario for me to be able to tell you exactly what to do in a given situation. One thing that

I can definitely do is to tell you what you shouldn't do. Some of this is knowledge that I learned the hard way and some of it is knowledge I gained by observing the mistakes made by others. It is never possible to say, which perfect certainty "If this happens, do X" but it is indeed possible to say "Well, if this happens, best thing is to not do Y". Remember, never make the problem worse than it already is. Therefore, the "what not to do" section of each scenario, which were popular in the first edition, contain some of the most valuable lessons in this book.

Searching Arenas and Stadiums

BACKGROUND

Major events will include numerous events at large venues such as stadiums, arena, and large music venues. General security provisions, regardless of CBRN/HAZMAT considerations, will likely include searches of these venues before events in order to ensure that they are free of conventional hazards. This scenario assumes such searches are already planned and concentrates on how to integrate CBRN/HAZMAT considerations into security provisions at an event.

Many terrorist incidents are perpetrated by hiding hazardous devices or weapons inside a venue prior to an event. In addition, there could be situations that are hazardous to the public or VIP visitors that have not been discovered. In either case, CBRN/HAZMAT materials or situations could be uncovered. The traditional countermeasure is to conduct a search operation called a "sweep" in some settings. In such an operation, trained security personnel, often search technicians or EOD technicians, with or without the assistance of dog teams or detection technology, physically search the entire venue involved. Due to size of venues and time available, search operations can take a long period of time and tie up a lot of personnel. Managing such searches is not easy. How to do a search properly is enough of a subject for another book, but I can recommend that training provided by the UK Police National Search Centre as an excellent resource for such operations.

Many police and security agencies understand the basic principle behind venue searches and have some experience in doing them with a view toward conventional threats. The purpose of this scenario is to try to integrate CBRN/HAZMAT concerns into such a search operation, rather than to re-invent the wheel.

CBRN and HAZMAT Incidents at Major Public Events: Planning and Response, Second Edition. Daniel J. Kaszeta.
© 2023 John Wiley & Sons, Inc. Published 2023 by John Wiley & Sons, Inc.

FIGURE A.1 Large venues, such as arenas, are labor-intensive to search.
Source: Used with permission from R. Mead.

THE SCENARIO

A major sporting event is being held at an arena in your city similar to that shown in Figure A.1 above. The arena holds basketball and hockey matches, as well as concerts. It can seat 20,000 people under normal conditions. This event in question is a championship match, which will be attended by a number of celebrities and several political leaders. It has been widely published that the various celebrities and politicians will be at the event. Significant amounts of money have been spent.

You have had several weeks of advanced notice to develop plans for the event. Within reason, you have the security personnel and resources that you have requested. No specific threat intelligence has been received, but the city has had extremist terrorist acts within the last five years.

Episode 1

You are tasked with planning for the security search of the arena. What is your basic scheme for searching the building? Discuss your concept, approach, and basic tactics. How much time will it need? How many personnel will you require? Will you need additional resources or equipment?

Episode 2

Your search team is experienced in conducting conventional searches and has never really considered CBRN/HAZMAT issues in searches before. How do you want them to approach the issue differently? Is there particular training or equipment that you want them to have? What will you do in your search operation differently to include CBRN/HAZMAT?

Episode 3

Your superiors have reviewed your plan and tell you that resources and time are not sufficient. You will have 20% fewer search staff and half a day less time than your original plan due to factors beyond your control. What do you do? What about if you had 50% less time? Or 50% fewer personnel?

Episode 4

You are conducting your search. While you thought that the building is secured, a large shipment of boxes had been delivered and placed inside the loading dock. While dealing with this situation, you encounter five employees who should not be at work, who had somehow gained entry through a back entrance that was supposed to be locked. How do you handle the situation?

Episode 5

During the search, the following situations are discovered:

- A vehicle in the underground loading dock area is marked with HAZMAT markers and contains a number of containers of industrial chemical.
- A search team notices an unknown but irritating smell in the air in a restroom adjacent to a lounge area designated for use by VIPs during the event.
- The search team discovers that the door to a room containing air-handling and ventilation equipment was left unlocked. There are signs that people have been in the room recently, such as footprints in the dust on the floor.

How do you think you will deal with these situations?

Episode 6

A search team member has found what appears to be a terrorist device located in a storage room in a basement level in the arena. It consists of a plastic drum containing several gallons of liquid and a possible explosive charge. It is near the end of your search schedule and you are within an hour of opening the venue for staff. How do you proceed?

DISCUSSION AND TIPS FOR SUCCESS

Planning

Keep your plan simple and realistic. Allow adequate time for your search and based your planning on actual walk-throughs and reconnaissance of the facility. Walk through every room of the building ahead of time. Take a whole day if you have to. Work with building management to learn the building. Be honest about time and personnel requirements. If you do not know how long it will take, do a practice run on a portion of the venue or a similar type of building. Use the practice run as a basis for calculating the larger search.

Search Scope and Search Integrity

Maintaining the integrity of the search is an important concept. Access to the search area must be controlled during the search. There is no point trying to do a search if the venue is not kept secure during and after the search. Otherwise, threats may be brought into the area after the search has been performed. If something is discovered or (worse) activated by the search team, then people may be in danger. You have a duty of care to the people who work there, so you should keep them away while you are looking for hazards.

Some terrorism acts have been perpetrated by employees at venues. Others have used deception, coercion, or blackmail to force employees to bring items into a venue. Because of this letting people into the site after the site has been searched should only occur through a screening process, otherwise the search is pointless. Once you've searched the place, everyone coming into it needs to be searched.

Scope of Search

The objective of this type of search operation is to systematically exclude the presence of hazards that are already present within the perimeter. Almost inevitably, resource and time constraints will preclude you from a perfect search. You will have the manpower, but not the time; or the time, but not enough manpower; or extenuating circumstances like late-running construction and maintenance activities that curtail your available time at the last minute; or an incident somewhere else that takes some of your personnel away. Some corners are going to have to be cut somewhere.

Likewise, the search will generally have to use nondestructive techniques to look for hazards. It is not a wartime environment where doors can be broken and walls knocked down. You do not go through locker rooms blasting open the lockers. Because of this, your search techniques will, by necessity, be more labor-intensive.

A search can and should be tailored in scope. Searching for items the size of a golf ball will be much more intensive than searches for items the size of a rifle or suitcase. If you have 30 people and 4 hours to search a convention center, finding small items that are well hidden is probably beyond the reasonable scope of the search.

The level of thoroughness of the search can be optimized or calibrated. A search should be roughly comparable to the thoroughness of the security screening. Searching for small amounts of liquid in the secured area is all well and good, but if the public or guests or staff are allowed to bring unscreened items into the venue that are equal or greater in size than the scope of the initial search, then the search has less value. For example, if guests bring in bottles and cans without the contents being screened, then searching for items of can or bottle size is of limited value. It could be done, as a small bottle of a hazardous substance hidden beneath the seats in an auditorium is certainly worth looking for. But bear in mind that the search will not systematically exclude the possibility of such an item being used in an incident at the venue.

Control Measures

One tactic for rooms, storage containers utility areas, closets, lockable doors, covers for manholes, and related spaces is to apply control measures, typically security seals. There are numerous types of tamper-evident or tamper-resistant security seals available. These can be applied to indicate that the space has been searched. Seals are useful as a security tool, if they are managed properly.

Seals need to be periodically inspected to ensure their integrity. At some large events, a coordinator and even an entire squad of personnel may be used to number, issue, and periodically re-inspect seals. Technology has advanced considerably in this area, and there are now such things as electronically readable RFID security seals. Proper use of such technology can aid a venue search.

Sector Search

If you do not have the resources to search an entire venue in a timely manner, consider using a "sector" approach if you have the security staff to manage it. Divide the venue into sectors and search them in a logical order, opening up each sector as it is searched. This also may ease the burden on security personnel as the numbers of security staff to keep the integrity of the search scales up over time. This tactic has been used to great effect in large venues.

A sector search could be used to allow gradually opening of a facility to essential event staff, if not necessarily the public. If you use this approach, you should still consider the safety of other sectors. Consider what would happen if a device were triggered. Leave some sectors as a safety buffer. But if you plan a search methodically, you could re-open search sectors that have already been searched, provided you leave some sort of safety buffer. This is, of course, a calculated risk and should be done in the broader context of an overall risk assessment.

Safety

A search effort could cause release of CBRN/HAZMAT materials. Search efforts could cause a terrorist device to function. Always have a contingency plan in place for this eventuality. There is not much point in holding a search operation if you are unwilling or unable to deal with a hazard if you find it. Part of a search plan is a response plan for dealing with hazards that are discovered by the search team. You need to have necessary support, such as medical and fire service support, if not on standby, at least close at hand.

"Find or Function"

The simplest, quickest, and most effective search tactic is also the most dangerous. "Find or function" is literally what it sounds like. It means that search teams conduct their searches in ways that are likely to trigger devices. Light switches are switched. Doors are open and shut. Toilets are flushed. In short, anything that might be connected to a hazardous device is operated to see if anything happens. Whether or not your search operation is a "find or function" operation is a management decision not entirely without risk.

Detection Technology

Detection technology may be useful in search operations, but given the state of the art, it is by no means a solution to all problems. Radiation detection is very easily integrated into search operations. Indeed, it is my recommendation that you should use radiation detection, if at all possible, in search operations. Chemical detection is not very easily adaptable to a search operation, as most likely sources of chemical warfare agents (such as improvised chemical devices) would be sealed to keep their payload from leaking away or provoking a reaction before they could function as intended. Biological detection generally does not work within a timeframe that is useful for search purposes.

Canine Searches

Do not underestimate the value of explosive detection canines in search operations. Even though dogs are not directly used to search for CBRN materials, they are excellent in detecting explosive materials that may be used to disperse CBRN materials. Never rely on dogs alone for a search and always take advice from professional dog handlers on their proper employment. Never use canines around cleaning chemicals, as this could damage their sense of smell.

WHAT NOT TO DO

Search Without Securing the Area After the Search

If somebody can sneak something into the building after the search was done, then the search effort was a waste of time and represents manpower that could have been used for something else.

Use Security Seals or Similar Control Measures Without a Means to Check Them

Security seals are useful tools in some conditions, but they never prevent entry. They only give indication that an entry has occurred, and this indication is only discovered if someone checks the seals frequently. As mentioned above, electronic means, such as scannable RFID or readable bar code seals can help.

Allow Cleaners, Staff, Media, etc. into the Venue while the Search Is Ongoing

A thorough search can set off a bomb or release toxic materials. Therefore, only people doing the search or securing the site should be in the proximity of searches. You do not want the responsibility for the death or injury of a cleaner.

Not Minding the Void

Rooms that can't be searched because nobody has the key or lockers that are not opened represent unsearched voids. An unsearched void is a failure in a counterterrorist search operation. If you don't look everywhere, then there's little value in the search. A good reconnaissance and planning will help you to identify where these voids are likely to be and allow you to have countermeasures available to you, such as master keys, a locksmith, a fiber-scope, or similar options.

Allowing too Little Time or Too Few People to Do It Properly

Many search operations give little or no protective value because of too few search technicians, too little time, or a combination of both. Be realistic and practical during the planning stages.

Neglect Manual Searches for Other Means

A good search will require manual inspection. Technical means should be used to supplement a manual search. Nor should dogs replace human searches. There is no substitute for hands and eyes.

Not Having a Response Plan for a "Find"

There's no point in having a good search plan and then not knowing what to do when you find something. Response procedures and "render safe" plans need to be in effect for the beginning of the first search, not the start of the event.

Ignore the Advice of Dog Handlers

The dog handler knows the dog's capability and temperament more than you ever will. Often, the handler is with the dog 24 hours a day and knows the dog better than some of his own family members. Dogs can also be dangerous. I've seen what an angry explosive detection dog can do, and it is not a pretty sight. One of the few times I actually got to apply my emergency medical technician (EMT) training was to patch up someone injured by a frustrated detection dog.

Use Chemical Detectors as Sniffers

Well-hidden chemical devices are not likely to be discovered by dragging photoionization detector (PIDs) and military chemical warfare detectors through a building as part of their search efforts. A PID is going to detect every cleaning product in the building. Some military surface contamination monitors may work at 1 second per square inch of surface area, but not at a faster rate. How many square inches of surface area are there in this sporting venue in this scenario? Sometimes chemical detectors are used in order to "establish a background," a technique I have seen in the United Kingdom. I think that this is a waste of time. Little value will be gained from such a tactic.

Screening People, Goods, and Vehicles

BACKGROUND

Searching and securing venues and areas will be an important part of a balanced, all-hazards approach to keep an event safe and secure. Well before CBRN issues became part of everyone's security horizon, searches for firearms, explosive devices, or knives were not uncommon. Nor has it been uncommon to search vehicles and incoming shipments for explosive devices. Further, if you are going to take great efforts to search an area or building or zone for hazards, you do not want this effort to go to waste by allowing people to enter the secured area and bring unwanted items into it.

There are necessarily trade-offs to be made in screening the public. There is a balance to be struck between thoroughness and speed. Where that balance point is to be found will be different at every event and may vary from venue to venue within a particular major event speed. Rushing people through and not screening them adequately creates unnecessary vulnerabilities. Being so thorough that the process is highly unpleasant to the public and creates long queues is also counterproductive and might even make the long queues of congregating people a target in itself.

The challenge is to adapt existing security screening processes to newer threats such as CBRN terrorism. While such measures are no safeguard against external release scenarios—a threat originating outside the perimeter, they are a reasonable countermeasure against internal release scenarios. Good screening techniques can deter or prevent terrorists from importing threatening materials and devices into an area, thus forcing them to use less efficient or more costly external attacks. Radiation screening technology is now available for checkpoint screening, as shown in Figure B.1.

CBRN and HAZMAT Incidents at Major Public Events: Planning and Response, Second Edition. Daniel J. Kaszeta.
© 2023 John Wiley & Sons, Inc. Published 2023 by John Wiley & Sons, Inc.

FIGURE B.1 Radiation detection portals have a role in some scenarios. *Source: J. Monde/Thermo Fisher Scientific.*

THE SCENARIO

It is a high-profile awards ceremony in the entertainment industry. The awards ceremony is a four-hour event, preceded and followed by numerous parties. The venue itself is a large theater complex in a major American city. Literally hundreds of famous names in entertainment will be descending on the venue. Both the venue and the date have been publicized for many months prior to the event. Several of the entertainers are security concerns as they have received threats. One of the award nominees is highly controversial in several countries in the Middle East.

Security and safety planning for the event has been underway for a year, and a reasonably good conventional security plan has been developed for the theater complex. Large numbers of contract security personnel will be used to operate X-ray machines and metal detectors. A "no bottled liquids" rule has been in effect for several years at the venue, but in practice small bottles and jars, such as medicines, personal hygiene items, and baby food, are routinely let in as the rule is largely focused on restricting alcoholic drinks. Vehicle and cargo screening took place several times in previous years for high-profile events and, while not routine, the venue knows how to accommodate such screening. The issue now is how to upgrade security to include CBRN concerns.

Episode 1

It is 6 months before the awards ceremony. You've been asked to develop and execute a training program for the security staff to implement screening for CBRN threats. Some of the security staff have never dealt with such concerns before.

What sort of procedures would you recommend? What kind of equipment do you need? Do you have adequate staff?

Episode 2

One area of concern is radiation detection. As part of the overall security posture, the primary security contractor has procured small radiation detection pagers from a reliable vendor. The concept is for at least one security screener per entry lane to have a radiation detection pager. Security staff have been using the radiation detectors for familiarization purposes. Staff have had numerous alarms that have turned out to be customers or staff who have had various medical procedures involving radioisotopes.

How will you ensure that disruptions do not occur based on such detection events on the night of the awards ceremony?

Episode 3

It is 2 weeks before the awards ceremony. Security staff have been practicing the security screening processes for liquids through the security checkpoints. Basically, nearly everyone has a bottle of water. The staff have asked "is there a way we can let bottled water through the checkpoints? It's all kicking off!" Is there a way you can screen bottles of water to make sure that they are, in fact, water?

Episode 4

It is the night of the event. The event starts shortly. You have already admitted some people and most of the staff through checkpoints. You have implemented all of the processes and technologies you have suggested above. However, a serious electrical fault has cut off power to part of the security screening checkpoints. Approximately half of the equipment that uses power sockets (e.g., X-ray machines, walk-through metal detectors) is not operable. Discuss how you will manage screening while maintaining the level of security that the event requires.

DISCUSSION AND TIPS FOR SUCCESS

Keep Your Expectations Reasonable

Every venue is different and the threat is widely varied. "Perfect" security is not possible and you will have to treat security screening as merely one component of a broad spectrum of plans and countermeasures. No screening technique is going to be universally effective against CBRN threats.

Screen the Staff

A common shortcoming is to allow event staff, press, or VIPs to circumvent security screening. Many staff may be temporary or contract employees and in the past terrorists have gained entry to venues through employment or use of fake identity to enter venues. In other situations, hazardous devices or substances could be smuggled in by staff. An employee could even be the perpetrator of an attack. There is greatly reduced value in screening the public if hundred or even thousands of staff are let in without scrutiny. By all means, set up staff-only and expedited

VIP lanes and checkpoints. This will help immensely. Known VIPS are rarely the problem, but even then, one US President was assassinated by a famous actor, John Wilkes Booth.

Perimeters and Standoff

Vehicle and cargo screening is different from personnel screening. You can effectively operate a system with outer and inner perimeters. The outer perimeter is for large items like vehicles and cargo. The inner perimeter is for personnel. Keeping larger quantities of potential threats further away from your protected core areas provides a degree of mitigation. If you are going to check vehicles and cargo, consider the possibility of locating your screening area so that a building is in between the screening area and the areas which you are most concerned with protecting. This intervening building could provide a buffer or shield in the event of a hostile device functioning.

Liquid Scanning Technology

Threats to civil aviation, particularly alleged plots in the early 2000s to use liquid explosives, lead to liquid bans on airliners. However, this same threat served as impetus for science and industry to develop liquid identification technologies. There are now devices that can scan liquids to determine if they are a hazard. Such technologies do not work in every type of container, but these technologies now have a place in screening at major public events. You will need to understand how a particular screening technology and product work and then develop a rational concept of operations to employ it.

WHAT **NOT** TO DO

Mismanage Queues

Operations to screen people and their possessions will lead to queues. Lengthy queues are not in anyone's interest. Part of the reason that you are providing security is to protect people, and if people are in lengthy queues outside your secured area, then these queues of people have become a terrorist's target in their own right and you've done part of the terrorist's job for him. Lengthy queues also can make management of an event difficult in other ways. People who wait for unnecessarily long periods of time can become more difficult to handle and can act in ways that tie up security staff who should be doing better things than handling angry people kicking off in the line. You want to have enough checkpoints, equipment, and staff to be able to effectively process people into your event.

Misuse Dogs

Explosive detection dogs have a place in security operations. But using them in and around the general public can cause problems. Even the most placid dog can have a bad day. Dogs can bite the public and interaction with the public can distract a dog from its duties. Some people have a fear of dogs. Employ dogs only after careful consultation with the specialists who are trained to work with them.

Ignore or Mistreat the Venue Staff

There are many scenarios where security screening is a temporary measure, implemented only on special occasions. The normal day-to-day employees at a venue may not be used to being screened. I have seen problems where temporary security staff, such as contract guards, try to do their job but end up handling the staff as brusquely as the public. While screening employees is as important as screening the public, it needs to be done with some respect and forethought. Employees are likely to have ideas about routes into the venue that might bypass security. It is also possible that an employee who is made disgruntled by security staff can cause problems. My suggestion is that management needs to communicate well ahead of time to the staff and manage expectations. Special entrances and/or dedicated lanes for staff are also an excellent practice.

Allow "Tailgating" or Other Bypassing

It is relatively easy in the confusion and bustle of a major event for a determined individual to exploit the situation. Be on the lookout for both accidental and deliberate attempts by people to literally slip into the event on the coat-tails of dignitaries or staff.

Threats and Hoaxes

BACKGROUND

Not every CBRN incident has to actually have CBRN materials present in order to cause chaos, confusion, and/or disruption. As mentioned in Chapter 1, the threat of use or hoax situations can cause many problems. The threat of CBRN materials is often enough to cause an incident in itself.

The history of threats and hoaxes well pre-dates CBRN terrorism. Telephonic "bomb threats" are nearly as the telephone. In written form, the communicated threat is even older. Threats can be simple or terribly complex in nature. Communicated threats can occur for many reasons. Some threats will be immature pranks. Others may be the result of actions by mentally ill persons. Some threats can be deliberate to confuse, distract, or dilute operational response assets or to create a general air of fear or disorder, possibly in preparation for an actual incident. In some instances, threats are "legitimate" in that real threat materials are actually present, albeit possibly not in a way that will harm anyone.

Hoaxes are also a problem. Fake devices, letters with powders, or vials of unidentified liquid can be used, often alongside some sort of communication, could easily be real threats until proven otherwise. Anthrax hoaxes were around for years before the first actual anthrax spores were used in mail threats.[1, 2]

Threats and hoaxes involving CBRN materials add a layer of complexity over the traditional bomb threat. The scope of potential harm is higher, but so is the technical difficulty of perpetrating the attack. Evaluating such threats requires technical knowledge of the materials involved.

THE SCENARIO

A large annual sporting championship is being held in your city. Tensions are high because the two competing teams are bitter rivals. In the past, matches between the two teams have resulted in civil disorder. This time, security issues are even more paramount because the vice president is going to attend the match, being a long-time fan of one of the teams. Several groups who are concerned about a current political issue are planning to hold a demonstration, due to the vice president's visit. Security and emergency preparedness plans have focused on VIP protection and civil disturbances. However, you are charged with CBRN/HAZMAT preparedness as part of the overall effort.

Episode 1

It is two days before the game. Several reporters from the main local newspaper receive SMS messages claiming that a radioactive device has been hidden in the stadium. The messages were sent from an otherwise unused "burner" phone.

  Discuss the various courses of action available to you and explain your thought process.

Episode 2

It is 24 hours before the start of the game. A courier package arrives at the home of the owner of one of the two teams. The package contains a threatening letter and a vial of clear unidentified liquid. The threatening letter demands cancellation of the game and claims that more of the liquid will be released at the game.

  As the emergency response evolves at the owner's residence, one police officer notes a person in a nearby park watching the evolving response with binoculars. This person flees before they can be questioned.

  Discuss the various courses of action available to you and explain your thought process.

Episode 3

It is 12 hours before the game. Final searches and sweeps are underway at the stadium. One of the search technicians receives an alert from a radiation detector. Searching closely, he finds a box. The radiation readings are coming from the box.

  Discuss your actions. How does this information change your analysis of the situation?

DISCUSSION AND TIPS FOR SUCCESS

Even ad young and inexperienced emergency planner will realize that always ignoring communicated threats is not an ideal practice. But doing a full "bomb scare" response to every threat is not a good policy either. It is best to find the right course to steer in between.

Consider Distraction and Diversion as Motives

Emergency responders need to consider that communicated threats may not just be nuisances. It is entirely possible that a phone or email threat may be designed to distract or divert responders. Tying up CBRN/HAZMAT and EOD resources at one end of a city can mean that a response to a real incident on the other end of the city could be delayed, thus increasing the effect of a terrorist attack. It is also possible that a threat message may be intended to provoke a response so that terrorists can observe response measures. This is what may be happening in Episode 2 above.

A balanced perspective is needed when responding to threats. For all but the most highly credible threats, response measures to search for a device should be discrete and kept out of the eye of the media to the greatest extent possible.

Credibility Assessment

An important part of any response to the communicated threat is a credibility assessment. A formal credibility assessment process should be developed as part of major event security plans. It is likely that such a process will be part of a general "bomb threat" antiterrorism plan. Bomb threat checklists to assist initial report, as shown in Figure C.1, are a good starting point. Explosives experts have long experience assessing whether a device described over the telephone is feasible or not. However, planners should endeavor to ensure that CBRN concerns are integrated into such a plan.

Credibility assessment has both objective and subjective aspects. How credible is the threat? Some threats are clearly not credible at face value, while others are so vague that they are hard to evaluate at all. But many threats have enough substance to analyze. Call center personnel are usually trained to keep such persons on the line as long as possible. In a few cases, there is some limited opportunity for interaction with the person making the threat. In other cases, a recording of the threat may be available for analysis. From an objective standpoint, threats should be analyzed using the following three criteria:

Psychological/Behavioral: Analysis of the psychological and behavioral characteristics of the perpetrator by analyzing the words and tone of the message. There is a small but growing field of "forensic linguistics" that shows some promise in being able to analyze the content of a message. Competent behavioral specialists should assess threats. An expert may be able to give insight on credibility, intent, and the mental health status of the perpetrator.

Technical Credibility: Particularly in situations involving CBRN materials, it is necessary to examine the technical credibility of the communicated threat. If the perpetrator uses wording that portrays a general lack of specialized knowledge, then the credibility of the threat drops. For example, a person who claims to have grown "bubonic plague virus" is not so credible, as anyone culturing it would know that it is a bacteria. If the threat makes reference to materials or methods of dispersion that are technically difficult, unlikely, or impossible, then the credibility also drops. A person who claims to have stolen a tanker truck of chlorine or phosgene may not be credible if no such thefts have been reported. And such products cost money and the truck has value, so legitimate industrial users will report such a loss.

Operational and Logistical Credibility: The operational and logistical credibility of a threat should be assessed. A threat may contain detailed and believable explanations of a technically feasible device or materials. Are the operational and logistical aspects of the threat feasible? Are the details consistent with the rest of the story? Does this threat match any claims in popular fiction?

BOMB THREAT CALL PROCEDURES

Most bomb threats are received by phone. Bomb threats are serious until proven otherwise. Act quickly, but remain calm and obtain information with the checklist on the reverse of this card.

If a bomb threat is received by phone:

1. Remain calm. Keep the caller on the line for as long as possible. DO NOT HANG UP, even if the caller does.
2. Listen carefully. Be polite and show interest.
3. Try to keep the caller talking to learn more information.
4. If possible, write a note to a colleague to call the authorities or, as soon as the caller hangs up, immediately notify them yourself.
5. If your phone has a display, copy the number and/or letters on the window display.
6. Complete the Bomb Threat Checklist (reverse side) immediately. Write down as much detail as you can remember. Try to get exact words.
7. Immediately upon termination of the call, do not hang up, but from a different phone, contact FPS immediately with information and await instructions.

If a bomb threat is received by handwritten note:

- Call _____
- Handle note as minimally as possible.

If a bomb threat is received by email:

- Call _____
- Do not delete the message.

Signs of a suspicious package:

- No return address
- Excessive postage
- Stains
- Strange odor
- Strange sounds
- Unexpected delivery
- Poorly handwritten
- Misspelled words
- Incorrect titles
- Foreign postage
- Restrictive notes

DO NOT:

- Use two-way radios or cellular phone; radio signals have the potential to detonate a bomb.
- Evacuate the building until police arrive and evaluate the threat.
- Activate the fire alarm.
- Touch or move a suspicious package.

WHO TO CONTACT (select one)

- Follow your local guidelines
- Federal Protective Service (FPS) Police 1-877-4-FPS-411 (1-877-437-7411)
- 911

BOMB THREAT CHECKLIST

Date: _____   Time: _____

Time Caller Hung Up: _____   Phone Number Where Call Received: _____

Ask Caller:

- Where is the bomb located? (Building, Floor, Room, etc.)
- When will it go off?
- What does it look like?
- What kind of bomb is it?
- What will make it explode?
- Did you place the bomb?   Yes   No
- Why?
- What is your name?

Exact Words of Threat:

Information About Caller:

- Where is the caller located? (Background and level of noise)
- Estimated age:
- Is voice familiar? If so, who does it sound like?
- Other points:

Caller's Voice	Background Sounds:	Threat Language:
❑ Accent	❑ Animal Noises	❑ Incoherent
❑ Angry	❑ House Noises	❑ Message read
❑ Calm	❑ Kitchen Noises	❑ Taped
❑ Clearing throat	❑ Street Noises	❑ Irrational
❑ Coughing	❑ Booth	❑ Profane
❑ Cracking voice	❑ PA system	❑ Well-spoken
❑ Crying	❑ Conversation	
❑ Deep	❑ Music	
❑ Deep breathing	❑ Motor	
❑ Disguised	❑ Clear	
❑ Distinct	❑ Static	
❑ Excited	❑ Office machinery	
❑ **Female**	❑ Factory machinery	
❑ Laughter	❑ Local	
❑ Lisp	❑ Long distance	
❑ Loud		
❑ **Male**	**Other Information:**	
❑ Nasal		
❑ Normal		
❑ Ragged		
❑ Rapid		
❑ Raspy		
❑ Slow		
❑ Slurred		
❑ Soft		
❑ Stutter		

Homeland Security

FIGURE C.1 Bomb threat forms can provide guidelines for people receiving threatening communications.
Source: Photo: US FEMA, public domain image.

WHAT <u>NOT</u> TO DO

Ignore or Fail to Circulate Intelligence and Warning

There have been circumstances where useful intelligence has not gotten to the people who really need it. If the person who assesses the credibility of a threat does not know about the stolen gas cylinders, then that circumstance represents a failure of the intelligence system.

Ignore Every Threat

Most threats are cries for attention or to cause diversion and chaos with relatively little expenditure. Most threats are not followed up by acts of violence. This may lead one to think that the best policy is to ignore threats. While some threats clearly can be ignored, there needs to be a formal process to evaluate them. It is important to remember that some terrorist groups deliberately provide advanced warning, albeit often in a vague or misleading way, before detonating a device.

Take Highly Visible Responses Immediately

It is possible that threats seek to force evacuations of buildings immediately. Is it the terrorist's intent to force everyone out onto the street where they are vulnerable to some other means of attack? Also, the entire intent may be disruption of normal business. A broad assessment of individual threats needs to occur. There may be a time to evacuate everyone in a hurry, but there needs to be a process that gets you to that decision, not a knee-jerk reaction to a phone call or email.

Amateur or Slapdash Credibility Assessment

The day when you get the threatening message is not the point at which to look in the telephone directory for a consulting psychologist. Nor should you rely on "well, did he sound like a nutter" as a means of making serious decisions. Build your assessment techniques and procedures early.

REFERENCES

1. Cole, L.A. (1999). Anthrax hoaxes: hot new hobby? *Bulletin of the Atomic Scientists* 55 (4): 7.
2. Monterey WMD-Terrorism Database Staff (2005). Anthrax Hoaxes: case studies and discussion. *Encyclopedia of Bioterrorism Defense* 29–31.

Unattended Items and Vehicles

BACKGROUND

Baggage and personal items, such as purses, suitcases, backpacks, or shopping bags, can contain dangerous devices or materials. A common terrorist tactic is to leave a dangerous item in a location where it can do damage, concealed in some kind of container. Usually, such an attack uses an object, one that appears ordinary for the setting. For example, terrorist attacks have used common luggage in and around hotels, train platforms, and airports or shopping bags near retail premises. Vehicles can also contain hazards, and given the size of vehicles, the potential quantities of hazardous substances will be larger.

Unknown, unattended, and suspicious objects and vehicles could easily be a conventional or CBRN device such as shown in Figure D.1. Devices constructed to spread CBRN materials can have a wider area of adverse effects than some conventional devices. Suspicious items are normally EOD problems. Determining if they are also CBRN problems can be difficult, depending on local policies.

THE SCENARIO

An important person has died. Their funeral is widely publicized and is taking place in a well-known cathedral in your city. A number of celebrities and several prominent political figures will be attending the funeral service. The time and place of the funeral is widely reported across the media, and it is now well-known on social media that various celebrities are staying at particular hotels.

Your team is the "hazard assessment team" for the event, and you have people from various response disciplines on your team, including EOD specialists. One of your jobs during the

CBRN and HAZMAT Incidents at Major Public Events: Planning and Response, Second Edition. Daniel J. Kaszeta.
© 2023 John Wiley & Sons, Inc. Published 2023 by John Wiley & Sons, Inc.

FIGURE D.1 Suspicious packages form a high percentage of CBRN incidents. *Source: Photo: US Dept of Defense, public domain image.*

event is to serve as the "eyes and ears" of the overall security command structure. You are called to the following situations. Discuss and make recommendations as to a course of action for each of the following. In particular, discuss whether or not the incident has CBRN concerns or is "merely" a conventional EOD issue.

Episode 1

It is the day before the funeral. The Grand Hotel is the temporary residence of two major celebrities who have arrived to attend the funeral the following day. A large suitcase has been reported in front of the hotel. A witness says that they saw someone leave the suitcase and walk away. Discuss your actions.

Episode 2

It is late that same evening, a car is found illegally parked next to the cathedral. The license plates do not match the description of the car. Both are clearly stolen. The car appears to be heavily weighted down in the rear. A pool of liquid is seen under the car.

Episode 3

It is three hours before the start of the funeral. Security sweeps have begun. A large, old suitcase is discovered in the alley beside the cathedral building, without any nametag on it. It appears to be abandoned. A review of CCTV footage shows a person leaving it there during hours of darkness.

DISCUSSION AND TIPS FOR SUCCESS

Standardize your Terminology

You want to have terminology and processes that are flexible and which reflect that every situation is a bit different. Not every unattended item is necessarily a "suspicious" item. The distinction depends very much on the context. A piece of luggage with a nametag on it, in a venue and context where luggage is commonplace, without additional information to make it threatening, can be an issue of concern. But it will not be the highest alert right off the bat. It could easily be resolved by a shout of "hey, who left the bag here" or a look at surveillance footage to see how the bag got there.

The circumstances of discovery, witnesses to the event, whether or not the contents can be seen without difficulty, available threat intelligence, and the overall security posture of the location where it was found all have a direct bearing on where to draw the line between merely unattended and suspicious. For example, a suitcase in a bus station is sometimes inherently less suspicious than the same suitcase placed against the wall of a courthouse.

A bag or object or vehicle that was deposited in a furtive or overtly suspicious manner would require a much higher index of suspicion right away. The same suitcase as above that was dropped by someone who immediately ran away would obviously be more suspicious.

Use Your Assessment Methodologies

All of the situations above are scenarios that call for judicious application of assessment schemes. No injuries have occurred, at least so far, and the incidents seem easily controllable. An assessment scheme should have basic guidance as to how and when to escalate a response to unattended items.

CBRN and EOD Need to Work Together

Any of the scenarios above might be a conventional explosive device or a CBRN device. EOD and CBRN experts will have to pool their expertise in the major event environment in order to adequately assess threats. It will never be easy to figure out a clear boundary between EOD and CBRN. There will be many situations where the two spheres overlap. By definition, an improvised explosive device designed to disperse CBRN materials is both a CBRN and an EOD issue. Equipment for protecting EOD operators against liquid and respiratory hazards now exists, thus making it easier to equip an EOD technician to operate in a possible CBRN environment. In practice, taking an EOD technician and giving him some CBRN skills and equipment is far easier, quicker, and cheaper than trying to take a CBRN technician and turning him into an EOD operator. In the situations above, the EOD expertise is needed to rule out the presence of energetic materials. CBRN expertise will be needed to rule out the presence of CBRN hazards.

Indicators That CBRN Is Involved

Usual practice in many places with suspicious packages and parcels is to treat them as conventional EOD concerns unless there is some indication that CBRN materials are involved. Again, this is a highly subjective area in which it is useful to hold discussions during the planning phases on exactly where you want to draw the line or if you want to draw the line at all. After all, unknown liquids could easily be chemical warfare agent, but could also be liquid explosives or something completely harmless. The liquid in Episode 2 could easily be a hazardous substance. But it could also be engine coolant or motor oil.

WHAT **NOT** TO DO

Use Dogs the Wrong Way

Explosive detection dogs can be a very useful tool in searches. However, I have seen situations where they have been used incorrectly. Once something has been declared as suspicious, do not use a dog to try to "clear" the item. Just because a particular dog doesn't alert on a bag or car does not necessarily mean that this item, which you have ALREADY thought was suspicious, is somehow safe. You might gain additional information if the dog alerts (i.e. a strong possibility of the presence of explosives), but you don't gain any additional information from the dog NOT alerting. Likewise, sending a second dog in and sending the same dog in a second time are pointless tactics. Finally, don't use a dog on loose powder or liquid. There's every chance you can damage the dog's sense of smell if they are exposed to many types of chemicals.

Treat Every Unattended Item as "Suspicious"

There may indeed be states of heightened alert when you might want to treat every single item as suspicious. But years of experience at protected sites have taught me that a two-tier system of "unattended" and "suspicious" is more manageable during most operations.

Fail to Keep Records

Document your responses. Take photographs of unattended and suspicious items. I have known more than one situation where an investigation was hindered because an EOD team disrupted a suspicious item and nobody had bothered to take any photographs of it.

Get "Tunnel Vision"

It is entirely possible that something suspicious has been placed for the purposes of observing your response procedures or providing a distraction to keep security personnel and attention diverted away from some other event. All too often, the unattended bag or vehicle becomes the sole focus of all the security personnel in the area. Interesting things may be happening in the background. Who is watching your reactions? Who's minding the perimeter? Has every CCTV camera now focused on the unattended box? Bear such factors in mind.

Be Wary of Secondary Devices

It is not an unknown terrorist tactic to deliberately use so-called secondary devices to specifically target emergency response personnel. A real or hoax device could be placed in order to gather response assets in a predictable location. Tunnel vision may make you vulnerable to such attacks. Factor the risk of such attacks into your plans and processes.

Suspicious Powders and Crime Scene Issues

BACKGROUND

Incidents involving suspicious powders purporting to be dangerous substances have been a phenomenon since the mid-1990s. The anthrax incidents in 2001 in the USA (the so-called Amerithrax attacks) served as a stark reminder that not every suspicious powder is a hoax. Because many dangerous substances could be sent in powder form and there is an understanding of this among emergency responders that these could be dangerous situations. As such, people with hostile intent understand that they can cause serious disruption to events simply by sending a powder and a message through the postal system.

THE SCENARIO

A large and highly prominent annual shareholder meeting of Corporation X is happening in your city. A number of high net-worth individuals will be present at the convention. Several prominent public figures are speaking at various events related to this meeting. The corporation is making much-publicized announcements of new products. The meeting is being held at a large conference center at a major hotel only a short distance away from the main corporate headquarters complex.

Some individuals and groups are using the meetings to highlight their issues. An environmentalist group, known for publicity stunts, has announced that they will hold daily protests about climate change. Some smaller groups, loosely affiliated with the broader movement, are suspected to be plotting more aggressive disruption tactics. Finally, several corporate executives who are attending the meeting are known to have received anonymous threats.

CBRN and HAZMAT Incidents at Major Public Events: Planning and Response, Second Edition. Daniel J. Kaszeta.
© 2023 John Wiley & Sons, Inc. Published 2023 by John Wiley & Sons, Inc.

FIGURE E.1   Police effecting entry in possible CBRN environments must proceed carefully.
Source: Used with permission from R. Mead.

Episode 1

It is one day before the meeting. The main mailroom at corporate headquarters receives an envelope addressed to the Chief Financial Officer (CFO). In accordance with current practice, this envelope is sent upstairs to the CFO's administrative staff. A relatively junior member of the staff opens the letter. Some light gray powder falls out of the envelope and there is a threatening message printed on a piece of normal paper. Along with threats, the printed letter claims that the powder is anthrax. There are two other employees in the room.

What is the best way for people to react in this situation? If you are called in as a responder, how would you manage the situation?

Episode 2

A group of protestors is outside across the street from the parking garage entrance to the conference center. The group has been generally peaceful but several demonstrators have been arrested after blocking traffic. The police are clearing the pathway for two expensive cars that evidently contain attendees visiting the meeting. One of the protesters pulls a paper bag out of his pocket and throws it at one of the two cars. The bag bursts revealing a rust-colored powder, which is scattered on the hood of the car. The person who threw the bag flees through the protesting crowd and police are unable to find him.

How do you react to this incident?

Episode 3

A major investor (Mr. A) is attending the shareholder's meeting. He is staying at a luxury hotel down the street from the meeting venue. He has been there for nearly a week as he has meeting with Company X. He has been seen receiving various letters and courier envelopes at the hotel's front desk. A courier envelope is given to him on the morning of the shareholder's meeting by the front desk staff. The investor dashes off to the shareholder meeting with the envelope under his arm.

Later, during a lengthy presentation, Mr. A. is sitting in the main auditorium. There are 600 people in the room. He opens the envelope, and it contains a threatening letter, a castor bean, and a quantity of powder. The letter alleges that the powder is ricin, a toxin extracted from castor beans. Mr. A stands up, flees the room, and takes refuge in a nearby restroom and calls for help on his mobile phone. However, the powder and the envelope were left in the auditorium and nearby people may have been exposed to the powder.

What is the best way to manage this situation?

Episode 4

Law enforcement sources have made enquires with the courier company involved in Episode 3. They have identified a suspect who is likely to have sent the envelope containing the castor bean. Investigators have surveilled him and located his place of residence. While the subject is at work, a search warrant has been obtained for his residence.

Explain how you would conduct a search. What sort of evidence are you looking for? What sort of processes, procedures, and equipment will you use? Will it look like Figure E.1?

DISCUSSION AND TIPS FOR SUCCESS

Build Mitigations into Your Plans Beforehand

The single biggest thing you can do to protect people in Episode 1 is to have sound and safe processes for opening mail and parcels. One way to ensure that hazards are mitigated is to have a separate, segregated facility for receiving parcels and mails. Such a facility could even be in a physically separate building, or it could have separate air-handling systems from the main building. Even a fume hood or glovebox to open parcels and envelopes is a useful protective measure. Episode 1 can also be mitigated by some basic awareness training for the office staff.

Assume Someone Is Going to Walk Away

In Episodes 1, 2, and 3 it is quite possible, and indeed probable, that someone is going to walk away from the incident site before proper perimeter controls are set up. From a public health perspective, if it turns out that dangerous substances really were used, it will be necessary to track people down. Make an effort to record the names and a means of contacting them. In Episode 2, it might not be possible to get everyone's name and the demonstrators may not be cooperative. It might be better to have medical staff or firefighter do the recording in that scenario, as they may be seen to be less confrontational than police. Offer to record their first name or even a pseudonym and explain that it is for medical reasons. In extremis, get a group leader's contacts and say "I'll call you if there really is a hazardous substance—then you can reach out to your people."

Enlist the Help of the Public

Episodes 2 and 3 are excellent example of a situation where crowd behavior becomes important. Try to apply some of the knowledge from Chapter 3. Large groups of people will behave in ways that are more helpful to emergency responders if you can find ways of getting the public to feel as if they are part of the emergency response, not just passive objects or recipients of the operation. Particularly in the early stages of response, you may be short of labor. You might be able to enlist a fair bit of public help for routine non-sophisticated tasks such as "can you write down the name of everyone that was at your table" or "can you ask your group if anyone has some video footage of what just happened and let me know in about ten minutes when I come back."

WHAT <u>NOT</u> TO DO

Over-react

In all of these scenarios there is possible threats to life and safety, but there are no immediate threats. Nobody is sick or ill. But in Episodes 2 and 3, responders are greatly outnumbered by members of the public. A heavy-handed or ill-thought response could alienate, annoy, or otherwise get the public to act in ways that may not be helpful to the overall response effort. Episode 2 could turn into a riot and Episode 3 could turn into a panic. Be very careful how you proceed. As mentioned above, an overly confrontational approach with protestors in Episode 2 could be counterproductive.

Destroy Evidence

In all of these scenarios, some type or degree of emergency decontamination is probably a very good thing to do. As raised earlier in the book, decontamination can destroy evidence. However, some effort should be made to preserve some of the evidence. I've been to too many training exercises where all of the useful evidence was flushed down drains. Some concerns could be addressed by very simple standing procedures. The office staff in Episode 1 could easily have been trained to put the offending envelope in a plastic bag. Episode 3 is more problematic, of course, but not unmanageable. It is also worth mentioning that the most useful bits of evidence may not be the actual threat materials but other items, like envelopes.

Industrial Chemical Accidents

BACKGROUND

Modern technological society relies heavily on the use of commercial chemicals. A vast industry exists to produce and transport chemicals. By necessity, much of this production, traffic, and commerce is in urban areas, often near locations where we hold major public events. The presence of large quantities of dangerous materials may be taken for granted or not even noted unless something goes wrong, but the major event planner must take into account the presence of large amounts of industrial materials in and around an event.

While terrorism planning often focuses on chemical and biological terrorism scenarios involving warfare agents, it should be noted that some industrial accidents involving chemicals have been truly horrific catastrophes. Many industrial materials are actually far more dangerous in many respects than most of the so-called warfare agents. Industry, commerce, and transportation can account for large quantities of such materials that dwarf the amount of chemical warfare agents in even large-scale incidents as illustrated in Figure F.1. It is imperative that industrial accident scenarios, where relevant, form part of your emergency planning and your threat basis.

THE SCENARIO

A major one-day sporting event is happening in your city. Over 60,000 tickets have been sold. The vice president is attending as well as the state governor and numerous other political figures. Several major entertainment celebrities are going to provide entertainment before, during, and after the event. Dozens of hotels are fully booked, as are dining venues.

CBRN and HAZMAT Incidents at Major Public Events: Planning and Response, Second Edition. Daniel J. Kaszeta.
© 2023 John Wiley & Sons, Inc. Published 2023 by John Wiley & Sons, Inc.

FIGURE F.1 Chemical factories can provide scope for accidents that affect major events.
Source: Photo: US Chemical Safety Board, public domain image.

The match itself is happening at a large outdoor stadium routinely used for such games. The stadium is on the edge of the city's central business district. It is located near a large interstate highway (600 meters away from the stadium at the closest) and a railway line used for significant freight traffic is 700 meters away from the stadium. On the other side of the railway line is an industrial district with numerous warehouse and light industrial operations.

Note that Episodes 2, 3, and 4 are separate and distinct.

Episode 1

It is six months before the sporting event. Many of the details are unclear, but the event is firmly in the diary and it is expected that some dignitaries will be attending. You have been asked to survey the city for possible industrial and commercial hazardous materials. How do you begin such a task? What are some of the resources you can use?

Episode 2

It is four hours before the event. People are beginning to arrive at the venue and a "tailgate party" scene is starting to evolve.

Your operations center received report of a traffic accident on the highway, about 700 meters away from the event. A semitrailer truck with numerous hazmat placards on it is involved in a serious collision with a car. A number of barrels of liquid are loose on the highway and several of them are reported to be leaking. None of the chemicals have been identified yet, although the hazmat team is going to make an entry soon to try to retrieve the shipping papers.

Describe your responses to safeguard your event. How do you assess if this situation is a potential hazard to your event?

Episode 3

The sporting event starts in two hours. Pre-game entertainment has begun and about 50% of the ticketed public have arrived. Various dignitaries are on their way to the event. A large "tailgate party" scene is underway in the large parking area east of the stadium.

The event has begun. A threatening email has been received by several local news stations and the local FBI field office. The threatening email claims that an explosive device has been planted on a railway car on the rail line near the sporting event. There is, indeed, a large freight rail train halted on the line. There are over one hundred rail cars, many of them with hazmat placards. The operations center is talking to the railway operator to get the full "consist"—the list of cars.

How do you evaluate the risk to your event? Do you consider cancellation?

Episode 4

Several reports have been received in your operations enter and have been confirmed by your fire department liaison officer. A large fire has broken out in a warehouse containing agricultural chemicals. This warehouse is about 3000 meters away from the stadium. Winds are not blowing in the direction to take the smoke directly to the stadium, but only a slight change in windspeed and direction could take the smoke to the event. A number of people have been taken to hospital with a variety of signs and symptoms of chemical exposure. The flames coming from the warehouse are strangely colored. A major hazmat incident has been declared.

What is your reaction? What measures should you consider?

DISCUSSION AND TIPS FOR SUCCESS

Tap into Existing Information and Expertise

Hazardous chemicals can be found in every major city and are transported on highways and railroads every day. The fire departments and various regulatory agencies have to deal with response to or oversight of such chemicals on a routine basis. These chemicals are not invisible to the people and agencies who work with them, regulate their transport or use, or respond to emergencies involving them.

Your advanced planning processes, as discussed in Episode 1, should investigate the current industrial and transportation situation and make maximum use of both existing documentation, liaison with existing response and regulatory agencies, and outreach with industry. Your planning should give you at least a basic understanding of the existing hazards within your area of interest.

Make Sure you have an Assessment Scheme

Episodes 2, 3, and 4 are good examples of situations where an assessment team such as a JHAT, could be very useful. As the manager of safety and security for the event, you will likely want your own eyes and ears at or near the scene of the problem to be able to provide some assessment as to how it may affect your own decisions. In a complex urban environment, there will likely be a number of different emergency response assets on the scene of these problems illustrated in the episodes, as well as news media coverage and social media posts.

Event Cancellation and Evacuation Decisions

The time to figure out how to make decisions on event cancellation and evacuation is NOT during the emergency. You will want to have at least thought about it ahead of time. Having a matrix, template, or checklist and having some firm guidelines about who actually makes the decisions is helpful, and these are not the sort of things you write down on the spot. You want them ahead of time.

WHAT <u>NOT</u> TO DO

Botch an Evacuation

While there might indeed be a need to evacuate people from the event venue, one must consider the overall situation. Does dumping 60,000 people out onto the streets, many of them in their own vehicles, actually make them safer? Or does it clog up traffic and make everyone less safe? On the other hand, Episodes 2 and 4 are things that are likely to be prominently seen on social media. Will people evacuate anyway? The most important principle is to try not to push people from a place of relative safety into a more hazardous environment. Many variables will have to be examined. There may be combinations of circumstances whereby keeping the event running and keeping people in the stadium might be the better option, since you have developed significant security, safety, and medical support at the stadium.

Overreact to the Physics of the Situation

If you use the GEDAPER process, you estimate the potential harm. Is there actually a mechanism by which the materials can travel to the event and harm people in Episode 2, the likelihood of leaking single drums of liquid, even highly dangerous liquid, being a hazard at 700 meters distance, even with the wind blowing toward the stadium is minimal. Realistically, you're talking about gallons or tens of gallons of liquid product on a highway nearly half a mile away. The mechanics of the situation just don't support a decision that this liquid is a threat to the event or justify an evacuation. Not only that, with the major highway blocked, an evacuation might complicate response efforts to deal with the incident or protect people who are closer to the environment.

If, however, the truck catches on fire and there is a lot of toxic smoke (as with Episode 4), the situation is entirely different. This is a hazard that can travel quickly, and given the unknown and likely highly mixed nature of the hazards, this is a dangerous situation. Three thousand meters sounds like a lot, but it will depend on a lot of variables.

Mystery Smells and Illnesses

BACKGROUND

Not every CBRN or HAZMAT incident is going to manifest itself with a big bang, a cloud of gas, or special effects out of a Hollywood film. Terrorist CBRN devices that function in a loud, obvious, and scary way may actually been less hazardous to the public than a slow, pernicious release of material. This is because a crowd's natural inclination will be to flee rather than stay put to absorb toxic materials. Some situations will unfold slowly over a period of time. Some incidents, ranging from nearly benign to truly catastrophic, may manifest themselves at first as suspicious odors or clusters of unexplained illnesses. Also, a terrorist device may leak or not function as designed, leaving suspect smells as perhaps the only indicator of an incident.

There are some historical trends of note. One is that there is some history of mass poisonings and mass illnesses related to unidentified smells. A number of such incidents occurred in Afghanistan, principally in girls' schools in the 2013–2017 period.[1] These incidents seem to bulk out historical reports of CBRN incidents, but follow-up on the incidents was poor. Given limited resources by responders, no perpetrators or toxic substances were isolated or identified. However, similar incidents have occurred even in the USA where sophisticated instrumentation was available. In addition, nuisance attacks by "stink bombs" or other offensive smelling material have a long history of use by people seeking to disrupt operations. Note the "Furries Convention" incident mentioned early in the book as an example.

As a general rule, strange smells reported by more than one person are worthy of investigation. Sudden onset of illnesses in clusters are also something that need investigating.

CBRN and HAZMAT Incidents at Major Public Events: Planning and Response, Second Edition. Daniel J. Kaszeta.
© 2023 John Wiley & Sons, Inc. Published 2023 by John Wiley & Sons, Inc.

FIGURE G.1 Photoionization detectors can help locate sources of chemical vapors. *Source: Photo: US Dept. of Defense, public domain image.*

THE SCENARIO

It is a major political party convention. The convention itself is being held in a large sporting arena in your city. Delegates and temporary party offices are housed in a number of large and medium-sized hotels. Although the sporting arena is the main venue, subsidiary events are being held at a number of other venues around the city. A large number of congressmen, senators, and state governors, as well as other officials from the political party are attending the event.

In this scenario, you are members of a multidisciplinary assessment team. You have CBRN, HAZMAT, EOD, and medical expertise on your team. For each of the following episodes, discuss whether or not you would deploy the assessment team, whether the incident is a CBRN situation, and what actions the assessment team should take. Discuss what additional support or responses would be needed.

Episode 1

It is one week before the convention is due to begin. You've just stood up your assessment team because security perimeters and searches are beginning. You receive a phone call from the emergency operations center. EMS personnel are treating several people in a temporary press office set up in a hotel conference room across the street from the main venue. An unidentified but offensive odor is reported and several people are reporting serious nausea.

Episode 2

It is the morning of the first day of the convention. A group of 500 protesters is holding a demonstration, having given prior notice, south of the security perimeter in a large plaza. The demonstration gets a bit agitated and some members of the protest end up throwing several apparent water balloons onto a media camera crew. The camera crew leaves the protest and enters the convention site through a dedicated security checkpoint for accredited media personnel.

The security personnel at the checkpoint have called for assistance. The equipment and clothing of camera crew, which had been at the demonstration, is emitting an unusual odor. One of the security staff and one of the camera crew are reporting eye irritation. Another member of the camera crew, who had been hit by a water balloon, is reporting skin irritation.

Episode 3

It is the evening on the first day of the conference. Venue security and medics are summoned to one of the VIP suites, where one of the delegations is hosting a number of officials and party supporters. Eleven people are suffering from headaches and gastrointestinal distress. One of the 11 victims is suffering from respiratory distress as well. There are 15 other people in the VIP suite who are apparently unaffected. Large amounts of food and drink have been consumed.

Episode 4

It is day two of the convention. Security and medical staff have been summoned to respond to an emergency situation. A strong noxious odor is emanating from one of the men's restrooms off of the convention floor. Several people are vomiting outside the restroom. Several others report respiratory distress and irritated eyes. The situation inside the restroom is unknown, but several people are reported to still be in there. Fire personnel in SCBA are about to do an emergency entry to try to rescue any persons left in there.

DISCUSSION AND TIPS FOR SUCCESS

Photoionization Detectors are Useful

The photoionization detector (PID) is good for discovering the presence of unknown dangerous gases and vapors. Although they cannot detect everything, PIDs are a very good generic detector for thousands of volatile organic compounds. Although a PID cannot identify specific compounds, it can help locate where gases and vapors are located and provide relative indications of concentration. In a room where a suspicious smell is reported, a PID can help you determine the presence of such a volatile organic chemical and, by seeing where the measurements are stronger or weaker, possibly isolate or identify the source. Figure G.1 illustrates use of a PID.

"Sick Building" Syndrome and "Worried Well" are Real Phenomena

There have been enough historical incidences of situations where people got sick in the same room or building, but without a clear cause, that the phrase "sick building syndrome"[2] was coined to cover such instances. Likewise, as discussed elsewhere in this book, the "worried well" phenomenon is a possibility. There can be people physically unaffected by something, who see or hear of signs and symptoms and then begin to think that they are affected as well. The two phenomena are likely related and you can have terribly confusing situations where you have people who appear to be sick but no easily discernible cause. Many instances of "sick building syndrome" have been declared to by psychogenic—in effect "worried well." However, psychogenic illness can be highly stressing and people will need help. Nor is it good policy to just assume "it's all in their head."

From a standpoint of management of security and safety of the overall event, the best thing you can do in many of these cases is to assess if there is acute danger to life, and if there is not, isolate the situation and let specialists follow-on responders deal with it. Get the victims off-site to clinics or hospitals for an assessment if needed and get an environmental health assessment of the location by specialists, freeing up your emergency responders for other assignments.

Understand Airflow in and Around Your Event Site

Part of managing this type of problem is understanding if the problem is contained or not. Problems of this type that happen in rooms, halls, or other enclosed spaces will require some attention to the HVAC situation. Knowing how the air-handling systems work, where air flows in a particular part of the building, and how the systems are "zoned" may be of critical importance to containing a problem and isolating an affected area.

Fresh Paint and Fresh Carpets

The old joke is that Queen Elizabeth thinks every room in Britain smells like fresh paint. However, planned visits by VIPs generate a lot of activity. Rooms do get painted afresh. New carpet gets put down. New furniture, which is full of synthetic materials, gets unwrapped. I once discovered at a trade show that a particular chemical warfare detector alerted on a type of vapor coming from still-drying carpet glue. Bear such things in mind when investigating an incident.

WHAT <u>NOT</u> TO DO

Search for Zebras Instead of Horses

Just because you are geared up for CBRN response does not mean that every incident that you are called into is a CBRN incident. I have seen CBRN teams that really want to do their job, so they inadvertently or subconsciously push hard to interpret incidents as being CBRN incidents. Episode 3 is a classic example. Yes, it COULD be some sort of dreadful terrorist attack. But given the parameters of the incident, it is much more likely that the incident is one of food poisoning. And if it is the food that is responsible, deliberate tampering of food is far less common than "normal" food poisoning. If you try to force-fit a diagnosis of CBRN terrorism into a crate of bad oysters, you will create much fuss and diversion.

Assume That Odors are at Nose Level

A lot of gases and vapors with strong odors will be heavier than air. A few might be lighter than air. If you are using detection methods to find a possible gas or vapor in one of these scenarios, you may need to get down to floor level and up to ceiling level inside a room in order to have a good chance of finding the culprit.

Field an Assessment Team Without Medical Expertise

Some sort of medical knowledge is very helpful in the types of scenarios highlighted above. Sick people are a form of CBRN detection. However, it takes competent medically-trained personnel to assess victims and process the information available at the incident scene. Having an assessment team without medical expertise deprives you of useful detection and identification methods.

Ignore the Possibility of Diversions

With reports of sick people or odd odors, a team could easily get distracted and lose sight of other events and occurrences. Do not preclude the possibility that a simple "stink bomb" nuisance device could be a ruse. An incident triggering a robust emergency response at one location could be used to draw attention and absorb resources from other areas.

REFERENCES

1. Hamdard, F. Afghan girls' school feared hit by poison gas. *Reuters*, 21 April 2013.
2. Joshi, S.M. (2008). The sick building syndrome. *Indian Journal of Occupational and Environmental Medicine* 12 (2): 61–64.

Chemical Warfare Agent Terrorism on Public Transport

BACKGROUND

Terrorist attacks with actual chemical warfare agents have been, blessedly, rare. However, the Aum cult in Japan demonstrated that such attacks are not impossible. State-directed attacks using nerve agents have been used to intimidate or silence enemies of the Russian and North Korean states. Furthermore, the use of actual chemical warfare weapons and military-style delivery systems have occurred in the Syrian civil war. It is difficult for a terrorist group to conduct terrorism with chemical warfare agents, but it is not impossible. This scenario explores several different ways in which such attacks might transpire.

THE SCENARIO

It is the Olympic Games and teams from all over the world are converging on your city for the majority of the events. Approximately 80% of the events are occurring in the greater metropolitan area around your city, with the other events in outlying locations. Please note, that for purposes of training exercises, this could be either the Summer or Winter Olympic games.

Your organization has had literally years to plan for these events, and a significant amount of planning has gone into overall safety and security efforts, with major efforts devoted to CBRN and HAZMAT concerns.

An extremist religious group, with a perverse and poorly understood ideology, has been plotting for some time to disrupt the games. In particular, they have a grievance against Country X and hope to attack that country's delegation to the games. The group operates in

CBRN and HAZMAT Incidents at Major Public Events: Planning and Response, Second Edition. Daniel J. Kaszeta.
© 2023 John Wiley & Sons, Inc. Published 2023 by John Wiley & Sons, Inc.

FIGURE H.1 Equipment such as the Smiths HAZMATID can rapidly identify toxic chemical.
Source: Used with permission from R. Mead.

secrecy. Although it has been monitored by law enforcement, few members of the group are known and little information about their activities is available.

Unknown to any security or emergency services personnel, the group has dabbled in chemical warfare agents. After extensive expenditure of funds and recruitment of several skilled chemists, the group has developed a limited capability to make single batches (less than one liter) of impure Tabun, which is a nerve agent.

Episode 1

Intelligence services have intercepted email communications indicating that this extremist group may be attempting to place "a chemical device" onto a bus or train. Additional details are not forthcoming, but this threat has been vouched for as "credible" by intelligence analysts.

Based on such scant information, are there ways in which you could increase security and preparedness measures?

Episode 2

Without the knowledge of police or security services, the extremist group has managed to place some of its members into jobs with a bus charter company that is providing transportation services to the games. The group has been using one of its members, a driver, to conduct reconnaissance against possible targets. The group is gauging the feasibility of leaving an improvised chemical device, containing about half a liter of impure Tabun and a very small explosive charge, on a bus. They are, of course, worried about whether the device would be discovered. They decide to conduct a test run, with an innocuous rucksack, to see if unidentified bags are discovered.

One member of the group, employed as a bus driver, is assigned to drive a team from their quarters to a practice facility. He carries a rucksack, filled with an unmarked bottle full of liquid, some clothing items, and a portable radio. The driver, at the end of his shift, removes his nametag (required by security policy) and hides the rucksack under one of the seats on the bus.

The rucksack is discovered an hour later by a routine security check. How do you react? Given the intelligence from Episode 1, would you react in a different way?

Episode 3

The extremist group uses their access to buses to replace a first-aid kit in one of the overhead compartments with a small Tabun device with a digital timer. Their driver leaves the device on the bus at shift change. Later, the bus is shuttling a press delegation from Country X to one of the event venues. While removing a bag from the overhead compartment, the fake first-aid kit is accidentally knocked to the floor of the bus. It is knocked open, and it is obvious that it is not a first-aid kit. The security staff evacuate the bus and isolate it. The situation is reported.

Describe your response procedures in this situation. Start with what kind of questions should be asked by the call center who is taking the initial report.

Episode 4

In this episode, the device in Episode 3 is not discovered and it functions as intended. The device explodes while the bus is driving, with its passenger load of journalists and camera crew. One person, a camera man, is immediately seriously injured, both by the explosion and by being covered in Tabun liquid. A number of other people on the bus are experiencing varying degrees of exposure to nerve agents. By the time responders are on the scene, a number of people have fled the bus. Some are displaying signs and symptoms and others are not. It is not known how many people are left on the bus.

DISCUSSION AND TIPS FOR SUCCESS

This scenario lays out a set of escalating episodes around a general theme. Episodes 1 and 2 are far more commonplace than Episodes 3 and 4. Most of the responses that you might consider for them would likely fall in the category of general antiterrorism and security processes rather than anything specifically tied to CBRN terrorism.

Detecting Hostile Reconnaissance

An astute investigator might look at Episode 2 and deduce the presence of a hostile reconnaissance effort. In a tabletop exercise, it would easily come up. But an exercise already has people thinking in the right mindset. Would it really work out that way in reality? After all, if bags were all screened anyway, would anyone care about a lone rucksack in that scenario. But on the other hand, this represents an opportunity to detect reconnaissance, tie this reconnaissance to a specific employee, and possibly disrupt a plot.

The Bus Provides Containment and Flexibility

The fact that the suspicious devices and bags are on a bus, which is a moveable object, presents some possible opportunities. The bus itself provides a degree of containment. Depending on an assessment of the situation, there may be some value in moving the bus in Episodes 2 and 3 to a location, which is safer and away from any hostile surveillance. It is a tricky judgment call in Episode 3 and will rely on an assessment of the device's condition. In Episode 4, moving the bus once victims are off may be the safest way to exploit the evidence, depending on where the bus is located.

There Is Ample Evidence

Forensic exploitation of the bag, first-aid kit box, devices, and any residue thereof could mean the difference between discovering and prosecuting the perpetrators and them getting away with the deed. Episodes 3 and 4 are prime CBRN forensic situations. In Episode 4, some of the evidence will be on the body and clothing of at least one victim.

Dermal Exposure Takes Time

Tabun is less volatile than Sarin, and it absorbs through the skin. But it can take hours for signs and symptoms to appear after dermal exposure to nerve agents. In Episode 3, it is possible that someone has got some level of dermal exposure to the Tabun. This is a classic case where medical surveillance and follow-up comes into play. You may have taken witness statements and released the passengers while someone else is working on the device. The discovery that it is Tabun may be rather late in the process. The person who knocked the first-aid kit over might only start to get ill two hours later.

WHAT NOT TO DO

Use a Dog Incorrectly

Every discovered bag or box in this scenario is already suspicious. There might be some value in use of an explosive detection dog in Episode 2 to rule out the presence of explosives. But in Episode 3 there's not much point as the device is clearly present and suspicious, and a dog and handler might be endangered.

Await an Exact Identification of the Agent

Tabun exposure presents just like any other nerve agent. The fact that it is Tabun and not Sarin, the pesticide parathion, or some other nerve agent makes little difference in the early stages of this response. Expending a lot of effort to try to narrow down the type of nerve agent does not really change how you treat patients and help people. Get on with helping people and let the exact identification, which may need sophisticated instrumentation or laboratory analysis catch up with you. It will not actually change any decision on medical care. Some identification instruments might be able to rapidly provide information at the scene, as pictured in Figure H.1, but do not delay a rescue.

Delay Rescuing People Off the Bus

I can certainly envisage some response protocols whereby a careful site assessment is done, a decontamination line is established, Level A PPE is carefully donned a safe distance away from the bus, and then, 25 minutes after arrival on site, a responder final boards the bus, only to be confronted with four dead bodies. But even a very high exposure to Tabun is medically treatable. Episode 4 is clearly a situation where an emergency entry, as soon as possible, in turn-out gear and SCBA, to make an emergency rescue attempt, followed by an emergency decontamination washdown, is tactically and morally justified.

Large-scale Chemical Terrorism

BACKGROUND

Chemical terrorism on a large scale has been rare in history. Individual poisonings are as old as human civilization itself, but large acts of violence involving chemical warfare agents or other toxic industrial chemicals are blessedly uncommon. Given the potential for both injury and disruption, such events, while low probability overall, must play some role in effective CBRN security planning.

Large quantities of the higher-grade military chemical warfare agents are likely off the agenda for nearly all terrorist groups without significant state support. But the Aum Shin Rikyo incidents in Japan in the mid-1990s saw a non-state group amass liters of the nerve agent Sarin. It was only their lack of an efficient dissemination mechanism that kept the number of dead in the low double-digit figures.

THE SCENARIO

A major international summit meeting involving 15 different finance and/or treasury ministers is being held in your city. This meeting has been planned for over a year. The summit itself is being held at a large hotel complex with an attached conference in the middle of the downtown. The summit will last three days. It is winter, and the weather is cold.

As has become customary during such events, various special interest groups are mounting protests and demonstrations before and during the event in areas near the summit. Much security planning has gone into monitoring the demonstrations and protests. However, a small violent extremist group has decided to target the event with a major terrorist attack. This group

CBRN and HAZMAT Incidents at Major Public Events: Planning and Response, Second Edition. Daniel J. Kaszeta.
© 2023 John Wiley & Sons, Inc. Published 2023 by John Wiley & Sons, Inc.

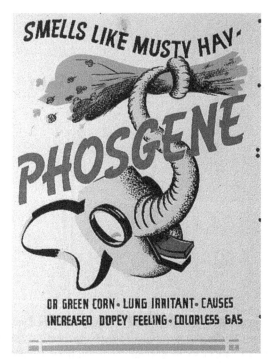

FIGURE I.1 Phosgene was infamous as a chemical weapon.
Source: Photo: US Government. Public Domain.

obtained toxic industrial gases in cylinders from several industrial sources in Mexico and managed to move them over the US border disguised as other less harmful materials.

Please consider these episodes as separate and distinct incidents.

Episode 1

The extremist group has been plotting an attack on the summit for a long time. The group set up a small front company and has rented an office on the second floor of an office building 200 meters outside the security perimeter of the summit meeting. Over the previous month, the group has secretly moved dozens of industrial gas cylinders containing chlorine gas into the office. They have fitted manifolds and hoses so that the gas will discharge out of the window of the office when the valves are opened.

The group had been careful to move their gas cylinders only late at night and had attempted to disguise the gas cylinders by rolling them in carpet. However, an astute security supervisor, while reviewing the previous evening's CCTV footage from the parking garage, noticed that two people were struggling with the weight of a roll of carpet and that a bit of a gas cylinder was visible in the end of the carpet roll. Considering this highly suspicious, he reported it to the building management. Reviewing further archived video footage revealed similar activities. As the rented office is nominally a firm of financial advisors, the presence of gas cylinders is highly suspicious. The police are contacted. It is one day before the security perimeters go into effect.

The next morning, two police detectives and the building manager arrive to investigate. The discover that the office has been re-keyed in violation of the lease agreement.

Based on this information, how do you proceed?

Episode 2

The extremist group has spent months infiltrating one of the protest groups, which is an environmental campaigning group known for disruptive stunts. One of the campaign group's disruption tactics, adopted from protests in other cities, is to get an old van or camper-van, block a major road, and let the air out of the tires to hinder moving the vehicle. The two extremists (Mr. X and Mr. Y) who have infiltrated the group have offered up an old camper-van for the disruption campaign.

Mr. X and Mr. Y. have put the toxic gas phosgene into several RV-size cooking gas canisters, which do not look out of place in a camper van. They've rigged the van so that hoses from both gas cylinders lead out to the undercarriage of the camper van.

On the second day of the summit meeting, Mr. X and Mr. Y are accompanied by four protestors in the camper van. The group parks the van in the middle of an intersection two blocks away from the summit's security perimeter. The four protestors let the air out of the tires, climb up a ladder to the roof of the camper van, and unfurl protest banners. However, Mr. X and Mr. Y. open the valves of the gas canisters and flee into the crowd of pedestrians along the side of the street. Moments later, a strong smell of newly mown hay is present. Police at the scene report a hissing sound from the camper van.

How do you react?

Episode 3

As in Episode 1, the extremists have been stockpiling gas cylinders in the office. However, their efforts are not discovered and disrupted.

It is the second day of the summit meeting. The attackers have been monitoring wind direction for many hours. Finally, the wind is blowing in the direction of the summit. The attacker opens the office window slightly, places the hoses out the gap in the window, dons a protective mask, and opens the valves on all of the cylinders. A high concentration of chlorine starts to build up on the street. At first, nothing is visible but people on the street start to small chlorine. Eventually, the gas cloud becomes visible and a yellow/green mist starts to drift toward the event. The first people affected are a group of demonstrators at street level. They start to flee in several directions as respiratory irritation and eye discomfort increase. Some, in the erroneous belief that police have used tear gas on them, start attacking the police. The security operations center is receiving dozens of reports, and CCTV footage clearly shows a yellowish mist approaching the building.

Discuss what actions you should take in this scenario.

DISCUSSION AND TIPS FOR SUCCESS

Not Every Casualty will be Immediate

The injuries caused by toxic chemicals often do not appear right away. Episode 2 involves the chemical phosgene (see Figure I.1), which is particularly notorious for the fact that the serious effects of phosgene exposure do not appear for many hours. You are going to have to assume that anyone who smelled the phosgene may, in fact, be a victim. This raises the interesting conundrum of a mass casualty incident where few, if any, of the victims feel unwell. Some will want to go home claiming nothing is wrong with them. Episode 3 is a bit less deceptive, in that chlorine provides immediate irritation. But even with chlorine, there are case histories of delayed onset of symptoms and mild symptoms getting far worse over time.

Not Every Chemical Weapon Is Like a Military Munition

The chemical weapon systems in this scenario are not artillery shells or bombs. These attacks harken back to the oldest chemical warfare attacks in the earlier stages of World War 1, when industrial gas cylinders were dragged to the front trenches. This scenario is a reminder that chemical weapons do not necessarily have to look like military munitions.

Understanding Airflow

You will be able to react with more precision and less danger to life and health if you have an understanding of how the environment around the summit site behaves, both in theory and in actual practice on the day of the incident. In Episode 2, some of your responses may depend on wind speed and direction. In Episode 3, the wind is in the direction of the secured area. In that case an understanding of how the gas will behave in and around your event site will be very useful.

WHAT NOT TO DO

Evacuate the Building into the Problem

Chlorine and phosgene are both heavier than air. The attendees of the summit are in a relatively modern building. It is highly possible that the air intakes for the building's ventilation system are on the roof, a fair distance above ground level. This is always something you should check during your planning. In all of these situations, the potential or actual harm is highest outside of the building. The building itself is a barrier to hazards in the air outside. If you are worried about exposure to toxic substances in the air, one of the worst things you can do in Episodes 2 and 3 is to evacuate the summit meeting into the open air near the release point. A "shelter in place" tactic is much better. Keeping them in the building, moving them higher in the building, and turning off air intakes are all far better than emptying the occupants of the building out into the open air.

Decontaminate When Unnecessary

The role of personnel decontamination in this scenario is minimal. In Episode 1, a forensic entry of the office will need some provision for emergency decontamination, as investigators do not know what sort of situation they will be entering or what hazards are present. The exposure to phosgene gas in Episode 2 and the chlorine gas in Episode 3 may require little or no decontamination, depending on individual exposure. Such decisions will require some decision-making and identification of the hazards, but the scenario uses commonplace well-known industrial hazards.

Decontaminate with Cold Water on a Cold Day

This scenario occurs in cold weather. Flushing people down with near freezing water on a cold day may kill some of the people you are trying to help. Yet I have seen this in the plan at some events. The legal aspects are troublesome. If someone wasn't going to die and you do something to them and then they die, a lawyer or a jury may call it manslaughter or negligent homicide. Build warm or at least tepid water into the plan.

Attacks with Biological Warfare Agents

BACKGROUND

Biological warfare agents have a checkered record as wartime weapons. Aside from the Japanese attacking the Chinese in the Second World War, there are few incidents of biological weapons being used in warfare. There have been some specific directed uses of biological warfare agents in acts of terrorism, such as a 1980s attempt to influence a local election in Oregon and various failed efforts by the Aum cult in Japan. The actual track record of biological weapons is checkered, due to the fact that results are highly dependent on environmental variables that can change quickly, such as sunlight and wind. However, due to the potential of mass casualties, major event planning does need to consider them as part of a threat spectrum.

THE SCENARIO

A very large annual music festival is being planned. Over 100,000 attendees are expected over the course of 5 days. Figure J.1 illustrates a typical festival of this type. This festival is well-known and a number of celebrities will be both playing at and attending the festival. The festival is being held in a rural area on a farm. A large percentage of the attendees are camping at the festival.

However, the international situation is heating up, and there is a crisis brewing in the Middle East. Intelligence agencies have issued general warnings about heightened concerns about terrorism, but no specific warnings have been applied to the music festival. Consider these three episodes as separate, distinct exercises.

CBRN and HAZMAT Incidents at Major Public Events: Planning and Response, Second Edition. Daniel J. Kaszeta.
© 2023 John Wiley & Sons, Inc. Published 2023 by John Wiley & Sons, Inc.

FIGURE J.1 Music festivals can be demanding operational environments for emergency responders.
Source: Paul Holloway/Wikimedia Commons/CC BY-SA 2.0.

Episode 1

It is one week before the event begins. Final preparations are underway for the event. The administrative offices for the music festival are actually located four miles away in the nearest town. A letter is delivered through the postal service to the administrative offices. No particular screening processes are in place for routine correspondence or deliveries. The envelope contains a threatening letter and some powder.

Emergency response protocols are followed. However, laboratory tests reveal that the powder in the letter contained a small number of actual viable anthrax spores. Based on sampling, it appears that the size of the particles is too large to constitute the highest category of threat. The small number of people who worked in the office are on prophylactic antibiotics. Nobody is ill. Both the conference organizers and the authorities have not made any public statement about the incident.

Do you go public with this news? Do you try to keep it under wraps? What happens if word leaks out? Do you consider canceling the event?

Episode 2

An extremist group has targeted the event. During the extensive run-up to the music festival, two members of the extremist group manage to get temporary jobs with two different food vendors at the event. A biologist who is a member of the group manages to culture the correct bacteria to produce Staphylococcal enterotoxin B (SEB). This toxin is most infamous for food poisoning and causes serious gastroenteritis. On day 2, the two group members who infiltrated the event secretly put the toxin in a number of different foods. Onset is rapid and people start becoming ill four hours after eating. Fatalities are rare with SEB, but the sanitary facilities and medical services for the event are quickly becoming burdened as several hundred people are seriously affected, and hundreds more have milder but still distressing symptoms.

This would have been passed off as an unfortunate but not atypical food poisoning event. However, the extremist group has posted a video on various social media sites taking credit for the outbreak.

How do you react?

Episode 3

It is 3 am on day 4. Most of the attendees are finally asleep in their tents, while a few are partying on into the night. The extremist group has managed to infiltrate the event, rather easily, by buying a ticket and turning up with camping gear. They leave a small explosive device in a rucksack. The explosive charge was small, only an ounce of TNT, but it was loud and injured several people nearby with flying debris. The explosion disseminated a quantity of powder and dozens of people nearby were covered in a very fine light-colored powder.

The explosion immediately caused a major self-evacuation of people and it took some time to evacuate the area, establish control of the scene, and identify what had actually happened. Six hours after the explosion, various social media posts purporting to be affiliated with the extremist group claim that the powder was anthrax spores. You do not know yet if this is true. Given the explosive means of dissemination, some of the spores, if they existed, would have been destroyed. This could be a hoax claim meant to provoke panic.

Discuss your response measures.

DISCUSSION AND TIPS FOR SUCCESS

Understand Crowd Behavior

With a crowd as large as this, crowd dynamics will be very important. Different types of crowds will behave in different manners. The same crowd may behave differently at different times of day or under different conditions. Weather, intoxicating substances, and the demographics of the audience may all have an effect on how a crowd will behave in a stressful situation. This is well understood in concert events management. A crowd of older people listening to Handel and Bach may behave in ways differently from a younger crowd listening to drill music. From an emergency plans perspective, liaise with those actually producing the event, who will advise you based on their experiences.

Leverage the Public

As with Scenario E, you will likely have more jobs to do than people to do them. This scenario is a classic example where you will want to enlist people from the event to do very routine tasks to free up labor for more specialized responders. Even if it is "here, do you mind wearing this vest and directing people to move that way toward the exit" or "hey, mind asking your group to stick together until we can get a handle on this" are small things but not inconsequential.

Understand the Topography in Outdoor Events

An outdoor event in a rural area poses different issues than an urban venue. Outdoor events are far more vulnerable to weather. Wet weather may make part of the venue muddy or basically unpleasant, both formally and informally pushing attendees into other areas. Also, large outdoor events tend to have fewer official and unofficial barriers between zones and areas. Control measures designed to corral attendees to await decontamination, for example, may be a moot point if the outer perimeter is basically a row of bicycle-rack fencing and people can flee in every direction. Decontamination efforts may result in water pooling in areas where you do not want it.

Understand the Diversity of the Event

In an event this large, it is statistically inevitable that you will have subsets of the crowd with special needs. You will have people with both seen and unseen disabilities. You will have people with medical conditions that make them more vulnerable to biological threats. There will be people addled by drink or drugs. In hot weather, there will be people who are already on the brink of heat injuries. There will be people who have little or no ability in the local language. Emergency planning will need to accommodate the full scope of human needs based on the actual audience and staff composition.

WHAT <u>NOT</u> TO DO

Invest Biological Weapons with Mythical Properties

Yes, anthrax is a potentially lethal bacterial disease. However, it is not contagious from person to person, nor do anthrax spores defy the laws of physics. In Episode 1, the problem, although potentially serious, is contained in a small area. That situation does not have scope for the anthrax spores to travel to the actual event. There's basically zero risk of these anthrax spores having a direct effect on the major event, being confined to a back office some distance away. However, the incident also shows that the extremist group has the technical capacity to deploy actual anthrax spores.

Likewise, the food-borne threat in Episode 2, while serious, will be constrained to people who ate food from the associated vendors. It will not magically affect people who did not eat the contaminated food. However, the full scope may not be understood at the time. Episode 3 is somewhat more problematic, but an ounce of TNT is going to do a lot to destroy anthrax spores. Depending on the exact circumstances, the TNT may cause more harm.

Try Too Hard to Corral People

If you are dealing with tens of thousands of people in a rural environment with basically tape or cheap fencing as the perimeter, you are not going to have enough people to enforce any sort of containment strategy, particularly if people are in fear of their lives.

Get Hung Up on the "Worst-Case" Scenarios

If you have an event with 100,000 people, getting fixated on how you are going to decontaminate all of them. You won't and you can't. A more reasoned approach is to try to figure out how to identify the hundreds, likely not thousands, who had the highest degree of exposure to the potentially dangerous substances.

Pandemic Illness

BACKGROUND

The contract to write this second edition was concluded shortly before the onset of the worldwide COVID-19 pandemic, which saw massive cancellation of public events for most of 2020 and well into 2021. No book on this subject could be considered to be a full discussion without consideration of how pandemics can affect major public events. It is still too early to learn the full lessons from the COVID-19 pandemic, but we can discuss ways to think about pandemic disease.

Pandemic disease outbreaks are natural. Despite conspiracy theories, the COVID-19 pandemic was not manmade.[1] Pandemics and epidemics do happen as naturally occurring biological events. Global travel means that an outbreak from one part of the world can travel easily to another part of the world. Many types of major events serve quite effectively to collect people from other parts of the world and concentrated them at close quarters. Therefore, event planners and emergency managers may have to make difficult decisions about mitigation measures, postponements, or cancellations.

Major events can become "super spreader" events, causing widespread damage to health, the economy, and reputations. For example, a racing event in Cheltenham, in England, was held at the very beginning of the pandemic in the United Kingdom. Difficult decisions were avoided and the event went ahead, possibly contributing to the rise of the disease in the United Kingdom.[2]

THE SCENARIO

It is the Olympics. Winter or Summer, take your pick, as the details do not matter. The games have been planned for 7 years. Significant planning across all aspects of logistics, security, and

CBRN and HAZMAT Incidents at Major Public Events: Planning and Response, Second Edition. Daniel J. Kaszeta.
© 2023 John Wiley & Sons, Inc. Published 2023 by John Wiley & Sons, Inc.

emergency planning have been underway for years. Many thousands of people will be descending on the city and the venues. Athletes from all over the world will be attending.

In this scenario, the pandemic disease in question is a variant of avian flu. It has emerged among poultry farmers in Southeast Asia. The incubation period ranges from 1 to 10 days, with an average of 5 days. This variant of the flu causes serious illness, with many people requiring hospitalization. The lethality rate is relatively high for an influenza virus, and this particular outbreak is seeing a number of fatalities.

Unlike other scenarios in this book, this scenario does not unfold in episodes. Rather, it has three variants. The various versions of this scenario unfold in different ways, and each way should be discussed.

Episode 1

It is 3 months before the start of the Olympics. This new strain of flew is emerging in Southeast Asia. Thousands are hospitalized and hundreds have died in the country where this disease has emerged. Dozens are hospitalized and a handful have died in other countries in the region. Small numbers have cropped up elsewhere around the world. Global and regional health authorities are concerned. The situation is generally similar to the situation how the COVID-19 outbreak was in early January 2020.

You are obviously under much pressure to continue the Olympics as so much has been invested, both economically and politically. However, from a safety and security standpoint, you must consider possible future developments. No cases have turned up in your city yet, but you must get in front of this potential problem.

Describe additional measures you will undertake and considerations you will examine at this point. At what point do you consider delaying or canceling the event? Are there mitigation measures you can plan and execute to keep the event safe? What logical criteria would you examine in order to make decisions about the event?

Episode 2

In this variant, it is now 10 days before the opening ceremony for the games. Athletes and staff are beginning to arrive in the city for the games. However, several people are now ill in hospital in your city with the new variant of bird flu. One of the sick people is a technician who has been installing equipment in several of the Olympic venues. Another is a sports journalist from Country X, which has reported a large number of cases. He had been screened upon arrival, but is now sick. Contact tracing shows that both of these people have been in contact with numerous individuals in and around the "Olympic Village."

How do you react? Do you cancel the event or continue? If you continue, what additional steps are you going to take?

Episode 3 (Hardest)

The games have been underway for 3 days. The emergency operations center receives several telephone calls late at night. Six athletes, all from the same country, are in hospital with serious flu symptoms. An hour later, 11 more athletes, housed in the same building, are showing symptoms and are being kept in quarantine in the building. A floor of one of the athlete's housing blocks is now sealed off under quarantine. However, given the incubation period and infectivity of this illness, it is highly likely that these are not going to be the only cases of the disease. Statistically, it is highly likely that the illness is now spreading at large in the community.

Now that the games are underway and the city is full of tens of thousands (or more) visitors, how are you going to react? Do you try to engage in containment measures, or is it too late? Do you cancel the events? But doesn't that just risk sending thousands of people home to spread the disease? There are numerous issues that you can discuss here.

DISCUSSION AND TIPS FOR SUCCESS

Work with Public Health Professionals

This scenario is more of a public health and epidemiology scenario than a CBRN/HAZMAT scenario. Although this scenario is clearly in the "B" category of CBRN, this is an area where you will likely be in a supporting role and will need to let the specialists take the lead. One lesson of COVID-19 has been that not enough attention was paid at the right points to the virologists and epidemiologists.

Have a Contingency Plan for Contact Tracing

Contact tracing is a core component of infectious disease response. One of the greatest lessons learned from COVID-19 is that those cities and countries which were prepared to do robust contact tracing early on in the pandemic suffered to a lesser degree. Arguably, the COVID-19 pandemic got a lot worse early on because contact tracing failed.

A rigorous contact tracing program will have a reasonable likelihood of mitigating harm in Episode 1 and may reduce harm in Episode 2. It might even, if you are both lucky and good, contain the problem in Episode 3 if implemented very quickly and thoroughly.

Don't Let your Responders Get Sick

Managing a major event, even if everyone stays healthy, requires a lot of labor. Losing people due to illness, even if it is just a relatively mild illness that keeps people at home instead of at work, can reduce your ability to maintain a safe major event. Use PPE early on and get your staff to self-isolate to protect the bulk of the responder workforce.

Reduce or Mitigate "Superspreader" Events

It is difficult to know what a "superspreader event" may be ahead of time. Mostly, such events are diagnosed after the fact. But we now know that things like ventilation, distancing, and placing events outdoors where possible may reduce the possibility of an event becoming a superspreader event.

Can You Help the World?

If the outbreak can be contained in the city where the games are being held, is it better to keep the affected people there and look after them in a setting where you already have infrastructure to do so? By canceling the events and sending people home, are you taking a local outbreak and turning it into a global one?

Know the Limits of Detection and Diagnosis Technologies

With infectious diseases, there are a range of diagnostic and detection techniques ranging from very good to abject quackery such as COVID-19 testing like shown in Figure K.1. With a new or rarely encountered disease, there may be little detection other than empirical observation of signs and symptoms. Or there may be laboratory-based techniques that take hours or days for results to come back. If you get an "old threat"—something already understood like H5N1 or COVID-19 coming back in a new outbreak, you will be in luck and have much diagnostic firepower that you can deploy. With a new pathogen, you may be completely out of luck.

At a planning stage, wherein your plans must by necessity be generic, do not make too many assumptions about what type of testing and detection capability you will have available to you.

Your Decontamination Plans May Be Helpful

Depending on the exact nature of the pandemic threat, it is quite possible that any advanced planning in the area of CBRN/HAZMAT decontamination. While, eventually, it turned out that COVID-19 was not primarily driven by surface contact, some other pathogens may spread this way. For example, equipment and ambulances may need decontamination.

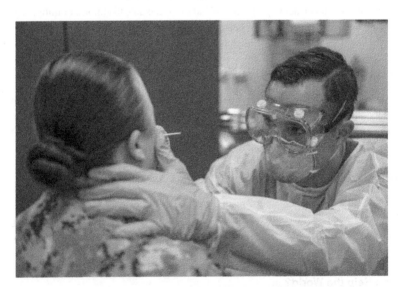

FIGURE K.1 The COVID-19 pandemic has yielded many lessons.
Source: U.S. Department of Defense.

WHAT <u>NOT</u> TO DO

Delay

Delay is your enemy. Complacency and optimism will incur a penalty. Act quickly to incorporate pandemic concerns into your planning. It is better to ratchet back on precautionary actions if it turns out you do not need them than to behind the curve.

Ignore Scientific Advice

One of the lessons we have seen from the COVID-19 pandemic is that ignoring scientific advice only to start following it a week or so later leads to problems. We now have great access to the world's finest minds and we should use that access.

Assume It Is Like COVID-19

Just like the generals who went into battle fighting the last war only to receive a great shock, we should not assume that the next pandemic is going to be like the last one. Even within the COVID-19 pandemic, different variants of concern acted differently.

REFERENCES

1. Kaszeta D. The coronavirus is not a man-made biological weapon. *The Washington Post*, 27 April 2020.
2. Humphries, W. Fears that Cheltenham festival may have spread Coronavirus throughout country, *The Times*, 3 April 2020.

SCENARIO L

Radiological Attacks

BACKGROUND

Attacks with radioactive substances are rare but not unknown. Covert emplacement of radioactive materials could injure people. Larger devices, so-called radiological dispersal devices (RDDs) or "dirty bombs" could be used to spread radioactive material over an area. The practicalities of health physics make it difficult, but not impossible for an RDD to pose an immediate threat of death or serious illness. What might constitute a dangerous amount of cobalt or cesium in one lump would be highly diluted if you spread that lump around a city block. However, such devices could easily serve as an area denial weapon, making it unsafe from both occupational and environmental health perspectives to for people to live, work, or transit through an area.

It is very important to understand that it is well established that much of public have a fear radiation which disproportionate to the scientifically validated risk. Ionizing radiation is invisible to our senses and has been subject to much misinformation in popular culture. Routinely and repeatedly, many people will look react to a particular situation involving radioactivity and overestimate the risk to themselves or family members. This is significant in the context of public events. A fear does not have to be rational or proportionate to cause problems. As the fear of radiation is well known, it is highly possible that this fear itself will be a motivator to use radiation sources to disrupt an event.

For purposes of protecting a major public event, it is nonsensical to disregard RDDs and covert emplacement of radiation sources due to their low likelihood of actually causing lethality. The point is that normal people fear radioactivity. Exposure to radiation or contamination by radioactive materials can force closure of venues and events.

CBRN and HAZMAT Incidents at Major Public Events: Planning and Response, Second Edition. Daniel J. Kaszeta.
© 2023 John Wiley & Sons, Inc. Published 2023 by John Wiley & Sons, Inc.

FIGURE L.1 Radiation detection and monitoring is critical in this scenario.
Source: Photo: US Dept. of Defense, public domain image.

THE SCENARIO

It is the annual State Fair, an event which goes on for weeks every summer. Many thousands of people attend the Fair every day during the season. It has been held at the same fairground every year. A wide variety of entertainment acts are booked, ensuring the presence of both major and minor celebrities. State officials, from the governor on down, attend the fair as a matter of tradition.

A small extremist group has political grievances against the state governor and the state government. Part of their beliefs is that the recent elections were stolen and that the current state government are not legitimate. Several members of the group are currently jailed or on bail awaiting trial for various acts of intimidation against state officials. One member of the group runs a construction business and has access to a highly radioactive industrial radiography source containing Cobalt 60.

Episode 1

A member of the extremist group is under periodic police surveillance. A grand jury is looking into his activities. Unexpectedly, this person is admitted to hospital with fairly severe radiation sickness, which is a highly unusual occurrence. A search of his home and workplace yields no radiation sources. However, significant materials about the state fair, including plans and diagrams are discovered in the person's home.

Episode 2

Part of the state fair is the annual rodeo. Habitually, the governor goes to the championship. The arena has a "Governor's" box. Several days before the championship event, the extremist group manages to use duct tape to hide a very small pellet of the Cobalt 60 under the floor of the governor's box, with the intent of harming the governor. Based on the events in Episode 1, some of the security staff at the state fair now have radiation detection devices, as does one member of the governor's protective detail from the state police. Several hours before the governor is due to arrive on site, one of the staff members reports very high radiation readings near the governor's box. A second staff member confirms this.

How would you go about verifying and localizing the source of the radiation? If you isolate the problem, can you proceed with the event? What would you do to further investigate the situation and improve safety at the event? Do you communicate with the public? If so, what do you say?

Episode 3

Upset by their failure in Episode 2, the extremist group develops a radiological dispersal device. They have ground their remaining Cobalt 60 to fine dust. They have mixed this dust with flour and placed it in a bag with a small explosive charge consisting of a small amount of commercial dynamite. They have fitted this with a simple digital clock and a detonator.

Although some efforts are being made to check bags and parcels coming into the fair, the group cleverly disguises the RDD and smuggles it onto the fairgrounds. Due to the danger in carrying the device, the various group members trade off carrying the bag containing the device to lower their individual exposure. A group member places the device in a trash container and leaves the fairgrounds.

After 1 hour, the device explodes, shattering the trash container and spreading the radioactive Cobalt 60 dust around a several thousand square foot area, although the extent of the contamination is not known until after the event. The explosion did not occur in a high traffic area of the fair. However, one person was immediately killed by the explosion and five people were injured seriously by flying debris. A number of other people were contaminated with dust. Because of the use of radiation monitors based on Episode 1, the emergency response effort almost immediately deduces that there is a radiological incident.

Explain how you proceed in this scenario.

DISCUSSION AND TIPS FOR SUCCESS

Incorporate Radiation Detection into Your Plans

All aspects of this scenario benefit greatly from the use of radiation detection instruments similar to those depicted in Figure L.1. Mitigating all of the hazards to health and safety in the various episodes above will require radiation detection. The vast majority of the time, the simplest possible detection methods will be good enough to alert you that there is some sort of problem. In an event such as this state fair scenario, you are better off with a proliferation of detection systems as a first layer of defense, backed up with a second-tier response with isotope identification.

Use Health Physics Expertise

The stolen cobalt used in this situation represents the higher end of the spectrum in possible industrial sources. This is a case where civilians or responders standing near a fragment of the pellet will start to gain significant radiation doses.

The professional discipline of health physics can provide a whole body of knowledge to help you understand what is safe and unsafe with regards to all of the radiological aspects of this scenario. For example, a competent health physicist can do some calculations in scenario two and estimate radiation exposure, thus letting you know whether or not anyone was exposed to levels of radiation that are unsafe. Based on such information, it may be possible to continue the event after the source has been contained.

One technique that may be useful for responder safety and regulatory compliance is the establishment of a "turnback dose." Teams could be instructed to retreat from the hot zone if their absorbed dose on a dosimetry instrument reaches a certain point. For example, if the occupational exposure limit is 5 rem (actual limits will depend on your regulatory environment), a "turn-back" dose of 1 rem or 1.5 rem could be enacted, thus providing a degree of assurance that the 5 rem limit will not be breached.

Remember That Exposure and Contamination are Not the Same Thing

In Episode 2, the radiation exposure is being caused by a single pellet of material. Safely remove this pellet and the hazard is gone. The gamma rays emitted by the pellet will not make other things radioactive. Episode 3 is entirely different, though. It poses a threat of widespread contamination as small particles of radioactive dust are scattered around. The particles could be spread by wind, carried away on the hair and clothing of people, or be washed away by rain.

WHAT NOT TO DO

Shut Down the Event for Every Radiation Alert

If you field a lot of radiation detectors, you will likely have positive radiation detections for benign reasons. Some small percentage of the public will have radioisotopes in their bodies for medical reasons. If a large state fair has tens of thousands (or more) visitors in a day, you will occasionally find medical isotopes. Some construction materials may include naturally occurring minerals that are more radioactive than the average background. You will need to have a second tier to your detection architecture that uses isotopic identification to weed out medical isotopes. There may even be building materials or geology that contains natural radiation sources. A general survey prior to opening up to the public would discover the background. However, the Cobalt 60 in this scenario will stand out like a sore thumb compared to natural or medical sources.

Ignore Radiation Safety Rules

You wouldn't let firefighters enter the building without SCBA and turnout gear, so you shouldn't let responders enter the dirty bomb scene without dosimetry and a way of recording dose. If you do not monitor accumulated dose of your responders, you may ruin them for future work at incidents. Monitor the dose closely and rotate teams to make sure people do not reach their exposure limits.

Focus on the Dirty and Not the Bomb

At the end of the day, a "dirty bomb" is more significant as a bomb than the fact that that it is radioactive. In this scenario, the likely lethality will come from blast and fragmentation from the explosion. Someone MIGHT get ill from radiation sickness if not decontaminated quickly. Furthermore, the evidence from the conventional components of the bomb and its container will be at least as useful as evidence as any radioactive fragments

Major Nuclear Power Plant Incident

BACKGROUND

In overall terms, commercial nuclear power is a safe method for generating electricity. Hundreds of commercial nuclear power plants have been operating around the world for decades. Modern reactors have significant safety features built into their designs, technology, and operating processes.

However, it is also clear that two of the largest CBRN incidents in modern history involved nuclear power plants. The 1985 Chernobyl and 2011 Fukushima accidents both demonstrated that, given the right circumstances, serious release of radioactive materials can occur, causing possible threats to public safety and health.

It is very important to note that nuclear power plant incidents are different in many ways from nuclear explosions and radiological dispersal devices. The types of radioactive material that can be released from a nuclear reactor are quite variable in their characteristic but are, at the end of the day, well understood by science. Some of the materials released in a catastrophic scenario might be gases, such as radioactive iodine or xenon. Others will be solids, such as cesium. A full treatment of the mechanics and mechanisms of radiation released from such accidents is beyond the scope of this scenario, but references are available on this subject.[1]

It is also important to note that direct fatalities from such incidents have been rare and generally limited to emergency responders on site at the nuclear incident, as opposed to civilians outside the site perimeter.

CBRN and HAZMAT Incidents at Major Public Events: Planning and Response, Second Edition. Daniel J. Kaszeta.
© 2023 John Wiley & Sons, Inc. Published 2023 by John Wiley & Sons, Inc.

FIGURE M.1 Nuclear power plant accidents are rare but have been significant in the past.
Source: Photo: US Nuclear Regulatory Commission, public domain image.

THE SCENARIO

A major international sporting event is being held in your region. Many thousands of visitors, but domestic and foreign have descended on the region. Hotels, restaurants, and public transport are at full capacity. This event has been expected for a long time, so that there is significant planning and additional resources in place.

A large nuclear powerplant, designed and built in the 1970s, but with a (so far) impeccable safety record. Figure M.1 illustrates an example of such a power plant. It is located north of your region. Like the vast majority of nuclear power plants, it is located on a water source for cooling. In this instance, it is on a major river, which is generally upstream from your sporting events. The power plant is located 25 miles (40 km) north of the major stadium used in the event.

Episode 1

Your command center receives an urgent notification of a serious security incident at the nuclear power plant involving a breach of the perimeter by armed intruders and a large truck. Additional security personnel and police are responding to the site. Weather reports show that winds are generally in the direction, at a speed of approximately 10 miles per hour. You have no further information at this time.

What do you do at this time? What questions are important for you to ask?

Episode 2

It is ten minutes after the previous episode. Both media reports and direct reports to your operations center report that the large truck used by terrorists to ram through the perimeter at the power plant was a vehicle-borne explosive device. The explosives have now detonated. It is estimated that at least ten tons of conventional explosives, possibly more, were in the truck. There is serious damage to the power plant, but the full extent of the damage is not available yet.

Gates at the stadium for the sporting event are supposed to start in two hours. Are you going to cancel the event or continue it? What measures are you going to take?

Episode 3

Several local politicians are holding a live television press conference. All of the major television networks are present, as are a number of radio stations and newspapers. The mayor of the city was asked a question about the dose rate of radiation near the nuclear plant. Referring to a handwritten note, he makes a serious mistake and uses the wrong unit to describe the level of radiation, greatly over-stating the risk. This causes some distress. Hundreds if not thousands of people are making calculations and looking things up online at home based on the wrong unit of measurements and wrong numbers.

How do you manage such a situation?

Episode 4

(At this point, you have likely decided either to continue the event or cancel the event. Regardless of which decision you made in Episodes 2 and 3, carry on based on the decision you made earlier on.)

It is announced that some radioactive iodine and radioactive xenon have both escaped from the power plant. These are radioactive gases and thus can spread with the wind. Winds are definitely in the direction to carry such gases from the power plant site to the city where your event in being hosted.

It is likely no longer a question of event cancellation, but now of emergency management and population protection. Explain what actions you are going to take.

DISCUSSION AND TIPS FOR SUCCESS

Leverage Existing Planning

The general risks associated with nuclear power plants are well understood. Modern countries have numerous safety and security measures built in to their facilities and procedures. Regulations and laws generally make it a requirement for nuclear power plants and the areas around them to have well-developed emergency plans. If you are planning an event that is anywhere near a nuclear power plant, then the planning processes should incorporate such plans. Likewise, if such an incident were to happen during a major event, the existing plans should be a framework on which you stand.

Understand Weather and Hazard Modeling

Assuming that your major event is not literally in the shadow of a nuclear plant, it will likely be weather that is the mechanism that moves many of the most significant hazards from the power plant to your location. Part of your decision process will involve technical expertise from nuclear engineers, part of it will rely upon scientific instruments monitoring the levels of radiation, and part of it will rely on understanding the weather.

In the USA, there are technical capabilities developed by the federal government, which are specifically designed to perform modeling of radiological releases in the event of nuclear power accidents. They use the latest in atmospheric science, meteorological data, and super-computers. Even if you are outside the USA, this capability, the National Atmospheric Release Advisory Capability, may be able to assist.[2]

Understand Units of Measurement and Do Not Make Mathematical Errors

Radiation and health physics are saturated with units of measurement (Gray, rads, Roentgen, Curies, Sieverts, Rem, etc.) and various prefixes to denominate them (milli-, micro-, centi-, etc.). Not only that, there are different things being measured, such as radioactivity, dose, and dose rate. Furthermore, there are different schemes (SI versus non-SI) that do not always easily convert from one to another. It's very easy to get confused on the units and measurements. Units usually come with prefixes, and if you are not used to them, it is easy to confuse milli (1/1000) with micro (1/1000000).

Understand Human Psychology

Nuclear radiation is an area where there are both irrational phobias and significant levels of ignorance among the public. Hazards that are invisible, tasteless, and odorless are harder for the human mind to process. The Covid-19 pandemic has taught us that some people will clearly deny threats that are invisible. But we also know that other people will make incorrect assumptions and assume a level of threat higher than what the actual situation warrants. Radiation incidents have been known to provoke a level of anxiety disproportionate to the actual risk.

Use the Buffer of Time and Distance to Your Advantage

This scenario actually gives you a considerable safety margin. There is 25 miles and easily at least 2–3 hours of time between the incident at the power station and the possible onset of radiation at the major event venue and surrounding area. Not only that, there are numerous other variables at play, and it is actually possible that the hazards may not reach you at all. Rushed or even panicked reactions early on in the incident can result in events and situations that are more hazardous to the public safety and health than doing nothing.

WHAT **NOT** TO DO

Hide Information from the Public

One of the lessons learned from the Fukushima event is that relatively cheap radiation detection technology can be combined with internet access to allow individuals to quickly and easily provide live radiation measurement online for the entire world to see. It is quite likely that, in the event of a major nuclear power plant accident, a lot of radiation monitoring (of a wide range of accuracy and quality) will go online in areas near the incident, regardless of whatever official monitoring is being done. It would be pointless to fight such efforts. Rather, you want to give context so that people understand the numbers.

Force Unnecessary Evacuations

In a situation like this, it is very likely that many evacuations could put people into less safe situations. Evacuations increase the risk that transportation routes needed for other purposes. Sheltering in place may be more useful as a protective action than mass evacuations, which will tie up logistics and may put a lot of people in positions where their potential for radiation exposure is higher. An evacuation could put a lot of people onto the roads at a point where they need better protection than a car or bus provides.

Get Tunnel Vision and Focus Only on the Incident

In this situation, it is easy to worry about the nuclear plant and what might happen there. You should not let your guard down. You need to focus on the security at and around your venues, which are 25 miles away from the problem at the power plant. It is entirely possible that the power plant situation is part of a bigger campaign of attacks and it could draw your attention away from other, more direct threats. Is it possible that the nuclear power plant incident can draw specialty resources away from your event plans? Certainly. It is also possible that such an incident could be completely unrelated to your event and that you could trigger unnecessary responses.

Let Your Imagination Give Radiation Mythical Properties

Nuclear incidents and radiation follow basic principles of physics. However, not everyone understands these principles. Radioactive materials released from a nuclear power plant do not assume mythical properties. Gases such as iodine and xenon do not travel faster than wind speeds. Solids like cesium need a mechanism to be able to travel, such as smoke from a fire or water flowing downstream. Rivers will not flow upstream and gases will not move very far upwind. Yet I have been in tabletop exercises where otherwise educated people thought that a cloud of fallout could travel 30 miles in half an hour on a calm day. Even on a bad day, the laws of physics apply.

REFERENCES

1. Fukushima Daiichi Accident. *World Nuclear Association.* May 2022.https://www.world-nuclear.org/information-library/safety-and-security/safety-of-plants/fukushima-daiichi-accident.aspx
2. National Atmospheric Release Advisory Center. *Lawrence Livermore National Laboratory.* https://narac.llnl.gov/, accessed 20 July 2022.

Example Threat Basis and Planning Threshold

This appendix is a *notional* threat basis and planning threshold for an imaginary event, and it is for illustrative purposes only. This is not meant to replicate any actual circumstance or apply to any actual event.

EVENT AND VENUE DESCRIPTION

International summit meeting, 1–3 June 2024, Smith Convention Center, Anytown, USA. The convention center is in a typical city center area in an average North American city setting. There is an interstate highway that passes within 500 m of the main venue. There is a railroad line that passes within 500 m of the main venue as well. No commercial waterways are present. No industrial facilities are located within 1000 m of the main event, although one of the delegation hotels is 500 m from a warehouse that ships and receives moderate quantities of industrial chemicals. The major venue has an inner perimeter to keep out pedestrians and an outer perimeter, to keep out vehicles.

PLANNING ASSUMPTIONS

1. For planning purposes, this planning document assumes a maximum occupancy of 7000 people at the convention center (main venue) and a maximum occupancy of 1000 people at each of 5 designated delegation hotels. We assume a maximum occupancy of 300 at the media center, across the street from the main venue.

2. The established conventional physical security measures will operate largely as intended.

CBRN and HAZMAT Incidents at Major Public Events: Planning and Response, Second Edition. Daniel J. Kaszeta.
© 2023 John Wiley & Sons, Inc. Published 2023 by John Wiley & Sons, Inc.

3. Site surveys find no unusual vulnerabilities above and beyond those normally found that cannot be remediated before the event.

4. Authorities have suspended shipment of hazardous substances by rail on the railway line during the event. We assume that this measure will be effectively enforced. Transportation authorities advised that a similar measure on the interstate would be unwise, as it would divert HAZMAT traffic onto local streets, which may be closer to the venue and/or pose more possibilities for transportation accidents.

5. The committee did not rank any of the scenarios as more likely than another.

The CBRN/HAZMAT committee developed our threat basis in a meeting on 1 February 20XX, which considered the following factors:

1. Security measures in place for the event

2. Intelligence provided by local, state, and national agencies

3. HAZMAT in the local area, as determined by examining available statistics, consultation with regulatory agencies and survey of the area.

4. Terrorist scenarios suggested by the committee members, who discussed each scenario for credibility and relevance to the event.

5. National planning scenarios. Some scenarios, namely a wide area aerosol anthrax attack affecting the entire region and an improvised nuclear device detonation were deemed by the committee to be beyond the scope of tactical planning for this event.

THREAT BASIS

The CBRN/Hazmat Threat Basis for this plan is composed of the following scenarios, agreed upon by the CBRN/HAZMAT committee. The following scenarios will become the basis for preparedness and exercises for this summit event. It was felt by the committee that these 11 scenarios cover a wide range of materials and locations sufficient to form a useful planning basis.

Commercial/Industrial HAZMAT incidents, either by accident or deliberate release:

1. Chlorine tanker or an anhydrous ammonia tanker in an overturn accident on the highway half a mile from the main event.

2. Gasoline tanker truck in a fire adjacent to the outer perimeter.

3. Structure fire at the industrial warehouse near delegate hotel #5. A wide variety of unknown chemicals, none in quantities larger than 55 gallon drum are involved.

Chemical Terrorism:

4. Small quantity (<1 kg) of nerve agent in a parcel device detonated upon inspection at the main venue shipping and receiving department loading dock, or at the front desk of any of the delegate hotels.

5. Small quantity (100 g) of nerve agent is smuggled into the main venue and thrown onto the stage in a small water balloon-type device.

6. Medium quantity (<10 kg) of a choking agent or blood agent in an improvised device not larger than a backpack or suitcase, dispersed by spray in the main entrance hall of the main venue at the security checkpoint.

7. Large quantity (100 kg) of blister or nerve agent dispersed explosively in a car-bomb type device at the outer perimeter.

Biological Terrorism:

8. Contamination of food at the main catering kitchen at the main venue with a toxin or food-borne pathogen causing widespread illness and some deaths.

9. Letter containing a small quantity of anthrax spores is opened in one of the rooms in the media center or one of the delegate hotels.

10. Outbreak of a pandemic respiratory illness in the city one week before the event.

Radiological Terrorism

11. A cobalt-60 (4 curies) industrial radiography source is hidden in the venue in an area that will expose many people to harmful gamma radiation. The size of device was based on radiation sources available for theft or diversion in the city.

12. An explosive device containing a cesium 137 medical source (10 curie) is detonated in a large parcel in the area between the outer and inner perimeter. The size of device was based on radiation sources available for theft or diversion in the city.

PLANNING THRESHOLD

Based on the parameters of the event and planning scenarios above, the committee developed the planning thresholds for the event. Modeling and simulations were used to estimate potential dispersion scenarios that might be relevant for the time of day and month of the year, given historical weather data for the month of June.

MAIN VENUE

Casualty Estimate:
Prompt fatalities: 100
Serious injuries (defined as requiring transport to definitive care): 500
Minor injuries (requiring some treatment at scene or follow-up): 1500
Worried well (no actual injuries, but requiring reassurance) 5000
Decontamination Requirements:
Non-ambulatory: 200
Ambulatory: 3000
Special needs ambulatory: 200
Responder decon: 200

HOTEL VENUES (EACH)

Casualty Estimate:
   Prompt fatalities: 25
   Serious injuries (defined as requiring transport to definitive care): 150
   Minor injuries (requiring some treatment at scene or follow-up): 400
   Worried well (no actual injuries, but requiring reassurance) 450
Decontamination Requirements:
   Non-ambulatory: 50
   Ambulatory: 500
   Special needs ambulatory: 30
   Responder decon: 100

Template for a CBRN/HAZMAT Site Survey

INTRODUCTION

No single template can accommodate every combination or permutation of possibilities for a major event. Venues can range from a golf course to a football stadium, with every conceivable type of area and structure in between. The example here is a very minimal template; an elaborate survey can contain far more information.

A guiding principle that should underpin the preparation of the CBRN/HAZMAT survey is the following question: "What would responders and incident commanders want to know if they had to respond to a CBRN/HAZMAT incident at this venue?" If you approach the survey from the standpoint of thoroughly answering that one question, you will end up with a product that has merit, even if it looks nothing at all like the template described below.

Write the survey in the simplest possible terms. Assume that the reader is not from the local area and is not familiar with the site at all, even if it a famous venue. Remember that when someone reads the survey, they may be in a very stressful situation. Short, declarative sentences using simple terms are the best. Assume the reader is a bit of an idiot. It is better to insult the reader's intelligence a bit than to have an on-scene commander say: "What the hell does that mean?"

Sometimes some precise information may not be available at the time the survey is prepared, such as exact event schedules. When exact information is unavailable, state so. Provide estimates when you can, but always make sure that you state clearly that it is an estimate.

Because a site survey document such as this one can contain a lot of information that would be of use to a terrorist, consider it to be a confidential document once it is completed.

CBRN and HAZMAT Incidents at Major Public Events: Planning and Response, Second Edition. Daniel J. Kaszeta.
© 2023 John Wiley & Sons, Inc. Published 2023 by John Wiley & Sons, Inc.

GENERAL INFORMATION

Site Name and Location

This may seem obvious, but if there are 120 sites on the master list, then you need to be specific. Include a map or diagram showing the site's location. Remember, the responder may be a military commander or a firefighter from a different city who is not familiar with the area.

When did you conduct the survey and who did you speak with?

Again, this seems obvious, but some of the readers will want to know how current the information is and how thorough the survey effort was.

Events to Occur at the Venue

What is going to happen here? When is it going to happen? Include a full schedule of events, to include preparation efforts. Event start times are important, but so are arrival times of dignitaries, security screening start times, and related scheduled events. In some major events, this information can be substantial, as events could go on for days or even weeks, such as at an Olympic venue.

Number and Nature of Participants

Who is attending this event? How many people?

Number of Staff at the Venue

Maximum/peak number of people expected on site:

    Are dignitaries/VIPs expected to attend? Provide details, to include any estimate on the presence of security details.

Contact Information

Who is responsible for what and how do we find them on the day of the event? Include names, contact numbers, and email addresses. This is useful for responders in many intermediate scenarios or for someone to do a last-minute update of the survey.

STRUCTURAL AND ENGINEERING FEATURES

General Description of the Site

The preparer should provide a good description of the site and include floor plans and diagrams.

Description of Building Mechanical Systems

How does the HVAC system work?

    An overview of the functioning of the HVAC system is important, in case adjustments need to be made after an incident.

    Where are the air intakes?

    Provide the location of the air intakes. Note their level relative to the ground. If possible, include a sketch or diagram.

How does one operate the HVAC system?

Where are the controls to the system? Is it possible to conduct an emergency shutdown? How would this be accomplished in an emergency scenario?

Where are the water, electrical, and gas supplies?

Location and Description of Known HAZMAT

Are there hazardous substances routinely used or stored in the venue? What about adjacent or nearby facilities? Take a bit of a tour around the neighborhood and see what type of commercial, industrial, and transportation infrastructure exists around the venue. Take particular note of highways, railroads, and commercial waterways, as these are often routes of travel for commercial HAZMAT shipments. It may be possible to speak with local regulators or government transportation agencies to get an idea of how much HAZMAT is routinely shipped in the vicinity of the venue.

Location of Any Potential "Safe Refuge" or "Shelter-in-Place" Locations

Bear in mind that most likely threats are heavier than air, so shelter-in-place locations are better if they are not in low-lying areas.

EMERGENCY MANAGEMENT INFORMATION

History of Incidents

Have major safety and security incidents occurred at the venue in the past? Has the venue ever had to do an emergency evacuation? How well did it go?

Existing Emergency Plans

What plans, procedures, and equipment are already in place for fire, medical emergencies, bomb threats, and other types of emergencies?

Known, Suspected, or Assessed Vulnerabilities to CBRN Attack

Given your "threat basis," how would you attack the facility if you were the terrorist?

SUGGESTED COURSE OF ACTION FOR KEY SCENARIOS

Evacuation Routes

What are the existing evacuation routes from the building? Are they adequate or sufficient for CBRN/HAZMAT scenarios?

Routes and Locations

Suggestions for emergency access/egress routes, staging of response assets, and locations for mass decontamination. Consider water supply for decontamination.

Example Task Lists, Capability Survey, and Capacity Survey

As described in Chapter 5, a methodical approach to analyzing one's capability and capacity is preferred. This appendix provides a few basic examples of the planning tools referred to in that chapter. For purposes of this appendix, we will analyze a fire department engine company that may be tasked to support an incident scene. This appendix is completely notional for discussion purpose and is not meant to represent any actual fire department. The planning scenario in question is one of the chemical attack scenarios from Part V.

SAMPLE MISSION STATEMENT

"Our mission at XYZ Engine Company is to respond to a CBRN/HAZMAT incident site with two fire engines and a minimum 8 firefighters, assist any emergency rescue that can safely be done, and establish decontamination (emergency, mass, and responder) as soon as possible. Once HAZMAT responders arrive, we will support them as required."

EXAMPLE COLLECTIVE TASK LIST

1. Respond quickly and safely to an incident scene
2. Conduct emergency rescue, if requested by incident commander, in accordance with SOP
3. Establish emergency decon within 5 minutes of arrival
4. Provide water and personnel to support mass decon

CBRN and HAZMAT Incidents at Major Public Events: Planning and Response, Second Edition. Daniel J. Kaszeta.
© 2023 John Wiley & Sons, Inc. Published 2023 by John Wiley & Sons, Inc.

5. Provide water and personnel to HAZMAT team or other specialists to support responder decon

6. Maintain assigned equipment in working order

7. Operate safely at a chemical or radiological incident scene

Example Individual Task List (Partial)

8. Wear assigned PPE (turnout gear and SCBA) correctly

9. Drive fire apparatus safely

10. Operate fire apparatus in accordance with set procedures

11. Perform emergency decontamination

12. Perform mass decontamination

13. etc.

SUMMARY OF CAPABILITY SURVEY

Equipment

"The engine company generally has the correct equipment for this mission. Some consideration should be given to providing handheld monitoring equipment to support the 'emergency rescue' mission task. The engine company cannot perform mass decon without equipment provided by other units and team. However, their job is to support those units with water and labor, so this is not considered a shortfall in capability by management."

Skills/Training

"While the basic level of skills and training of company members appears adequate, a majority of company members have not had decontamination refresher training in at least one year. The survey team also feels that training for the 'emergency rescue' task is inadequate."

Plans and Procedures

"Adequate plans and procedures are in place for the decontamination tasks. However, the 'emergency rescue' task is ill-defined in present organizational procedures and guidelines."

Summary of Capacity Survey

Equipment

"Sufficient equipment is available for conducting the assigned missions. However, supply of replacement SCBA cylinders is a limiting factor in the endurance of the engine company's personnel."

Personnel

"The engine company has sufficient personnel for the mission. However, there are periods when sick leave and vacation leave reduce manning to a level where mission accomplishment may be borderline."

Time

"Emergency decontamination can be established within 5 minutes of arrival on site, provided a fire hydrant is available. If equipment is available from other units, the engine company is capable of supplying water and personnel to a mass decon operation within 5 minutes of arriving on site. Also, the team is capable of supporting the HAZMAT team with responder decon within 5 minutes. These figures were all validated in a recent departmental HAZMAT exercise."

Other Assessments of Capacity

"Based on a recent training exercise, the team is capable of providing emergency decon for 20 victims. It is possible that capacity may exceed this figure, but it may require a change of SCBA bottles in a clean area. The ability to perform 'emergency rescue' remains unverified and untested."

Synchronization Matrix—Simplified Example

This is a small, notional synchronization matrix. As a planning tool, such a matrix can be as small or as advanced as needed.

Note: For the purposes of this example, the scenario is a small, improvised chemical device detonated at a sporting event.

CBRN and HAZMAT Incidents at Major Public Events: Planning and Response, Second Edition. Daniel J. Kaszeta.
© 2023 John Wiley & Sons, Inc. Published 2023 by John Wiley & Sons, Inc.

Time or event	Before incident	Incident	Incident +1 minute	I+5	I+10	I+15
Response element						
Event security personnel	Normal operations		Security staff begin evacuating portion of stadium.	Evacuation on-going	Evacuation on-going	Evacuation on-going
Event staff (other)	Normal operations		Event is suspended. Event medical team alerted, liaises with ambulance crews.	First aid station reports over 100 ambulatory victims.	Event medical staff overwhelmed and no longer capable of support	
Incident command (IC) team	Not activated			Battalion chief arrives, assumes incident command. Starts GEDAPER process	Safety officer appointed, initial ICS roles assigned	Asst. Chief arrives on scene, assumes IC.
Operations center	Normal operations	Receives alert.	Notifications being made. CCTV of scene pulled up.	Notifications continue	Monitors ongoing developments	Essential additional personnel summoned for operations center duty
JHAT	1 team near main venue, 2d team 10 minutes away	Team moves on foot to assess incident	Observes incident from distance with binoculars, reports likely nerve agent.	Establishes contact with IC, conveys preliminary assessment		JHAT provides an assessment to IC
Fire department	Normal operating posture	Receives alert from op center	"HAZMAT Box" dispatched to site	First due engine and ladder teams arrive on site. Emergency decon set up.	Additional units arrive. Emergency decon established.	Establishing mass decon line.

Fire department HAZMAT team	1st due team at normal station, 10 minutes away. 2nd due team 15 minutes away. 2 more teams available by mutual aid 20 minutes away		Team 1 dispatched. Alert sent to team 2 and mutual aid	Team 1 is en route. 2d due team dispatched and 2 mutual aid teams summoned.	Team 1 arrives, reports to IC. Team members begin dressing in Level A and/or Level B	Team 1 ready for an initial entry Team 2 arrives, reports to IC. Team 2 prepares for entries
EMS	2 BLS ambulances and 2 ALS ambulances on scene		Additional units dispatched	2 more ambulances. Senior EMS official declares MCI. Triage area set up	EMS staff begins treating casualties at emergency decon line	
Hospitals	3 hospitals identified for emergency care for event, all operating on an alerted basis. Hospitals need 20 minutes to set up decon operations.			Hospitals receive alert from operations center	Based on information from the scene, MCI plan is enacted. Hospital decon area setup is initiated	
Military CBRN team	Staged with HAZMAT team, 10 minutes from site. Team is ready to roll with 10 minutes warning	Team is monitoring comms and "self-alerts"		CBRN starts movement towards incident scene		Initial team from CBRN team arrives at site, reports to IC
Other military assets (specify)	None identified					
EOD/bomb squad	1 team on site, others available		Dispatched to scene to evaluate situation			Advises IC on possible secondary devices

(Continued)

Time or event	Before incident	Incident	Incident +1 minute	I+5	I+10	I+15
Police	100 on site, with 8 sergeants, 2 lieutenants, and 1 captain		Police don CBRN survival equipment. Additional units dispatched.	Additional personnel arrive.	Mobile command unit dispatched to scene	Initial perimeter established inside venue. Police in PPE at decon line to assist with public order and evidence collection.
CBRN crime scene unit	Team members are on other duties, with 30 minute recall time			Team recall signaled. On duty members move to office, off duty members await instructions		Team begins to assemble at their designated assembly point
Laboratory	Normal operations			Lab is alerted by operations center.	Lab stops non-essential work and issues recall notice to essential employees who are off duty	
Public and media affairs	Normal operations. Police and fire PA personnel are at their offices				Public affairs staff notified	Duty PA officer begins movement to incident scene to establish Joint Information Center
Logistics	Normal operations					
Command and control	Normal operations					
Other (specify)	Normal operations					

Bibliography

Abbott, E.B. and Hetzel, O.J. (2018). *Homeland Security and Emergency Management: A Legal Guide for State and Local Governments*. American Bar Association: 3rd ed.

Agro, F., Frass, M., Benumof, J.L., and Krafft, P. (2002). Current status of the Combitube™: a review of the literature. *Journal of Clinical Anesthesia* 14 (4): 307–314.

American Hospital Association, Society for Healthcare Strategy and Market Development (2002). *Crisis Communications in Healthcare: Managing Difficult Times Effectively*. Chicago: Society for Healthcare Strategy and Market Development of the American Hospital Association.

Argonne National Laboratory. *Radiological and Chemical Fact Sheets to Support Health Risk Analyses for Contaminated Areas*. 2007.

Armed Forces Radiobiology Research Institute (2003). *Medical Management of Radiological Casualties*, 2nde. Bethesda (MD): US Government.

Baker, D.J., Jones, K.A., Mobbs, S.F. et al. (2009). Safe management of mass fatalities following chemical, biological, and radiological incidents. *Prehospital and Disaster Medicine* 24 (3): 180–188.

Barbera, J.A. and Macintyre, A.G. (2003). *Jane's Mass Casualty Handbook: Hospital: Emergency Preparedness and Response*. Surrey (UK): Jane's Information Group.

Barsky, L.E., Trainor, J.E., Torres, M.R., and Aguirre, B.E. (2007). Managing volunteers: FEMA's Urban Search and Rescue programme and interactions with unaffiliated responders in disaster response. *Disasters* 31 (4): 495–507.

Bevelacqua, A. (2001). *Hazardous Materials Chemistry*. Albany (NY): Delmar.

Bice, S. and Brown, S. (2008). Public health and crisis management. *Journal of Medical CBR Defense* 6.

Bidilă T., Pietraru, R., Ioniță, A., and Olteanu, A. Monitor Indoor Air Quality to Assess the Risk of COVID-19 Transmission. *2021 23rd International Conference on Control Systems and Computer Science (CSCS)*. IEEE, 2021.

CBRN and HAZMAT Incidents at Major Public Events: Planning and Response, Second Edition. Daniel J. Kaszeta.
© 2023 John Wiley & Sons, Inc. Published 2023 by John Wiley & Sons, Inc.

Bikkina, S., Manda, V.K., and Rao, U.A. (2021). Medical oxygen supply during COVID-19: a study with specific reference to State of Andhra Pradesh, India. *Materials Today: Proceedings* https://www.sciencedirect.com/science/article/pii/S2214785321002856.

Broder, J.F. and Tucker, G. (2011). *Risk Analysis and the Security Survey*. Elsevier.

Bulson, J., Bulson, T.C., and Vande Guchte, K.S. (2010). Hospital-based special needs patient decontamination: lessons from the shower. *American Journal of Disaster Medicine* 5 (6): 353–360.

Buncefield Major Incident Investigation Board (2007). *Recommendations on the Emergency Preparedness for, Response to and Recovery from Incidents*. London: UK Government.

Campbell, L.M. (2009). *A Technological Countermeasure for Chemical Terrorism Against Public Transportation Systems: A Case Study of the "Protect" Program*. Monterrey (CA): US Naval Postgraduate School.

Caponecchia, C. (2012). Relative risk perception for terrorism: Implications for preparedness and risk communication. *Risk Analysis: An International Journal* 32 (9): 1524–1534.

Carter, H., Drury, J., Rubin, G.J. et al. (2015). Applying crowd psychology to develop recommendations for themanagement of mass decontamination. *Health Security* 13 (1): 45–53.

Carter, H., Gauntlett, L., Rubin, G.J. et al. (2018). Psychosocial and behavioural aspects of early incident response: outcomes from an international workshop. *Global Security: Health, Science and Policy* 3 (1): 28–36.

Carter, H., Drury, J., Rubin, G.J. et al. (2012). Public experiences of mass casualty decontamination. *Biosecurity and Bioterrorism: Biodefense Strategy, Practice, and Science* 10 (3): 280–289.

Carter, H.E., Gauntlett, L., and Amlôt, R. (2021). Public perceptions of the "Remove, Remove, Remove" information campaign before and during a hazardous materials incident: a survey. *Health Security* 19 (1): 100–107.

Carter, H., Drury, J., and Amlôt, R. (2020). Recommendations for improving public engagement with pre-incident information materials for initial response to a chemical, biological, radiological or nuclear (CBRN) incident: a systematic review. *International Journal of Disaster Risk Reduction* 51: 101796.

Center for Counterproliferation Research, National Defense University (2002). *Anthrax in America: A Chronology and Analysis of the Fall 2001 Attacks*. Washington (DC): US Government.

Chilcott, R.P. and Wyke, S.M. (2016). "CBRN Incidents." *Health Emergency Preparedness and Response*, 166–180. Oxford (UK): CABI Publishing.

The Chlorine Institute (2006). *Guidance on Complying with EPA Requirements Under the Clean Air Act by Estimating the Area Affected by a Chlorine Release*. Arlington (VA): Chlorine Institute, Inc., Edition 4, rev 1.

Cole, D. (2000). *The Incident Command System: A 25-year Evaluation by California Practitioners*. Emmitsburg (MD): National Fire Academy.

Cole, L.A. (1999). Anthrax hoaxes: hot new hobby? *Bulletin of the Atomic Scientists* 55 (4): 7.

Craighead, G. (2009). *High-rise Security and Fire Life Safety*. Butterworth-Heinemann.

Dembek, Z. (2008). *Textbook of Military Medicine: Medical Aspects of Biological Warfare*. Washington (DC): Office of the Surgeon General, US Army.

Denman A.R., Parkinson S., Groves-Kirkby C.J. A comparative study of public perception of risks from a variety of radiation and societal risks. *Proceedings of the 11th International*

Congress of the International Radiation Protection Association, 23–28 May 2004, Madrid, Spain. 2004.

Dickson, E.F.G. (2012). *Personal Protective Equipment for Chemical, Biological, and Radiological Hazards: Design, Evaluation, and Selection*. John Wiley & Sons.

Doel, S. (2006). *Terror Tactics*. Police Magazine (UK): The Police Federation of England and Wales.

Drake, C. (1998). *Terrorists' Target Selection*. New York: St. Martins Press.

Drielak, S. (2004). *Hot Zone Forensics: Chemical, Biological, and Radiological Evidence Collection*. Springfield (IL): Charles C. Thomas.

Drielak, S. and Brandon, T. (2000). *Weapons of Mass Destruction: Response and Investigation*. Springfield (IL): Charles C. Thomas.

Drury J. Interview with the author, 18 January 2021.

Drury, J., Novelli, D., and Stott, C. (2013). Representing crowd behaviour in emergency planning guidance: 'mass panic' or collective resilience? *Resilience* 1 (1): 18–37.

Drury, J. (2018). The role of social identity processes in mass emergency behaviour: an integrative review. *European Review of Social Psychology* 29 (1): 38–81.

Dvorak P. Health officials vigilant for illness after sensors detect bacteria on mall. *Washington Post*. Washington (DC); 2005.

Eidson, M., Komar, N., Sorhage, F. et al. (2001). Crow deaths as a sentinel surveillance system for West Nile virus in the northeastern United States, 1999. *Emerging Infectious Diseases* 7 (4): 615.

Elliot, A. and Rehfisch, N. (2011). Mortuary provision in emergencies causing mass fatalities. *Journal of Business Continuity and Emergency Planning* 5 (1): 430–439.

Evrard, O., Patrick Laceby, J., and Nakao, A. (2019). Effectiveness of landscape decontamination following the Fukushima nuclear accident: a review. *Soil* 5 (2): 333–350.

Fearn-Banks, K. (2016). *Crisis Communications: A Casebook Approach*. Routledge.

Fenelly, L. (2004). *Effective Physical Security*. Burlington (MA): Elsevier.

Fischer, H., Stine, G., Brenda, L. et al. (1995). Evacuation behaviour: why do some evacuate, while others do not? A case study of the Ephrata, Pennsylvania (USA) evacuation. *Disaster Prevention and Management* 4 (4): 30–36.

Fish, J.T., Stout, R., and Wallace, E. (2010). *Practical Crime Scene Investigations for Hot Zones*. Boca Raton (FL): Taylor and Francis.

Franz, D. (1997). *Defense Against Toxin Weapons*. revised ed. Fort Detrick (MD): US Government.

Galatas B.G.I. (Army of Greece ret.) Interview. Conducted by Dan Kaszeta, 8 November 2011.

Gayev, Y. (2004). *Flow and Transport Processes with Complex Obstructions: Applications to Cities, Vegetative Canopies, and Industry*. Dordrecht (Netherlands): Springer.

Georgopoulos, P.G., Fedele, P., Shade, P. et al. (2004). Hospital response to chemical terrorism: personal protective equipment, training, and operations planning. *American Journal of Industrial Medicine* 46 (5): 432–445.

Gerace, E., Seganti, F., Luciano, C. et al. (2019). On-site identification of psychoactive drugs by portable Raman spectroscopy during drug-checking service in electronic music events. *Drug and Alcohol Review* 38 (1): 50–56.

Governments of the United States, Canada, and Mexico (2020). *Emergency Response Guidebook*. US Department of Transportation.

Halliday J. 'Stay Put' safety advice to come under scrutiny after grenfell tower fire. *The Guardian*, 2017.

Haslam, J.D., Russell, P., Hill, S. et al. (2022). Chemical, biological, radiological, and nuclear mass casualty medicine: a review of lessons from the Salisbury and Amesbury Novichok nerve agent incidents. *British Journal of Anaesthesia* 128 (2): e200–e205.

Hata, A., Meuchi, Y., Imai, S. et al. (2021). Detection of SARS-CoV-2 in wastewater in Japan during a COVID-19 outbreak. *Science of The Total Environment* 758: 143578.

Hawley, C. (2018). *Hazardous Materials Monitoring and Detection Devices*. Burlington (VT): Jones & Bartlett Learning.

Hawley, C. (2002). *Hazardous Materials Incidents*. Albany (NY): Delmar.

Hawley, C., Noll, G., and Hildebrand, M. (2009). The need for joint hazard assessment teams. *Fire Engineering* 162: 9.

Heide, E.A. (2004). Common misconceptions about disasters: Panic, the disaster syndrome, and looting. In: *The First 72 Hours: A Community Approach to Disaster Preparedness* (ed. M.R. O'Leary), 337. New York: iUniverse.

Hincal, F. and Erkekoglu, P. (2006). Toxic industrial chemicals – chemical warfare without chemical weapons. *FABAD Journal of Pharmaceutical Sciences* 31: 220–229.

Houghton, R. and Bennett, W. (2020). *Emergency Characterization of Unknown Materials*. Boca Raton (FL): Taylor & Francis.

Humphries W. Fears that cheltenham festival may have spread coronavirus throughout country. The Times, 2020, 3 April 2020.

International Atomic Energy Agency. *The Radiological Incident in Goiana*. Vienna; 1988.

Interpol (2018). *Guidelines for Dead Body Management and Victim Identification in CBRN Disasters*. Lyon: France.

Johnson S. *Does my face look big in this?* CBRNe World. 2012; 38–40.

Johnson S. *Plumes are for the birds*. CBRNe World. Autumn 2011; 86–89.

Joint Staff, US Department of Defense (2011). *Joint Publication 4-06, Mortuary Affairs*. Washington (DC): US Government.

Joint Staff, US Department of Defense (2019). *Joint Publication 4-0, Joint Logistics*. Washington (DC): US Government.

Joshi, S.M. (2008). The sick building syndrome. *Indian Journal of Occupational and Environmental Medicine* 12 (2): 61–64.

Kaplan D. Nuclear monitoring of Muslims done without search warrants. *US News and World Report*. 22 December 2005.

Kaszeta, D. (2016). Decontamination of buildings after an Anthrax attack. *Journal of Terrorism and Cyber Insurance* 1 (1): 47–60.

Kaszeta D. No, Coronavirus is not a biological weapon. *Washington Post*. 27 April 2020.

Kaszeta, D. (2019). Restrict use of riot-control chemicals. *Nature* 573 (7772): 27–30.

Kaszeta, D. (2020). *Toxic: A History of Nerve Agents*. C. Hurst & Co (UK).

Lai, F.Y., O'Brien, J.W., Thai, P.K. et al. (2016). Cocaine, MDMA and methamphetamine residues in wastewater: consumption trends (2009–2015) in South East Queensland, Australia. *Science of the Total Environment* 568: 803–809.

Laing, M.K., Tupper, K.W., and Fairbairn, N. (2018). Drug checking as a potential strategic overdose response in the fentanyl era. *International Journal of Drug Policy* 62: 59–66.

Lannan, T. (2004). Interagency coordination within the national security community: improving the response to terrorism. *Canadian Military Journal*, Autumn 5: 49–56.

Larsen, J. (ed.) (2013). *Responding to Catastrophic Events: Consequence Management and Policies*. Springer.

Lerner B. In a hospital hierarchy, speaking up is hard to do. *New York Times*; 2007, 17 April 2007.

Lesak, D. (1998). *Hazardous Material Strategies and Tactics*. New Jersey: Prentice Hall.

London Chamber of Commerce and Industry (2005). *The Economic Effects of Terrorism on London: Experiences of Firms in London's Business Community*. London: August.

London Emergency Services Liaison Panel (2007). *Major Incident Procedure Manual*, 7the. London: The Stationery Office.

Lynn, M. et al. (ed.) (2018). *Disasters and Mass Casualty Incidents: The Nuts and Bolts of Preparedness and Response to Protracted and Sudden Onset Emergencies*. Springer.

Majid, A.M. and Spiro, E.S. (2016). Crisis in a foreign language: Emergency Services and Limited English Populations. *Proceedings of the International ISCRAM Conference* https://idl.iscram.org/files/amirahmmajid/2016/1363_AmirahM.Majid+EmmaS.Spiro2016.pdf.

Maniscalco, P. and Christen, H. (2011). *Homeland Security: Principles and Practice of Terrorism Response*. Burlington (MA): Jones and Bartlett.

Manoj, B.S. and Baker, A.H. (2007). Communication challenges in emergency response. *Communications of the ACM* 50 (3): 51–53.

Manto, S.E. (1999). *Weapons of Mass Destruction and Domestic Force Protection: Basic Response Capability for Military, Police & Security Forces*. Army War College, Carlisle, PA USA. Ft. Belvoir Defense Technical Information Center.

Marrs, T., Maynard, R., and Sidell, F. (2007). *Chemical Warfare agents: Toxicology and Treatment*, 2nde. Chichester (UK): Wiley.

McDonald, J., Coursey, M., and Carter, M. (2004). Detecting illicit radioactive sources. *Physics Today* 57 (11): 36–41.

McEntire, D.A. (2015). *Disaster Response and Recovery: Strategies and Tactics for Resilience*. John Wiley & Sons.

McGeary, J. (2007). *Applying Goldwater-Nichols Reforms to Foster Interagency Cooperation Between Public Safety agencies in New York [Thesis]*. Monterey (CA): US Naval Postgraduate School.

McIsaac, J.H. (2010). *Hospital Preparation for Bioterror: A Medical and Biomedical Systems Approach*, 7–9. Elsevier.

Medema, J. (2010). *Principles of Chemical Defense*. Netherlands: Self published.

Monterey WMD-Terrorism Database Staff (2005). Anthrax hoaxes: case studies and discussion. In: *Encyclopedia of Bioterrorism Defense* (ed. R. Katz and R.A. Zilinskas). Hoboken (NJ): Wiley-Liss.

Moran, C.G. and Webb, C. (2008). Lessons in planning from mass casualty events in UK. *World Health* 359: 3–8.

National Commission on Terrorist Attack upon the United States. *The 9/11 Commission Report*. Washington (DC): 2004. 242.

National Fire Protection Association (2002). *NFPA 472 Standard for Professional Competence of Responders to Hazardous Materials Incidents*. Quincy (MA): National Fire Protection Association.

National Fire Protection Association (2016). *Emergency Evacuation Planning Guide for People with Disabilities*. National Fire Protection Association.

National Health Service Scotland. *Hospital lockdown: a framework for NHS Scotland*. 2010. https://www.sehd.scot.nhs.uk/emergencyplanning/Documents/FinalLockdown Guidanceforweb.pdf

National Institute for Occupational Safety and Health, US Department of Health and Human Services (2002). *Guidance for Protecting Building Environments from Airborne Chemical, Biological, or Radiological Attacks: NIOSH Publication 2002-139*. Washington (DC): US Government.

National Research Council (2011). *BioWatch and Public Health Surveillance: Evaluating Systems for the Early Detection of Biological Threats: Abbreviated Version*. National Academies Press.

Noll, G. and Yvorra, J. (1995). *Hazardous Materials: Managing the Incident*. Stillwater (OK): Oklahoma State University Press.

North Atlantic Treaty Organization. *Reporting nuclear detonations, chemical and biological attacks, and predicting and warning of associated hazards and hazard areas: operators manual ATP-45(B)*. Brussels; 2001.

Ogie, R., Rho, J.C., Clarke, R.J., and Moore, A. (2018). Disaster risk communication in culturally and linguistically diverse communities: the role of technology. *Multidisciplinary Digital Publishing Institute Proceedings* 2 (19): 1256.

Oliver M. Radiation found at 12 Sites in the Litvinenko case. *The Guardian*, London, UK; 2006, 30 November 2006.

Olympic Delivery Authority (UK). *Learning legacy Lessons learned from the London 2012 Games construction project*. 2011.

Palmer, I. (2004). The psychological dimension of chemical, biological, radiological, and nuclear terrorism. *Journal of the Royal Army Medical Corps* 150: 3–9.

Pavlin, J. (2007). Medical surveillance for biological terrorism agents. *Human and Ecological Risk Management* 11: 3.

Pearce, M. (2018). The most comprehensive look yet at how the Las Vegas concert massacre unfolded. *Los Angeles Times* 19. https://www.latimes.com/nation/la-na-las-vegas-timeline-20180119-story.html, accessed 19 January 2018.

Perlman, Y. and Yechiali, U. (2020). Reducing risk of infection–The COVID-19 queueing game. *Safety Science* 132: 104987.

Persily, A., Chapman, R.E., Emmerich, S.J. et al. (2007). *Building Retrofits for Increased Protection Against Airborne Chemical and Biological Releases*. Gaithersburg (MD): National Institute of Standards and Technology.

Petterson J. Perception vs. reality of radiological impact: the Goiania model. *Nuclear News*; 1988.

Price, P.N., Sohn, M.D., Gadgil, A.J. et al. (2003). *LBNL/PUB-51959. Protecting Buildings from a Biological or Chemical Attack Actions to Take Before or During a Release*. Berkeley (CA): Lawrence Berkeley National Laboratory.

Price, J. (2016). Hazardous area response teams: celebrating 10 years in the making and counting. *Journal of Paramedic Practice* 8 (8): 390–393.

Proceedings of the International Symposium on the use of Incident Command Systems in Fire Management, Inje University, Gimhae, South Korea, 2009. https://www.alnap.org/system/files/content/resource/files/main/pan-asia-proceeding-ics%28final%29.pdf

Reuters. Bomb suspect's brother mutilates himself. *New York Times*; 1998, 11 March 1998.

Richards, A., Russey, P., and Silke, A. (ed.) (2011). *Terrorism and the Olympics*, 203–204. London: Routledge.

Richmond, C. (2010). *Special Events medical Services*. Sudbury (MA): American Academy of Orthopedic Surgeons and Jones and Bartlett.

Schumacher, J., Weidelt, L., Gray, S., and Brinker, A. (2009). Evaluation of bag-valve-mask ventilation by paramedics in simulated chemical, biological, radiological, or nuclear environments. *Prehospital and Disaster Medicine* 24 (5): 398–401.

Sengupta K. Head of bomb detector company arrested in fraud investigation. *The Independent*. London; 2010, 23 January 2010.

Shanker T., Schmitt E. US military goes online to rebut extremism. *New York Times*; 2011, 17 November 2011.

Shelton D. Testing for anthrax has overwhelmed state labs: officials in Missouri and Illinois say they might have to start denying requests. *St. Louis Post-Dispatch*; 2001, 16 October 2001.

Shiwakoti, N. and Sarvi, M. (2013). Enhancing the panic escape of crowd through architectural design. *Transportation Research Part C: Emerging Technologies* 37: 260–267.

Sidell, F., Takafuji, E., and Franz, D. (ed.) (1997). *Textbook of Military Medicine: Medical Aspects of Chemical and Biological Warfare*. Washington (DC): Office of the Surgeon General, US Army.

Stanovich, Lt Col Mark (2006). Network-centric emergency response: The challenges of training for a new command and control paradigm. *Journal of Emergency Management* 4 (2): 57–64.

Stone, F. (2007). *The 'Worried Well' Response to CBRN Events: Analysis and Solutions, Counterproliferation Paper 40*. US Air Force: Air University.

Strom, K. and Eyerman, J. (2008). *Interagency coordination: a case study of the 2005 London train bombings. National Institute of Justice Journal* 260: 8–11.

Török, T.J., Tauxe, R.V., Wise, R.P. et al. (1997). A large community outbreak of salmonellosis caused by intentional contamination of restaurant salad bars. *Journal of the American Medical Association* 278 (5): 389–395.

Toronto Police Service. *G20 Summit Toronto Police Service after-action review*. Toronto; 2011.

Transportation Research Board of the National Academies (2008). *The Role of Transit in Emergency Evacuation*. Washington (DC): National Research Council.

United Kingdom Home Office (2010). *Audit and Review of Olympic and Paralympic Safety and security Planning: Summary*. London: UK Government.

United States Army (2000). *Field Manual 3–5: NBC Decontamination*. Washington (DC): US Government.

United States Army (2002). *Field Manual 3-06.11 Combined Arms Operations in Urban Terrain*. Washington (DC): US Government.

United States Army (1990). *Field Manual 3-9: Potential Military Chemical/Biological Agents and Compounds*. Washington (DC): US Government.

United States Army (2008). *Field Manual 3-11.21: Multiservice Tactics, Techniques, and Procedures for Chemical, Biological, Radiological, and Nuclear Consequence Management Operations*. Washington (DC): US Government.

United States Army (1997). *Field Manual 101-5 Staff Organization and Operations*. Washington (DC): US Government.

United States Army Center for Health and Preventive Medicine (1999). *Technical Guide 238: Radiological Sources of Potential Exposure and/or Contamination.* Edgewood (MD): US Army.

United States Army Soldier and Biological Chemical Command (1999). *Guidelines for Incident Commander's use of Firefighter Protective Ensemble with Self-contained Breathing Apparatus for Rescue Operations During a Terrorist Chemical Agent Incident.* Edgewood (MD): US Government.

United States Army Soldier and Biological Chemical Command (2000). *Guidelines for Mass Casualty Decontamination During a Terrorist Chemical Agent Incident.* Edgewood (MD): US Government.

United States Army Medical Research Institute of Chemical Defense (2007). *Medical Management of Chemical Casualties Handbook,* 3rde. Edgewood (MD): US Government.

United States Army Medical Research Institute of Chemical Defense (2007). *Medical Management of Chemical Casualties Handbook,* 4the. Edgewood (MD): US Government.

United States Army Medical Research Institute of Infectious Disease (2005). *Medical Management of Biological Casualties Handbook,* 6the. Frederick (MD): US Government.

United States Army Training and Doctrine Command (2007). *A Military Guide to terrorism in the Twenty First Century.* Washington (DC): US Government.

United States Department of Commerce (2011). *National Oceanic and Atmospheric Administration. CAMEO Fact Sheet.* Washington (DC): US Government.

United States Department of Defense (1993). *Directive 5230.16. Nuclear Accident and Incident Public Affairs Guidance.* Washington (DC): US Government.

United States Department of Defense (2013). *Medical Consequences of Radiological and Nuclear Weapons.* US Government Printing Office.

United States Department of Defense Center for Counterproliferation Research, National Defense University (2002). *Anthrax in America: A Chronology and Analysis of the Fall 2001 Attacks.* Washington (DC): US Government.

US Department of Defense. *Department of Defense Plan for Integrating National Guard and Reserve Component Support for Response to Attacks Using Weapons of Mass Destruction.* 1998.

United States Department of Health, Education, and Welfare (1959). *Effects of Biological Warfare Agents.* Washington (DC): US Government.

US Department of Homeland Security, Office of SAFETY Act Implementation. *Best Practices for Anti-Terrorism List for Commercial Office Buildings;* 2018. https://bpatsassessmenttool. nibs.org/pdfs/BPATSforCommericalFacilites_102018.pdf

US Department of Homeland Security, Command, Control and Interoperability Center forAdvanced Data Analysis. *Best Practices in Anti-Terrorism Security for Sporting and Entertainment Venues Resource Guide;* 2013. https://www.safetyact.gov/externalRes/ refdoc/CCICADA%20BPATS.pdf

United States Department of Homeland Security (2009). *Emergency Support Function 15 Standard Operating Procedures.* Washington (DC): US Government.

US Department of Homeland Security, Office of SAFETY Act Implementation. *Field Guide: Conducting BPATS Based Assessments of Commercial Facilities;* 2018. https:// bpatsassessmenttool.nibs.org/pdfs/FieldGuideforConductingBPATSBasedAssessme nts_092018.pdf

United States Department of Homeland Security (2007). *National Preparedness Guidelines.* Washington (DC): US Government.

United States Department of Justice (2000). *Crime Scene Investigation: A Guide for Law Enforcement.* Washington (DC): US Government.

United States Department of Justice (1999). *Death Investigation: A Guide for the Scene Investigator.* Washington (DC): US Government.

United States Department of Justice. *Managing Large-Scale Security Events: A Planning Primer for Local Law Enforcement Agencies.* United States of America; 2013.

United States Department of Justice (2007). *Planning and Managing Security for Major Special Events: Guidelines for Law Enforcement.* Washington (DC): US Government.

United States Department of Transportation (2011). *Transportation Planning for Planned Special Events.* Washington (DC): US Government.

United States Department of Transportation, Pipeline and Hazardous Materials Safety Administration (2011). *Top Consequence Hazardous Materials by Commodities and Failure Modes 2005–2009.* Washington (DC): US Government.

United States Environmental Protection Agency (2007). *Building Retrofits for Increased Protection Against Airborne chemical and Biological Releases.* Washington (DC): US Government.

United States Environmental Protection Agency (2000). *First Responders' Environmental Liability Due to Mass Decontamination Runoff.* Washington (DC): US Government.

United States Environmental Protection Agency (2001). *List of Lists - EPA 550-B-01-003.* Washington (DC): US Government.

US Environmental Protection Agency *Region III Fact Sheet: Quality Control Tools: Blanks* (rev.1, 2009). United States Environmental Protection Agency, 27 April 2009.

United States Federal Bureau of Investigation (2007). *FBI Laboratory Annual Report 2007.* Washington (DC): US Government.

US Federal Emergency Management Agency (2020). *National Incident Management System Basic Guidance for Public Information Officers.* US Government.

US Federal Emergency Management Agency (2019). *National Response Framework.* US Government.

US Federal Emergency Management Agency (2019). *Emergency Support Function Annexes.* US Government.

United States Federal Emergency Management Agency (2004). *Responding to Incidents of National Consequence: Recommendations for America's Fire and Emergency Services Based on the Events of September 11, 2001, and other Similar Incidents.* Washington (DC): US Government.

United States Fire Administration (1997). *Fire Department Response to Biological Threat at B'nai B'rith Headquarters.* Washington (DC). Washington (DC): US Government.

United States General Accounting Office (2003). *Bioterrorism: Public Health Response to Anthrax Incidents of 2001, Report GA-04-152.* Washington (DC): US Government.

United States Government Accountability Office, Timothy M. Persons, and Chris Currie. *Biosurveillance: DHS Should Not Pursue BioWatch Upgrades Or Enhancements Until System Capabilities are Established: Report to Congressional Requesters.* United States Government Accountability Office; 2015.

United States National Fire Academy (1999). *Emergency Response to Terrorism Self Study: FEMA/USFA/NFA-ERT:SS.* Emmitsburg (MD): United States Federal Emergency Management Agency.

United States Occupational Safety and Health Administration (2005). *OSHA Best Practices for Hospital-based First Receivers of Victims from Mass Casualty Incidents Involving the Release of Hazardous Substances.* Washington (DC): US Government.

United States Secret Service (2009). *Multi-agency Response to Concerns Raised by the Joint Congressional Committee on Inaugural Ceremonies for the 56th Presidential Inauguration [Redacted].* Washington (DC): US Government.

Venugopal, A., Ganesan, H., Raja, S.S.S. et al. (2020). Novel wastewater surveillance strategy for early detection of COVID-19 hotspots. *Current Opinion in Environmental Science & Health* 8–13.

Vouriot, C.V.M., Burridge, H.C., Noakes, C.J., and Linden, P.F. (2021). Seasonal variation in airborne infection risk in schools due to changes in ventilation inferred from monitored carbon dioxide. *Indoor Air* 31: 1154–1163.

Wagner, R. (2021). *Saving Our Own: Maximizing CBRN Urban Search and Rescue Capabilities to Support Civil Authorities.* [Diss.] Monterey (CA): Naval Postgraduate School.

Wahlert, M. and Tomashot, S. (2006). Protecting the superbowl: the terrorist threat to sporting events. In: *Homeland Security: Protecting America's targets,* vol. 2: public spaces and institutions (ed. J. Forest), 163–188. Westport (CT): Praeger.

Walker, D. (1968). *Rights in Conflict: The Violent Confrontation of Demonstrators and Police in the Parks and Streets of Chicago During the Week of the Democratic National Convention of 1968,* vol. 3852. New American Library.

Walker, R. and Cerveny, T. (1989). *Textbook of Military Medicine: Medical Consequences of Nuclear Warfare.* Washington DC: Office of the Surgeon General, US Army.

Wardle, C. and Derakhshan, H. (2017). Information disorder: toward an interdisciplinary framework for research and policy making. *Council of Europe* 27: 20–27.

Index

Printed and bound by CPI Group (UK) Ltd, Croydon, CR0 4YY

16/04/2025

14658347-0003